THE EVOLUTION OF
CHARLES DARWIN

Also by Diana Preston

Eight Days at Yalta

Paradise in Chains

A Higher Form of Killing

The Dark Defile

Cleopatra and Antony

Taj Mahal

Before the Fallout

A Pirate of Exquisite Mind

Lusitania

The Boxer Rebellion

A First Rate Tragedy

THE EVOLUTION OF CHARLES DARWIN

THE EPIC VOYAGE OF THE *BEAGLE* THAT FOREVER CHANGED OUR VIEW OF LIFE ON EARTH

DIANA PRESTON

GROVE PRESS
New York

Published simultaneously in Canada
Printed in the United States of America

First Grove Atlantic hardcover edition: October 2022
First Grove Atlantic paperback edition: September 2023

Library of Congress Cataloging-in-Publication data is available for this title.

ISBN 978-0-8021-6122-2
eISBN 978-0-8021-6019-5

Grove Press
an imprint of Grove Atlantic
154 West 14th Street
New York, NY 10011

Distributed by Publishers Group West

groveatlantic.com

23 24 25 26   10 9 8 7 6 5 4 3 2

CONTENTS

PART THREE
AFTER THE *BEAGLE*

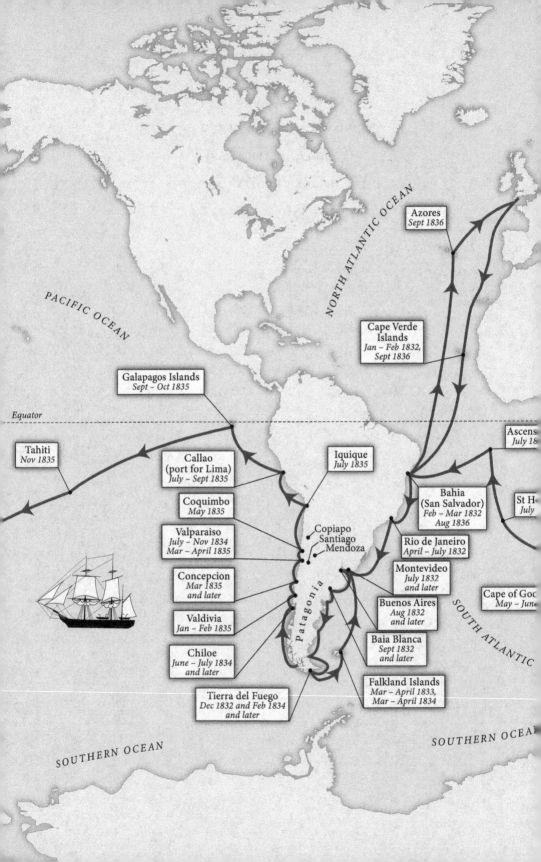

PACIFIC OCEAN

NORTH ATLANTIC OCEAN

Azores
Sept 1836

Cape Verde Islands
Jan – Feb 1832, Sept 1836

Equator

Galapagos Islands
Sept – Oct 1835

Tahiti
Nov 1835

Callao (port for Lima)
July – Sept 1835

Iquique
July 1835

Ascens
July 18

Coquimbo
May 1835

Bahia (San Salvador)
Feb – Mar 1832
Aug 1836

St H
July

Valparaiso
July – Nov 1834
Mar – April 1835

Copiapo
Santiago
Mendoza

Rio de Janeiro
April – July 1832

Concepcion
Mar 1835
and later

Montevideo
July 1832
and later

Patagonia

Cape of Goo
May – Jun

Valdivia
Jan – Feb 1835

Buenos Aires
Aug 1832
and later

SOUTH ATLANTIC

Chiloe
June – July 1834
and later

Baia Blanca
Sept 1832
and later

Tierra del Fuego
Dec 1832 and Feb 1834
and later

Falkland Islands
Mar – April 1833,
Mar – April 1834

SOUTHERN OCEAN

SOUTHERN OCEAN

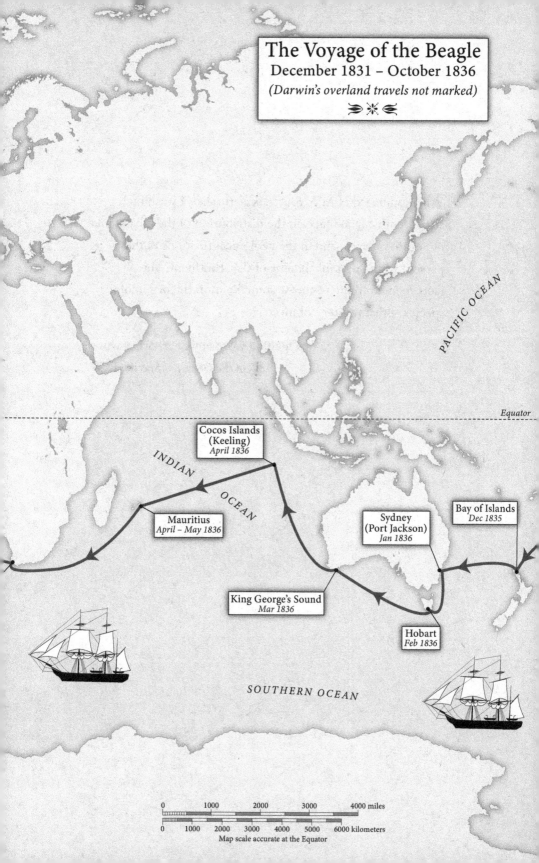

The Voyage of the Beagle
December 1831 – October 1836
(Darwin's overland travels not marked)

PACIFIC OCEAN

Equator

Cocos Islands (Keeling)
April 1836

Mauritius
April – May 1836

INDIAN OCEAN

Sydney (Port Jackson)
Jan 1836

Bay of Islands
Dec 1835

King George's Sound
Mar 1836

Hobart
Feb 1836

SOUTHERN OCEAN

0   1000   2000   3000   4000 miles
0   1000   2000   3000   4000   5000   6000 kilometers
Map scale accurate at the Equator

When on board H.M.S. *Beagle*, as naturalist, I was much struck with certain facts in the distribution of the inhabitants of South America, and in the geological relations of the present to the past inhabitants of that continent. These facts seemed to me to throw some light on the origin of species—that mystery of mysteries . . .

Opening words of Charles Darwin's
On the Origin of Species, 1859

Introduction

In the early afternoon of December 27, 1831, just beyond the breakwater of Plymouth harbor, a tall, ruddy-complexioned, gray-eyed young man transferred from a small yacht to HMS *Beagle* after enjoying a convivial last lunch ashore. Though the weather had recently calmed, the *Beagle*'s crew had labored for three hours to tack her into open water to begin her long voyage under her commander Robert FitzRoy. His demanding mission was to survey the coast of South America from Buenos Aires to Lima and complete a circumnavigation of the world, taking longitudinal measurements as well as returning to Tierra del Fuego three of its inhabitants he had virtually kidnapped on the *Beagle*'s previous voyage and brought to England.

The new arrival clambering aboard the *Beagle* was the expedition's supernumerary and self-financed "gentleman naturalist," twenty-two-year-old Charles Darwin. His excitement was tempered by justifiable apprehension. Could he withstand the physical and mental rigors of the voyage? Would he be seasick? Would he get along with his diverse companions in the ship's cramped confines and lack of privacy? Could he really bear parting from family and friends for so long? And, perhaps above all, did he know enough about the various fields of science he would be expected to cover? Writing in his diary two weeks earlier, he had tried to reassure himself that the voyage was a "great and uncommon . . . opportunity of improving myself" that must not be thrown away.

Because of its twin scientific and philosophical consequences for humanity, the voyage of HMS *Beagle* was to become one of the most important ever undertaken, arguably surpassing the expeditions of Leif Erikson, Ibn Battuta, Zheng He, Ferdinand Magellan, Christopher Columbus, and James Cook, and even the first moon landing. Yet when the *Beagle* departed England, little suggested the intellectual revolution to follow. Charles Darwin was a conventional young man, but as the voyage progressed, he began to develop unconventional ideas. The theories that grew from his research on the voyage would redefine perceptions of humanity and its relationship to other species, showing it had evolved from earlier life forms and was not the divinely created and ordained apex of an unchanging hierarchy. Darwin's thinking would consign the first chapter of Genesis, and with it Adam and Eve, to a mythological limbo, though he would never become a declared atheist himself.

At the time, Darwin would have appeared an unlikely radical thinker. From a wealthy middle-class background, he was, by his own admission, comfortably used to being the focus of family attention, not least from his devoted sisters. He had accepted his medical doctor father's prescription for his unfocused son of a quiet life as a country parson until unexpectedly offered the opportunity to sail on the *Beagle*. He anticipated the coming voyage would show him the glories of tropical places so vividly described by his hero, the explorer and natural scientist Alexander von Humboldt.

Sociable, usually good-natured, and eager to please, Darwin was well suited to fit in on a small naval vessel. His uncle Josiah Wedgwood II, son of the founder of the Wedgwood potteries, had already spotted that Darwin was "a man of enlarged curiosity." Yet to many who knew him, the likable young man, installed with his books and microscopes aboard the *Beagle*, did not seem out of the ordinary. Indeed, Adam Sedgwick, Cambridge University professor of geology, thought, "There was some risk of his turning out an idle man."

Like many of his contemporaries, Darwin was cheerfully and unashamedly chauvinistic, nationalistic, and sexist, as the diary he kept aboard the *Beagle* as well as his subsequent writings reveal. However, though far from radical, his political views were liberal for the time and deep-seated. He opposed slavery, and during the voyage his abhorrence was reinforced by seeing slave-owning societies firsthand. While he believed that different peoples—such as the indigenous Aboriginal peoples of Australia and the Fuegians subsisting near-naked in twig wigwams in chill Tierra del Fuego—might be at differing stages of "civilization," he never wavered from the belief that all humankind belongs to a single species.

By its eventual length—nearly five years—and the fact it circumnavigated the world, returning from South America across the Pacific via Polynesia, New Zealand, and Australia, the *Beagle* voyage gave Darwin a rare opportunity to gather a massive collection of data. These enabled him to examine with growing confidence geological strata and compare and contrast flora and fauna, whether on opposite sides of the Andes or within islands such as the Falklands, St. Helena, Ascension, and the Galapagos—although contrary to popular myth, he did not have a eureka moment about evolution in the latter. He was able to collect fossils of extinct species, prompting lines of thought about their connections to species still living. In Chile he felt the ground shudder beneath his feet— his first experience of an earthquake—seeming proof that the earth's crust was in perpetual motion.

The voyage marked an evolution in Darwin himself. The more facts he gathered—and he was, throughout his life, an inveterate list maker—the more ideas came into his head. Many of these would have seemed heretical to the embryo clergyman he had been when he sailed, not doubting the literal truth of the biblical picture of Creation. For most of the voyage Darwin thought of himself as primarily a geologist. However, in its latter stages he turned increasingly to biology and zoology. As the *Beagle* finally headed for home, he was already making notes on how species changed, though it would be many years before he felt confident enough to reveal his ideas about evolution publicly and face the storm of hostility he knew they would provoke. Integral to his thinking was the interrelationship between living organisms and their environment, making him a pioneer of what we today call ecology.[1]

Eager to explore ashore at any opportunity—Darwin spent some three-fifths of the time on land—he plunged into remote, sometimes hazardous, hinterlands for weeks, even months, usually carrying his pistols

---

1. The German biologist Ernst Haeckel is thought to have invented the word *ecology* in 1869, basing it on the Greek word *oikos*, or *house*, seeing the natural environment as the house or home of all the animals and plants living in it.

with him. In Argentina, he rode hundreds of miles with only gauchos for company, relishing their independent way of life, cooking beef over fires of animal bones and sleeping on their saddles beneath the starry Southern Hemisphere skies. He became caught up in South America's frequent political upheavals. On one occasion he marched with an armed detachment of *Beagle* crew through the streets of Montevideo to intervene in a rising to help defend British property, even if he soon retired back to the *Beagle* with a headache.

During the voyage Darwin embraced new experiences, sampling his first banana in the Cape Verde islands and drinking tortoise urine in the Galapagos. Traveling rough, he sometimes became ill. Whether the poor health that harassed and hampered him in later years stemmed from this period, perhaps from bites from the notorious bugs of the South American pampas, has been much debated. As the voyage went on and as he collected ever more data, Darwin's curiosity about the natural world grew more focused and analytical. He developed from a diffident young man, conscious of his inexperience, into an assured, ambitious natural scientist, prepared to question the ideas of his mentors. He came ashore in Falmouth in October 1836 no longer planning life in a quiet country parsonage, but determined to establish himself in academic circles.

The *Beagle* voyage also shaped the lives of others on board, in particular Captain Robert FitzRoy and his Fuegian protégé, Jemmy Button. FitzRoy, whose desire to take a naturalist gave Darwin his chance, spent a very different five years from the young scientist he nicknamed his "Philos," or philosopher. Striving to fulfill his exacting orders from the Admiralty— even feeling compelled to spend his own money on the expedition—the perfectionist, possibly bipolar FitzRoy suffered a nervous breakdown. He also saw his attempts to establish a Christian mission in Tierra del Fuego fail. In later years, as a traditional Creationist, FitzRoy blamed himself for allowing Darwin the opportunity to develop his "heretical" theories. FitzRoy's own career endured peaks and troughs. Though he distinguished himself as a meteorologist, devising the world's first weather forecasts, his life ended tragically.

Of the three Fuegians FitzRoy brought to England as a kind of social experiment and then returned, Jemmy Button was the one most deeply affected by the *Beagle* voyage. Like other indigenous people plucked from their homes—such as the tragic Bennelong dispatched to Britain by Arthur Philip, governor of the first convict settlement in Sydney Cove, Australia—he was left stranded between two societies, with serious consequences.

Others aboard the *Beagle*, from erstwhile ship's "fiddler and boy to the poop cabin," Syms Covington, who became Darwin's personal assistant, to young naval officers Bartholomew Sulivan and John Wickham, who became Darwin's lifelong friends, thrived through their experiences during the expedition.

The voyage of the *Beagle* was about discovery in every sense, at every level, from self-discovery through detailed broadening of knowledge to the widest scientific revelation. This book belongs to all who sailed, but especially to Darwin, without whom the voyage would have been a footnote, albeit quite an important one, in the history of marine charting and meteorology. As he himself wrote, it was "by far the most important event in my life and . . . determined my whole career."

PART ONE

Prelude

CHAPTER ONE

The Selection of Darwin

Luck, at least as much as any defined personal ambition or readily apparent outstanding talent, gave Darwin the chance to sail as the *Beagle*'s gentleman naturalist. He was neither the first nor the obvious choice. Born on February 12, 1809 (the same day as Abraham Lincoln), he was the fifth of a family of six children. His father, Dr. Robert Darwin, was a shrewd investor, as well as a successful physician, wealthy enough to loan money to the great and good of Shropshire from his home, The Mount, a large, square redbrick house he'd had built overlooking the River Severn on the outskirts of Shrewsbury.[1]

Robert Darwin was the son of doctor and renowned natural philosopher Erasmus Darwin. Among his patients and clients was his brother-in-law, Josiah "Jos" Wedgwood II, son of the famous potter.

Looking back on his early life, Charles Darwin claimed, "All my recollections seem to be connected most closely with self." His mother, Dr. Darwin's wife Susannah Wedgwood, died aged fifty-two of an abdominal tumor when Charles was only eight and a half. He wrote that all he remembered of her was her deathbed, her black velvet gown, and her worktable. He believed this was "partly due to my sisters, owing to their great grief, never being able to speak about her or mention her name; and partly to her previous invalid state." However, he may have found his mother's loss so difficult to cope with that subconsciously he blotted it out. His elder sisters, Marianne, eleven years his senior, Susan, and Caroline took on the role of mothering this somewhat shy, sometimes nervous, and abstracted boy and his younger sister Catherine. Darwin

---

1. The Mount still stands as the Shrewsbury District Valuer's Office.

Charles Darwin, age six, with his sister Catherine.

on occasion told deliberate lies—like pretending to find "stolen" fruit that he himself had picked and hidden in the shrubbery—"for the sake of causing excitement" and securing his indulgent family's attention.

Dr. Darwin sent his son at the age of nine to board at Shrewsbury public school, where he did not excel. He recalled being "at that time very passionate (when I swore like a trooper) and quarrelsome." The traditional rote learning of the classics bored him, and whenever he could he dashed the mile or so back to the reassuring comfort of the Darwin family home. He read and reread a book titled *Wonders of the World*, which, he recalled, "first gave me a wish to travel in remote countries . . . ultimately fulfilled by the voyage of the *Beagle*." He became an enthusiastic collector of everything from coins to shells. A later hobby, in which he joined his older brother Erasmus, was chemical experiments creating gases and compounds from domestic substances in a makeshift laboratory they built for themselves in a shed in the garden of The Mount. These activities caused his schoolfellows to nickname him "Gas," but he believed they "showed me practically the meaning of experimental science."

Such a hobby needed money and—as he would into adulthood—Charles used his sisters as a conduit to obtain it from their formidable father, of whom he seems to have been a little afraid. Darwin recalled his father was easily angered and somewhat "unjust" to him in his youth. Dr. Darwin was physically

Dr. Robert Waring Darwin, Charles Darwin's father

imposing—"the largest man I ever saw"—broad-shouldered, six foot two, and weighing well over twenty-four stone (296 pounds). Though by other accounts inclined to be distant and given to intimidating brooding silences, he seems to have had his son's interests at heart and the transmitted requests for money usually succeeded.

In 1825, when Charles reached sixteen, Dr. Darwin realized that keeping him at Shrewsbury School was pointless. Worried that his easygoing son lacked focus, he decided that he should become a physician like himself and dispatched him that autumn to join his brother Erasmus at Edinburgh University, Britain's most prestigious medical school. Charles disliked his medical studies, finding most of the lectures "intolerably dull" and complaining that one of his tutors, Dr. Duncan, "is so very learned that his wisdom has left no room for his sense." He watched two surgical operations, including one on a child, but, in these days before chloroform, found them so grisly with their writhing, screaming, and gore, that he rushed away traumatized from both before they were over and would never attend another.

At first Darwin relied almost exclusively on the company of his brother, four years his senior. However, when Erasmus left Edinburgh the following year, Charles made efforts to broaden his circle. His interest in the natural sciences was growing, and he made friends with several like-minded young men. He also began attending lectures on zoology given by the thirty-three-year-old Dr. Robert Grant, a well-traveled polymath who, while originally qualifying as a medical doctor, now lectured on invertebrate animals. Grant was interested in the connections between plants and animals and, in particular, whether primitive organisms might have the characteristics of both.

Darwin's enthusiasm impressed Grant, a rather "dry and formal" bachelor who despite his deeply held socialist and anti-establishment views lectured in evening dress. Grant quickly became his mentor and took him to meetings of scientific societies usually closed to undergraduates. He invited him to join him on trips to Leith Harbour to gather marine specimens in chilly tidal pools. He also taught him how to dissect and lent him a single-lens microscope and asked him to examine some puzzling aspects of a species of flustrae—a seaweed-like animal found on tidal rocks. Thus Darwin made his first discovery: "that the so-called ova of Flustra had the power of independent movement by means of cilia and were in fact larvae." Grant shocked him by presenting the finding publicly as if it were his own. Darwin's first exposure to "the jealousy of scientific men" was a jaundicing lesson he never forgot.

However, thanks to Grant, Darwin got to hear the renowned American ornithologist John James Audubon lecture on North American birds. His talk inspired Darwin to take lessons from a taxidermist—"a negro . . . a very pleasant and intelligent man"—who made his living teaching university students at the cost of a guinea a term how to preserve dead animals. His lodgings were just a few doors from Darwin's. His name was John Edmonstone, and he was a former enslaved man, now freed—Edmonstone being the name of his former owner in Guyana. He had accompanied Edmonstone's friend, explorer and naturalist Charles Waterton, in his travels through the South American jungle. Waterton had taught him taxidermy.

Waterton's book *Wanderings in South America*, full of sensational stories of wrestling caimans and boa constrictors, had just been published and was widely reported in the newspapers and the talk of dinner tables. Accompanying Charles Edmonstone back to Britain as a free man, John had settled first in Glasgow and then Edinburgh. During his lessons, Darwin, who described himself as being on "intimate" terms with Edmonstone——"intimate" was a word he rarely used of his relationships with acquaintances—had ample opportunity to learn more about South America and its jungles and indeed about slavery. He later wrote that his friendship with Edmonstone showed him they shared "many little traits of character" and "how similar their [people of color's] minds were to ours." American visitors to Edinburgh and Britain in general at this time were appalled by the mixing between the races. A clergyman from Charleston found it "revolting" to see "well dressed young white men and women walking arm in arm with negroes in the streets of Edinburgh."

Darwin was growing increasingly interested in natural history, though he lamented his continuing inability to sketch specimens adequately. When he enrolled at a cost of four pounds and five shillings for the eminent Professor Robert Jameson's course, which ranged across zoology and botany through to paleontology and geology, he was disappointed. Darwin found Jameson an "old brown dry stick" and his lectures on geology so "incredibly dull" that he resolved "never as long as I lived to read a book on geology or in any way to study the science."

By 1828, Charles Darwin had also decided that medicine was not for him, a message "my father perceived or he heard from my sisters," Darwin later wrote, suggesting that once again he had used his sisters to intercede for him. Dr. Darwin, who had earlier accused his son of caring "for nothing but shooting, dogs, and rat-catching" and predicted "you will be a disgrace to yourself and all your family," was bitterly disappointed. To prevent him embracing a life of leisured indolence, Dr. Darwin decided his son should go to Cambridge University as a prelude to taking holy orders.

Darwin asked for time to consider but concluded he rather liked both the idea of Cambridge and that of becoming a country clergyman. As he knew from friends and relations in the church, a rural clergyman with a good living had plenty of time to pursue private pursuits and interests. He had no scruples on religious grounds, later writing that though not convinced by every dogma of the Church of England, "I did not then in the least doubt the strict and literal truth of every word in the Bible." Also, agreeing to study to become a clergyman would appease his querulous father, his only source of income. Dr. Darwin duly and obligingly gave him a generous allowance of £300 per year at a time when a young army officer received less than £200, a country parson £140, a butler around £40, and farm laborers and seamen around £30.

After cramming to polish up his Greek, which was a requirement for entry, Darwin went up to Christ's College Cambridge in January 1828, where he thoroughly enjoyed the camaraderie of university life. More confident than in Edinburgh, he went hunting and shooting with a sporting set with whom he spent convivial evenings when, he recalled, they "sometimes drank too much, with jolly singing and playing at cards." He begged his family to fund the purchase of a new double-barreled gun, pleading his present one was so old it might malfunction and injure him. He recalled, "When at Cambridge I used to practice throwing up my gun to my shoulder before a looking glass to see that I threw it up straight."

With friends Darwin founded the Glutton Club—a dining society whose eight members met weekly to sample animal flesh "unknown to the human palate," including squirrel, hawk, and bittern. It disbanded after a disagreeable tasting of a stringy brown owl. Darwin's college bill alone of more than seven hundred pounds for his three and a half years was testament to his enjoyment of university. He later claimed, "Upon the whole the three years which I spent at Cambridge were the most joyful in my happy life." However, he also recalled that he had frittered them away as comprehensively as he had his time at Edinburgh University. Even though his tutor at Christ's College agreed with Darwin's own recollection, remembering him "dawdling away his Cambridge days with his horse and his gun until thrown into a panic by the approaching examinations," this was not entirely true. At both universities he focused on extracurricular scientific subjects that interested him rather than those he was supposed to be studying. At Cambridge, he did do sufficient work—however last minute or panicked—on his formal syllabus to obtain a bachelor of arts degree in 1831, ranking tenth out of the 178 not reading for the more onerous honors degree.

During his years at Cambridge, Darwin's enthusiasm had become beetle hunting, then a popular pursuit, to which his second cousin and lifelong friend William Darwin Fox, also an undergraduate at Christ's College, introduced him. One day, peeling a strip of bark from a tree trunk, he uncovered two rare specimens and seized one in each hand. But then he spotted a third. Unable to contemplate letting it go, he popped the beetle he was clutching in his right hand into his mouth, upon which it secreted an acrid fluid, burning his tongue so badly he had to spit it out. Before long Darwin was so keen that he hired an assistant—a laborer—to scrape moss off old trees and gather debris from barges in which reeds were transported into the city from the fens. As a result Darwin acquired some "rare species." In 1829 he sent a beetle to Stephens's *Illustrations of British Insects* and was thrilled when the magazine published a drawing of it above "the magic words, 'captured by C. Darwin, Esq.'"—the first time his name had appeared in print. It is, however, perhaps symptomatic of where his strongest enthusiasms lay at Cambridge that Darwin, who was to become an inveterate list maker, kept few records of his beetle collecting but a meticulous, detailed inventory of the kinds and numbers of game he shot and when and where he shot them.

Also through his cousin William Darwin Fox, Darwin first met John Stevens Henslow, Professor of Botany, who would be key to securing him the chance to sail on the *Beagle*. Like most Cambridge academics, Henslow was a clergyman since college fellows were required, after a certain period, to take holy orders in the Church of England.

John Henslow

Though only thirteen years older than Darwin, Henslow was already respected in Cambridge as "a man who knew every branch of science." His lectures on botany were the only formal ones on natural science that Darwin attended at Cambridge, paying a guinea for the course to do so.

Henslow held weekly Friday evening soirees at his modest house for undergraduates and academics interested in science, to which Fox secured his cousin an invitation. Darwin relished these gatherings, becoming a regular attendee and, before long, Henslow's friend. Nearly every day he accompanied the professor on long rambles so that, as Darwin recalled in the autobiography he wrote for his family late in his life, he became known as "the man who walks with Henslow." What impressed him was Henslow's excellent judgment and intellectual breadth—his knowledge of botany, entomology, mineralogy, and geology was extensive—and the way he based his conclusions upon "long-continued minute observations," a practice Darwin himself would adopt.

During his final year at Cambridge, guided by Henslow, Darwin read voraciously about natural history. Two books particularly impressed him—Alexander von Humboldt's massive, multivolume *Personal Narrative*, detailing his scientific travels, especially in South America, and Sir John Hershel's *Introduction to the Study of Natural Philosophy*. Humboldt mingled vivid but accurate descriptions of the tropics with observations on how geography and climate influenced nature. His account of the jungles and volcanoes of Tenerife, where he observed a meteor shower, so excited Darwin that he tried to persuade Henslow and others to join him on a scientific expedition there and even began inquiring about chartering ships.

Meanwhile Henslow, who before the chair of botany had held that of mineralogy, encouraged Darwin to study geology. The subject that had seemed so unattractive in Edinburgh now caught Darwin's imagination. He was gratified when, in the summer of 1831, at the end of his final academic year, Henslow persuaded Cambridge's most eminent geologist, Reverend Adam Sedgwick, to let Darwin accompany him on a two-week study tour of North Wales designed to improve the geological record of the area. During their trip Sedgwick gave Darwin a grounding in field geology, teaching him about the stratifications of rocks, how to mark them on a map and how to scour rock specimens for fossils, while encouraging him to collect specimens and make observations of his own.

On his return home Darwin must have anticipated a discussion with his father about the next steps toward becoming a clergyman. However, waiting for him was a letter from Henslow, enclosing another from a fellow Cambridge academic, the mathematician Reverend George Peacock, that turned his life in an entirely unforeseen direction. Henslow explained that Captain Francis Beaufort—Admiralty Hydrographer since 1829 and creator of the Beaufort Scale of Wind Force—had asked Peacock to recommend a young man interested in the natural sciences to sail on a two-year naval survey expedition to Tierra del Fuego, returning home across the Pacific. Convinced this was a rare opportunity and a serious loss to natural science if not seized, Peacock consulted Henslow. The expedition's commander, Robert FitzRoy, appeared to be seeking someone "more as a companion than a mere collector and would not take any one however good a Naturalist who was not recommended to him likewise as a *gentleman*." The position would be unsalaried, the incumbent was expected to pay his own mess bills, and the expedition would depart in a month's time.

Henslow believed Darwin was the ideal candidate: "any thing you please may be done—You will have ample opportunities . . . —In short I suppose there never was a finer chance for a man of zeal and spirit." Anticipating that Darwin might harbor doubts about being adequately qualified, Henslow reassured him that "I consider you to be the best qualified person I know of who is likely to undertake such a situation—I state this not on the supposition of your being a *finished* Naturalist, but as amply qualified for collecting, observing, and noting any thing worthy to be noted in Natural History . . . you are the very man they are in search of."

Darwin was eager to accept. His father, however, objected strongly, pointing out a host of reasons why such a step was unthinkable for a young man about to enter the church. Yet seeing his son's evident disappointment, he told him, "If you can find any man of common sense, who advises you to go, I will give my consent." That night, dutiful son that he was, Darwin wrote back to Henslow declining the opportunity: "My father, although he does not decidedly refuse me, gives such strong advice against going that I should not be comfortable, if I did not follow it . . . My father's objections are these; the unfitting me to settle down as a clergyman.—my little habit of seafaring.—*the shortness of the time* and the chance of my not suiting Captain FitzRoy." But, he added, had it not been for his father, he would have seized the chance.

The next day, Darwin rode over to visit his Wedgwood relations— his uncle Jos and bevy of cousins—at Maer Hall, the late seventeenth-century house at the heart of their one-thousand-acre estate thirty miles away in Staffordshire. Given the closeness of the Darwin and Wedgwood families, Maer was his second home, especially in the shooting season— "Bliss Castle" he called it.[2] As Charles Darwin knew, his father regarded his brother-in-law Jos as "one of the most sensible men in the world." Indeed, Dr. Darwin had given his son a letter to him: "Charles will tell you of the offer he has had made to him of going for a voyage of discovery for 2 years.—I strongly object . . . but I will not detail my reasons that he may have your unbiased opinion . . . and if you think differently from me I shall wish him to follow your advice."

Where Robert Darwin perceived dangers and difficulties, Jos Wedgwood saw a glorious opportunity. He asked his nephew to list Dr. Darwin's objections, then wrote a long letter crisply demolishing each in turn: sailing on the *Beagle* would not be "in any degree disreputable" to his nephew's future as a clergyman; there was no reason to suppose it would make him "unsteady and unable to settle"; the Admiralty would never dispatch an unseaworthy ship on such a venture; although the voyage would be "useless" as regards Darwin's career in the church, "as a man of enlarged curiosity," it would give him an opportunity "of seeing men and things as happens to few."

Early the next morning, September 1, 1831, Wedgwood dispatched the letter to Dr. Darwin, together with one from Charles warning his father, "I am afraid I am going to make you again very uncomfortable" but

---

2. Maer Hall still stands.

adding that he was not so hell-bent on going "that I would for one *single moment* hesitate, if you thought . . . you should continue uncomfortable." Afterward, uncle and nephew went partridge shooting—the ostensible reason Darwin had ridden over to Maer—but by ten o'clock Jos was driving his nephew back to Shrewsbury to plead his case in person. He could have saved himself the effort. By the time they arrived, Dr. Darwin had already given in after reading Jos's letter. When his son assured him that though he had been extravagant at Cambridge, "I should be deuced clever to spend more than my allowance whilst on board the *Beagle*," he replied, "But they all tell me you are very clever."

Hoping he was not too late, a buoyant Charles wrote at once to Captain Beaufort accepting the position, packed a few clothes and rushed the one hundred and fifty miles to Cambridge to see Henslow. On his arrival, his mentor had disquieting news. A cousin of Captain FitzRoy, Alexander Wood, who happened to be one of Professor Peacock's undergraduates, had heard that Darwin had been offered the position on the *Beagle* and had written to FitzRoy lauding Darwin's suitability as a scientist. Less helpfully, he had mentioned that Darwin came from a Whig background. The FitzRoys were, by contrast, high Tories, and FitzRoy himself had stood unsuccessfully for Parliament as such in the general election only a year earlier. FitzRoy had replied saying he thought that the vacancy might well have been filled. Henslow, to whom Wood had passed FitzRoy's letter, was disappointed both on his own and Darwin's behalf and complained that Peacock had misled him. Darwin's first reaction was to abandon any hope of sailing on the *Beagle*. Then, reflecting further, he decided he would lose nothing by going to London to see FitzRoy.

Robert FitzRoy, on whom Darwin's hopes depended, was, at twenty-six, four years his senior and from an aristocratic family. Through his father, General Lord Charles FitzRoy, he was descended from the first Duke of Grafton, one of King Charles II's illegitimate sons by his flamboyant, grasping mistress Barbara Villiers. FitzRoy's mother, Lady Frances Stewart, his father's second wife, was the eldest daughter of the Marquis of Londonderry. She had died when FitzRoy was only five, leaving

Robert FitzRoy, commander of
HMS *Beagle*

him motherless even younger than Darwin. Soon afterward Lord Charles, who was a Tory member of Parliament for Bury St. Edmunds, moved his family to Wakefield Lodge, a Palladian hunting lodge near the village of Pottersbury in Northamptonshire. There Robert grew up with a half brother eight years his senior, a brother five years older, and a sister, Fanny, two years older.

In 1818, Lord Charles sent his son, aged twelve and a half, to the Royal Naval College at Portsmouth to prepare for a naval career. He showed an early aptitude for science and mathematics and won a gold medal for completing the syllabus—supposed to take three years—in under eighteen months. His first experience of life at sea came in 1819 when he sailed as a College Volunteer—as the young students were called—aboard HMS *Owen Glendower*, voyaging between Brazil and Northern Peru around Cape Horn. By 1820 he was a midshipman on the same ship, writing to his sister Fanny, "I am very happy and comfortable on board and am sure I shall like sea life very well." In 1824 he gained full marks in seamanship exams back at the Royal Naval College, graduating first of twenty-six candidates, and was soon promoted to lieutenant. In 1828, Admiral Sir Robert Otway, commander of the British navy's South American fleet, appointed him his flag lieutenant.

In November that year "a death vacancy," as the navy succinctly called such opportunities, gave twenty-three-year-old FitzRoy his first command when Otway appointed him to HMS *Beagle*. Together with a larger companion ship, HMS *Adventure*, the *Beagle* had spent the previous two years surveying Patagonia and the island of Tierra del Fuego under the overall command of Captain Philip Parker King, and her task was not yet complete. Delighted to be appointed to "a Discovery ship," FitzRoy enthused to his sister: "We are ordered to collect everything—animals—insects—flowers—fish—anything and everything we can find . . . Will it not be a most interesting employment? . . . it opens a road to credit and character, and further advancement in the Service. Providing I do not fail in my exertions."

Understandably, FitzRoy did not dwell on the tragedy behind the vacancy that had delivered him this prize. Worn down by the scale of his task, "the dreary sea" and "inhospitable shores," and the feeling that "the soul of man dies in him," the *Beagle*'s previous captain Pringle Stokes had grown morbidly depressed, locking himself in his cabin and leaving the survey work to others. Finally, in the Strait of Magellan, Stokes had taken his pistol and shot himself in the head, badly wounding himself. Gangrene had set into his brain and after ten agonizing days, he died.

FitzRoy took the *Beagle* with its shaken crew, many recovering from the seamen's scourge of scurvy and some convinced that Stokes's miserable ghost haunted the ship, from Montevideo back down the coast of South America to Tierra del Fuego to continue the survey work. Entering the Strait of Magellan, FitzRoy encountered his first Fuegians—an old woman, her daughter, and a child paddling a canoe made of wooden branches, covered with bark, and held together with seal intestines. The wildest people he had ever seen, they reminded him "of drawings of the Esquimaux, being rather below the middle size, wrapped in rough skins, with their hair hanging down on all sides, like old thatch, and their skins of a reddish brown colour, smeared over with oil, and very dirty . . ."

Over succeeding months FitzRoy's sometimes violent and usually mutually unsatisfactory encounters with the Fuegians convinced him "that so long as we were ignorant of the Fuegian language, and the natives were equally ignorant of ours, we should never know much about them . . . nor would there be the slightest chance of their being raised one step above the low place which they then held in our estimation." He conceived a plan to take some Fuegians to England to be educated in Christianity and British ways before being returned to their homeland as a catalyst for "civilizing" others. By a variety of methods he inveigled aboard a little girl around nine years old and three young men, quieting any twinges of conscience by attempting to make sure his social experimentees understood that one day they would return.

As he would on further occasions, FitzRoy acted both impulsively and entirely on his own authority—"I incurred a deep responsibility, but was fully aware of what I was undertaking." He hoped that a sympathetic Admiralty would defray some of the costs but, if not, he was prepared to fund the Fuegians' upkeep and education from his own pocket until their return. This was just as well because, when the *Beagle* returned to England in autumn 1830, the Admiralty refused any responsibility, financial or otherwise, for the young visitors. More seriously, as time passed FitzRoy learned that though the survey of South America was still incomplete, the Admiralty had no intention of resuming the work in the near future. There would, therefore, be no naval vessel to take the Fuegians back to Tierra del Fuego.

To add to FitzRoy's worries, one of the young men died of smallpox despite, like the others, having been vaccinated against the disease in Montevideo and again upon arriving in England.[3] "I . . . could not but feel how much I was implicated in shortening his existence," FitzRoy lamented. The death reinforced his determination to honor his commitment to take his surviving three "guests" home. In the end he saw no option but to pay to charter a small merchant ship, *John of London*, to carry him and his party to South America.

At the eleventh hour—though too late for FitzRoy to cancel the contract with the owner of the *John* and save himself nearly one thousand pounds[4]—his uncle, the Duke of Grafton, and his friend Captain Beaufort convinced the Admiralty to send FitzRoy back to South America to complete the survey work. The decision had little to do with repatriating

---

3. Thirty-five years before, in 1796, having observed that milkmaids who had contracted cowpox appeared immune from smallpox, Edward Jenner had created a vaccine using cowpox. Though people were initially skeptical, by the 1830s vaccination was well established in England.

4. One thousand pounds was a very substantial sum, equivalent to more than one hundred thousand pounds today.

the Fuegians, though it would provide their passage home. The spur was Britain's growing commercial ambitions. Spain's former colonies in South America had recently gained their independence, and their new freedom offered considerable opportunities for trade. Eager to steal a march on rivals, especially France, the British government realized British merchant shipping urgently needed more complete information about coastlines, currents, shoals, and other hazards. Initially, the Admiralty told FitzRoy he would be given the *Chanticleer*, a sister ship of the *Beagle*. However, only recently returned from South America, she was found to be entirely unseaworthy. Instead FitzRoy learned that he would again command the *Beagle* after a refit.

As FitzRoy made his preparations, perfectionist that he was, he considered how best to fulfill his remit from the Admiralty. His desire to take a naturalist was sincere. On the first *Beagle* voyage, perplexed by some sulfurous-smelling rock which he suspected might be metal bearing but had no way of confirming, he had written, "If ever I left England again on a similar expedition, I would endeavour to carry out a person qualified to examine the land; while the officers and myself would attend to hydrography." However, as Peacock had told Henslow, he also wanted a companion, a "gentleman" to talk with in a way he couldn't with his officers, to share his meals—captains usually dined alone—and generally to buoy his spirits during the voyage. Thoughts of Pringle Stokes's breakdown in the bleak and desolate regions to which he would soon return must have weighed on FitzRoy, who, Darwin later wrote to his sister Catherine, worried about his "hereditary predisposition" to mental health problems. In 1822 FitzRoy's uncle Lord Castlereagh, a former British foreign secretary and representative at the Congress of Vienna at the end of the Napoleonic Wars, had grown so paranoid that, falsely convinced he was being blackmailed for homosexuality, he had slashed his own throat.

FitzRoy therefore suggested to his friend Beaufort that "some well-educated and scientific person should be sought for who would willingly share such accommodations as I had to offer, in order to profit by the opportunity of visiting distant countries yet little known." Beaufort, who anticipated important scientific results from the voyage, had approached Professor Peacock, whom he knew well and who had duly sought Henslow's advice.

Henslow's first thought had been not Darwin, but his own brother-in-law, Leonard Jenyns, a young clergyman and naturalist and also a Cambridge friend of Darwin's, who later recalled that during their Cambridge years, Darwin was "a most zealous entomologist and attended but little—so far as I can remember—to any other branch of natural history." Jenyns went as far as starting to pack until qualms about deserting his parish overcame him. Henslow was also highly tempted to go himself. Longing to travel to further his research and of far more limited means than the privileged Darwin, who could blithely contemplate financing an expedition to Tenerife, he knew few such chances would come his way. However, Henslow too had a parish, his wife had recently given birth, and her forlorn expression when he talked of going decided him that he could not leave her. And so he had sent Darwin the letters he had found waiting on his return from geologizing in Wales with Sedgwick.

A letter dated September 1, 1831, from Beaufort alerted FitzRoy that his "Savant" had been found: "a Mr Darwin grandson of the well known philosopher and poet [Erasmus Darwin]—full of zeal and enterprize and having contemplated a voyage on his own account to S. America . . . Let me know how you like the idea . . ." By now, however, FitzRoy had begun to wonder whether he really wanted to live in close proximity with someone he did not know and probably of a very different political hue. Overcome by what he later called "a sudden horror of the chances of having somebody [I] should not like on board the vessel" he decided "to throw cold water on the scheme" and to write his cousin in Cambridge the letter that so dismayed Darwin.

Charles Darwin and Robert FitzRoy met for the first time in Beaufort's offices in the Admiralty on September 5, 1831. Darwin found the handsome, fine-featured, aquiline-nosed captain courtesy itself. No one could have been "more open and kind," he wrote the same day to his sister Susan. Though FitzRoy did not immediately say the vacancy had been filled, he was at first politely discouraging, pointing out every likely difficulty, danger, and discomfort of the coming voyage and warning Darwin that should he need more time to complete his South American survey he would not return across the Pacific to determine longitudes as currently planned.

FitzRoy was testing Darwin in various ways, some unknown to his visitor. FitzRoy believed in physiognomy and phrenology—"bumpology," he called them—a popular pseudoscience at the time whose disciples contended a man's character could be told by the bumps in his skull or the outline of his features. As Darwin later discovered, his rather broad, fleshy nose did not at first suggest to FitzRoy that he possessed sufficient energy, determination, and endurance for a long voyage. However, as they talked, he clearly passed FitzRoy's tests. Before long FitzRoy told him that entirely coincidentally, just five minutes before their meeting, the "friend" whom he had invited to accompany him had informed him that he was no longer available. With this doubtless fictional person neatly disposed of, FitzRoy assured Darwin there would therefore be room for him on the *Beagle*, provided he could "bear being told that I want the cabin to myself when I want to be alone.—if we treat each other this way, I hope we shall suit, if not, probably we should wish each other at the Devil." If Darwin became weary of life at sea he would be free to depart.

That night the two young men dined together. Delighted and relieved, Darwin reported back to Henslow, "You cannot imagine anything more pleasant, kind and open than Cap. FitzRoy's manners were to me . . . I think he really wishes to have me." That same evening FitzRoy wrote to Beaufort, "I like what I see . . . and I now request that you will apply for him to accompany me as a naturalist. I can and will make him comfortable on board, more so perhaps than you or he would expect, and I will contrive to stow away his goods and chattels of all kinds and give him a place for a workshop."

CHAPTER TWO

"A Birthday for the Rest of My Life"

During their first meetings, FitzRoy informed Darwin that the *Beagle's* departure was delayed until October 10, but that was still only a month away. Darwin stayed in London and set about his preparations enthusiastically, writing to his sister Susan to instruct their servant Nancy to make him some extra shirts, all to be marked "DARWIN." He also requested a long list of other items, from his carpet slippers and walking shoes to his Spanish books to help him converse with people in South America; his new microscope, which needed to have cotton stuffed inside to protect it; his geological compass; and a small book about taxidermy, presumably a relic of his studies with John Edmonstone, which he thought was in his bedroom. He asked her to seek their father's advice on whether he should take arsenic, the remedy Dr. Darwin prescribed for the eczema that had troubled him since soon after his mother died and that had now erupted on his hands, a possible sign of nervous excitement or apprehension about the voyage.

Darwin also had a more delicate mission for his sister—to raise the question of money with their father. He assured her that though FitzRoy was generally "all for Economy," he had urged Darwin to purchase a pair of good pistols for his protection "and never to go on shore anywhere without loaded ones." The cost would be around sixty pounds. Throughout his life, and not just within his family, Darwin preferred when faced with a troublesome topic to avoid personal confrontation and make his representations through a friendly intermediary. Dr. Darwin must have agreed, since on September 9, Darwin wrote again to his sister that "I am very much obliged to my Father and everybody else." The same day, FitzRoy drove the two of them around London in his gig on a shopping trip during which Darwin purchased a pair of pistols and a rifle for fifty pounds—"there is a saving," he reported triumphantly to Susan. He also bought for five pounds another telescope, with a compass, to carry ashore. FitzRoy himself spent a princely four hundred pounds on firearms. Contradicting his previous remarks about FitzRoy's economy, Darwin wrote home, "I never saw so . . . extravagant a man as regards himself . . . How he did order things."

The previous day, London's shops had been closed and the streets packed with people waiting to see the coronation procession of sixty-five-year-old William IV pass by. Darwin paid thirteen pence to secure a prime view of the gilded carriages bearing the king, his wife Adelaide, and the royal entourage to Westminster Abbey amid fluttering flags and new-fangled gas illuminations. He thought it a truly glittering spectacle, "like only what one sees in picture books of Eastern processions," but noted "very little enthusiasm" among the crowds. A reason for their subdued mood was tension over the Reform Bill that the recently elected Whig government under Lord Grey was attempting to push through Parliament against stiff Tory opposition.

The Bill's objectives—supported by the liberal Darwins and Wedgwoods but not the Tory FitzRoys—included lowering the property qualifications required to vote, redefining constituency boundaries, and reducing the number of so-called "rotten boroughs"—boroughs able to return a Member of Parliament despite having very few voters and usually controlled by one powerful family. A few weeks after Darwin's and FitzRoy's shopping expedition, the House of Lords threw the Bill out, causing seventy thousand angry people to march through London demanding electoral reform.

On September 11, FitzRoy took Darwin on a three-day voyage by steamer—the first sea-going steamship in Britain, the aptly named *Experiment*, had made her debut in 1813—from London around the south coast to the naval dockyard at Devonport in Plymouth to show him the *Beagle*. A brig-sloop, she was being extensively and expensively refitted. The final cost to the Admiralty was £7,583, only slightly less than that of originally building her a decade earlier at Woolwich Naval Dockyard on the Thames. Launched in May 1820, the third naval ship to be named

Beagle, her first mission had been to sail upriver, passing beneath the six-hundred-year-old and shortly to be replaced stone-arched Old London Bridge, to fire a gun salute at the coronation of George IV. Though built for war, the *Beagle* had never seen action and in 1825 was converted for survey work for which her small dimensions—just ninety feet long and twenty-four and a half feet wide—and maneuverability well suited her.

FitzRoy, as he wrote, was "resolved to spare neither expense nor trouble in making our little Expedition as complete . . . as my means and exertions would allow." In particular, he had the upper (main) deck raised eight inches aft and twelve inches forward, increasing the headroom below to a more comfortable six feet. Raising the upper deck without a compensating rise in the height of the ship's wooden sides would, FitzRoy claimed, keep the ship "much dryer on deck" because less water could collect there and thus the vessel's stability would be improved. This was a wise precaution since the *Beagle* belonged to the Cherokee class of brig-sloops nicknamed "coffin" brigs for their propensity to roll, even capsize, in rough seas. Another consequence of the change in deck height was that the size of the ports on the main deck through which the guns were fired was reduced. To withstand the conditions, FitzRoy decided to fit the square-rigged *Beagle* with extra strong crosstrees and heavier rigging "than is usual in a vessel of her tonnage. Our ropes, sails and spars were the best that could be procured."

The ship's stern was being rebuilt and much of the living space remodeled. For understandable reasons, FitzRoy had ordered the quarters on the lower deck where Pringle Stokes had shot himself to be converted for storage and a new captain's cabin constructed. The midshipmen's berths and the crew mess were also being refitted. The cook in the galley would have a new type of patent stove in which the fire, unlike in the previous open fireplace, did not have to be put out in rough weather, meaning hot food and drink would be available when it was most needed. A more modern design of rudder was being installed. The capstan was being replaced by a windlass, a type of winch that made raising and lowering the anchor easier and quicker. To protect the hull from marine parasites and rot, it was being resheathed first with two-inch-thick fir planking,

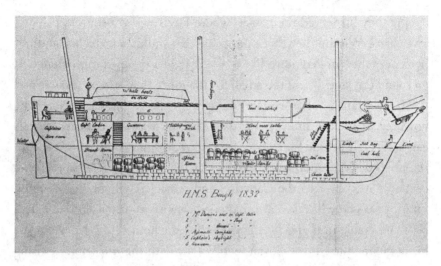

Longitudinal section of HMS *Beagle*, 1832, by Philip Gidley King, 1890

then a layer of felt, and finally new copper sheeting. The effect of all the modifications to the *Beagle* was to increase the vessel's "burden," or total carrying capacity, from 235 to 242 tons.

FitzRoy also ensured the *Beagle* had plenty of boats—important since, unlike her previous voyage, she would have no companion ship to aid her if damaged or sinking. These included four whaleboats—so named because they were built to the design used by whalers—two of which FitzRoy paid for from his own pocket. On the *Beagle*'s first voyage he had seen how effective for survey work these light, versatile craft were. They could be rowed but were also equipped with mast and sails and fitted with centerboards that could be lowered to serve as a keel in deeper waters and raised in shallower ones. In addition to the whaleboats—two twenty-eight feet long and the others twenty-five feet—the *Beagle* would carry a dinghy and two further boats—a two-masted twenty-six-foot yawl and a slightly smaller cutter to be stored within it.[1]

---

1. FitzRoy brought the word *dinghy*—an Indian word used on East India Company vessels—into Royal Naval usage to replace the term *jolly boat*. On the *Beagle* voyage, he pioneered the naval use of the merchant shipping term *port* instead of *larboard*, to avoid confusion with *starboard* when orders were being given.

As they inspected the vessel, where FitzRoy saw progress Darwin saw chaos. Without her three masts and her bulkheads, to his landlubber's eyes she "looked more like a wreck than a vessel commissioned to go round the world" and too small for the more than seventy people and the large quantity of stores and equipment she would carry. FitzRoy, however, reassured him. A few days later he was less concerned. His cabin, directly behind the ship's wheel, above the new rudder and beneath the poop deck (a small, elevated deck at the stern), was "a capital one, certainly next best to the Captain's and remarkably light . . . Captain FitzRoy says he will take care that one corner is so fitted up that I shall be comfortable in it and . . . consider [it] my home . . ." Nevertheless his home measured just eleven feet by ten, with the mizzenmast rising straight through it and a large survey table in the middle. During the day he would share it with nineteen-year-old mate and assistant surveyor John Lort Stokes, a midshipman on the *Beagle*'s previous voyage and now promoted, and fourteen-year-old midshipman Philip Gidley King, son of Captain Philip Parker King, overall commander of the previous expedition on which he had sailed aged only nine, and grandson of his namesake, the third governor of New South Wales.

Despite misgivings about coming discomforts—which, as FitzRoy forewarned this founder member of Cambridge's Glutton Club, would include "no wine and the plainest dinners"—Darwin's admiration for FitzRoy still overflowed. To his sister Susan he wrote: "I will give you one proof of FitzRoy being a good officer, all officers are the same as before, 2/3 of the crew and the eight marines, who went before all offered to come again; so the service cannot be so very bad." Though he had previously praised FitzRoy as "my beau ideal of a Captain . . . all that is quite a joke to what I now feel. Everybody praises him . . . and indeed judging from the little I have seen of him he well deserves it. Not that I suppose it is likely that such violent admiration as I feel for him can possible last; no man is a hero to his valet, as the old saying goes; and I certainly shall be in much the same predicament as one." So fulsome was Darwin that his sisters took to calling FitzRoy "Captain Wentworth," after the dashing naval officer in Jane Austen's *Persuasion*. Darwin's pen portrait of FitzRoy—"a slight figure . . . dark but handsome"—enhanced this romantic image.

Darwin was pleased FitzRoy was equipping his ship with every modern scientific convenience. He was taking no fewer than twenty-two marine chronometers—eleven provided by the Admiralty, six purchased at his own expense, and five loaned by their makers or owners—for the chain of longitude calculations he was to make. He was also bringing five sympiesometers—mercury-free marine barometers—that he prized for their accuracy. To reduce magnetic interference with his instruments, he insisted that the ship's seven cannons—the *Beagle* was officially a ten-gun vessel, but the Admiralty would not permit so many for a mere surveying voyage—be brass, not iron.

Beaufort agreed "few vessels will have ever left this country with a better set of chronometers . . ." By his painstaking development of marine chronometers in the previous century, John Harrison had made navigation far less hazardous. To sail with confidence, a captain had to know the exact position both of his ship and of his destination. That meant knowing their precise latitude and longitude. Latitude—the position northward or southward relative to the equator—could be measured by calculating the height of the sun or stars above the horizon. Longitude—the position eastward or westward—was more complex. It could only be calculated by measuring the difference between the time at the local position and Greenwich Mean Time—longitude was 0 degrees at Greenwich—with each hour's difference equating to 15 degrees of longitude. This meant using chronometers set to, and reliable enough not to deviate from, Greenwich Mean Time, which was what Harrison had achieved. Beaufort also wanted FitzRoy to cross-check the accuracy of the *Beagle* chronometers by measuring the moon's position in relation to certain stars, a more complex procedure known as the lunar distance method.

The *Beagle* was the first naval vessel to carry a copy of the Beaufort Scale, which denoted the force of the wind. Still used today, the scale extends from "0 Calm" through numbered intermediate strengths of breeze, gale, and storm to "12 Hurricane." FitzRoy also had a new code devised by Beaufort for recording other meteorological observations in a standardized way. Beaufort developed both in 1806 but only since being appointed Admiralty Hydrographer in 1829—a post he would fill until 1855—had he been able to put his innovations into practice.

The *Beagle* would pioneer new lightning conductors designed by William Snow Harris—copper strips fitted to the masts, then running to the keel, thus providing a clear path to the water beneath. Darwin attended a public demonstration by "Mr. Thunder-and-lightning Harris," as he called the inventor: "By means of making an Electric machine, a thunder cloud—a tub of water the sea, and a toy for a . . . battle ship he showed the whole process of it being struck by lightning and most satisfactorily proved how completely his plan protects the vessel from any bad consequences."

With so much still to be done, the *Beagle*'s departure was further postponed until October 20, so Darwin left Devonport for London by coach—a 250-mile journey accomplished in just twenty-four hours of continuous travel. Concerned about the ownership of specimens he collected during the voyage, he called at the Admiralty to see Beaufort, who explained that scientific specimens collected on naval expeditions—most often by the ship's surgeon, who was expected to double as a naturalist— were usually regarded as government property. However, to Darwin's relief, he continued that since the government would have the surgeon's collection from the *Beagle* voyage, Darwin would be free to retain or deposit in scientific institutions of his choice any specimens he collected.

Darwin next traveled to Cambridge, where Henslow agreed to take custody of the specimens he sent back during the voyage. He then went home to Shrewsbury to spend ten days packing and saying goodbye to his family. Back in London, Darwin discovered that the *Beagle*'s departure was again delayed—this time to November 4. He used the extra days to consult experts about how best to stuff, store, and preserve specimens. Robert Grant, who had recently moved from Edinburgh to take the chair of zoology at London University, advised him how to handle marine creatures. For example, he told Darwin to slit the gills and clean the abdomens of crabs before pickling them in neat alcohol. Darwin's mind remained fixed on departure. "What a glorious day the 4th of November will be to me.—My second life will then commence, and it shall be as a birthday for the rest of my life," he wrote to FitzRoy.

On October 24, a wet, cold day, Darwin arrived back in Devonport only to find that progress with the refitting was so slow that the *Beagle* would now not sail until late November. That same day, Darwin began keeping a diary, writing on faintly lined paper with a red margin in a format he would follow until November 7, 1836—1,841 days and 751 pages later. An early entry recorded that everything remained "in a state of bustle and confusion" on the *Beagle*.

Exploring the cluttered decks, he tried "to look as much like a sailor as ever I can," but confessed to Henslow, "I have no evidence of having taken in man, woman or child." He also thanked Henslow for all his help: "What an important epoch 1831 will be in my life, taking one degree, and starting for Patagonia . . . —And you have been most instrumental in getting them both . . . I often think of your good advice of taking all uncomfortable moments as matters of course, and not to be compared with all the lasting and solid advantages . . ." Henslow's other advice had included how to deal with the ship's captain—"bridle your tongue when it burns with some merited rebuke"—and the true, if prosaic, not to take too many things, "a mistake all young travellers fall into."

While Darwin waited, FitzRoy arranged for him to share lodgings ashore with John Lort Stokes, who taught him how to use his surveying and navigational instruments so that by mid-November, Darwin was, every morning, taking and comparing the differences in the barometers. Darwin enjoyed the good-natured Stokes's company but was less enthusiastic about other officers. Dinner in the gunroom convinced him that though superficially they were "all good friends," there was "a want of intimacy" that he attributed to their differences in rank—and they were dull. With a waspishness worthy of Jane Austen, whose novels he admired as much as did his sisters and indeed FitzRoy, Darwin thought it "quite surprising that the conversation of active intelligent men who have seen

so much . . . should be so entirely devoid of interest." They were "like
the freshest [university] freshmen," and he felt grateful that at sea he
would dine with FitzRoy.

The officers and midshipmen included thirty-three-year-old Scot John
Clements Wickham, the first lieutenant—a veteran of the *Beagle*'s previous voyage. The second lieutenant, tall twenty-one-year-old Bartholomew
Sulivan, like FitzRoy, had passed through the Royal Naval College, but,
unlike him, he was poor. The son of a talented but impoverished naval
officer and eventual rear admiral forced to reject a knighthood because
he could not afford the associated fees, Sulivan had secured one of the
free places offered to families with a naval tradition but without funds.
A cheerful man and a tremendous talker, he would become Darwin's
lifelong friend. They shared an interest in botany; Sulivan's father had
told him, "Pick and send home any strange plant you find." Darwin was
amused when the other young midshipman aboard introduced himself
thus: "I'm Arthur Mellersh of Midhurst, I have read Lord Byron and I
don't care a damn for anyone!"

   Darwin's relationship with the ship's surgeon, Robert McCormick,
was inherently uneasy because—at least in McCormick's eyes—they
were rivals. At thirty-one he was older than both Darwin and FitzRoy.
Indeed, with John Wickham and thirty-five-year-old purser George
Rowlett, he was among the oldest officers aboard. The son of a naval
surgeon, McCormick had hoped to enter the navy as a midshipman, but
while he was still a boy his father had been lost at sea. Lacking money—
or the connections of a young man like Sulivan—he had decided his
only route into the navy was as a medical officer. Early in his career he
contracted yellow fever in the West Indies, which left him subject to
bouts of ill health but did not prevent him from sailing as assistant surgeon on Captain William Parry's North Pole expedition, which reached
82°45′ N—a record unbeaten for more than forty years.

   In 1830, unable to find an appointment at sea, like Darwin,
McCormick had gone to Edinburgh University to study medicine further.
There, also like Darwin, he had attended Professor Jameson's geology

lectures. Though Darwin had found them dispiritingly dull, they inspired McCormick to seek to make a name in the natural sciences on voyages of scientific discovery. Therefore, unlike for many ship's surgeons who regarded gathering natural history specimens as an unwelcome addition to their primary duties, the presence on the *Beagle* of a "gentleman naturalist" enjoying special privileges was always going to be difficult for McCormick. As for Darwin, he told Henslow, "My friend the Doctor is an ass, but we jog on very amicably: at present he is in great tribulation, whether his cabin shall be painted French Grey or a dead white—I hear little excepting this subject from him."

Darwin also got to know a landscape artist, thirty-eight-year-old Augustus Earle. The son of an American painter, Earle had studied at London's Royal Academy before traveling the world to paint in the Mediterranean, North Africa, Brazil, Australia, and New Zealand, once becoming stranded for eight months on the tiny island of Tristan da Cunha in the mid–South Atlantic. FitzRoy had personally engaged him at a cost of three hundred pounds to produce a pictorial record of the voyage. The captain was also personally employing instrument-maker George James Stebbing to tend the marine chronometers and other delicate navigational apparatus. Even before these commitments, the money FitzRoy had spent on arms, instruments, equipment, the two extra whaleboats, and the abortive chartering of the *John* meant he was already thousands of pounds out of pocket.

On November 12, Darwin noted that the seeming confusion aboard the *Beagle* was finally yielding to order. The dockyard workers had finished painting her, and the decks were clear, with everything—boats, kegs, ropes, tarpaulins—neatly stowed. "For the first time I felt a fine naval fervour; nobody could look at her without admiration."

The next day he first glimpsed the three Fuegians accompanied by the twenty-one-year-old and totally inexperienced missionary Richard Matthews, who was intending to establish a mission station in Tierra del Fuego. Crewmen on the *Beagle*'s previous voyage had named the girl— now around ten or eleven years old—Fuegia Basket to commemorate the basket-type twig boat they built to regain the ship after Fuegians stole their whaleboat. Her name among her own people was Yokcushlu. Of the two surviving males, one was in his mid-twenties and from the same people—the Alakaluf of western Tierra del Fuego. His real name was Elleparu, but FitzRoy called him York Minster because he had taken him hostage near a rocky promontory named for the cathedral. The second was an adolescent named Orundellico, who was from the Yaghan, or Yamana, people, renamed Jeremy or "Jemmy" Button because FitzRoy had acquired him from a group of Fuegians near Cape Horn in return for "a large shining mother-of-pearl button." Darwin would become fascinated by the Fuegians, but his diary accorded their arrival only two brief sentences.

On November 14, FitzRoy's twenty-two chronometers were installed in a special compartment next to his cabin at the stern of the lower deck. That evening his formal instructions arrived from the Admiralty. Darwin, to whom FitzRoy must have shown them, noted in his diary that they were "in every respect most perfectly satisfactory, indeed what Cap. Fitz himself wished.—The orders merely contain a rough outline.—There could be not be a greater compliment paid to Cap FitzRoy than in so entirely leaving the plans to his own discretion."

The orders were indeed brief. FitzRoy was to sail via Madeira or Tenerife and the Cape Verde and Fernando Noronha islands to South America, there to undertake the operations and surveys as directed by Beaufort as Admiralty Hydrographer. If anything happened to FitzRoy, the officer assuming command was, if possible, to complete the part of the survey on which the *Beagle* was currently engaged but then bring the vessel home. The Admiralty also provided a note stating "it is their lordships' direction that no senior officer who may fall in with Commander Fitz-Roy, while he is employed in the above important duties, do divert him therefrom, or in any way interfere with him, or take from him, on any account, any of his instruments or chronometers."

The detail of FitzRoy's task was in a far longer accompanying memorandum from Beaufort. For example, in Rio de Janeiro, "as all our meridian distances in South America are measured from thence" and since its precise longitude was disputed, he was to calculate its position definitively. Next, he was to head south to Montevideo on the River Plate and there check his chronometer readings at precisely the same place as had Captain King on the *Beagle*'s first voyage. South of the Plate the main survey work would begin, including the Rio Negro and the eastern shores of the island of Tierra del Fuego. FitzRoy was also to call at the Falkland Islands to remedy "our present ignorance" about safe approaches and harbors useful to British ships seeking a haven in distress.

Entering the Pacific, FitzRoy was to complete and correct existing charts of the Chilean coast "and indefinitely to the northward . . . If he should reach Guayaquil, or even Callao, it would be desirable he should run for the Galapagos, and . . . survey that knot of islands." FitzRoy was then to make for Tahiti and check his chronometers at Point Venus—"a point . . . indisputably fixed by Captain Cook's and by many concurrent observations"—before calling at Port Jackson [Sydney] and, time and season permitting, other places such as Hobart and the Keeling Islands before finally heading home.

Beaufort additionally listed other demanding tasks, including measuring the Earth's magnetic field: "No day should pass at sea without a series of azimuths, and no port should be quitted without having ascertained not only the magnetic angle, but the dip, intensity and diurnal variation." FitzRoy was also to record tidal patterns, monsoon and trade winds, take detailed barometric readings, and note the wind and weather using Beaufort's scale.

He was to watch for "remarkable phenomena . . . highly interesting to astronomers." If a comet was sighted, "its position should be determined every night by observing its transit over the meridian . . ." Beaufort suggested the sheltered anchorage offered by coral reefs in the Pacific was ideal for astronomical observations and that "while these are quietly proceeding . . . a very interesting inquiry might be instituted respecting the formation of these coral reefs." Finally, Beaufort asked FitzRoy to report regularly to the Hydrographic Office in London "so that if any disaster should happen to the Beagle, the fruits of the expedition may not be altogether lost." Naval practice was to send such dispatches home on British naval vessels or sometimes, if documents were not militarily sensitive, on merchant ships.

As departure finally approached, Darwin enthused to Henslow that the *Beagle* now looked "most beautiful, even a landsman must admire her . . . one thing is certain no vessel has been fitted out so expensively, and with so much care.—Everything that can be made so is of Mahogany . . ." He outlined the ship's itinerary and then added, with a touch of anxiety, "I grieve to say time is unlimited, but yet I hope we shall not exceed the 4 years." Henslow had, of course, originally told Darwin the expedition would last two years. At his first meeting with Darwin, FitzRoy had suggested the voyage would extend to three years. By now, he had perhaps revised this estimate, given the scale of the task outlined for him by the Admiralty.

On November 21, a day that a drunk marine fell overboard, never to be seen again, Darwin brought aboard his own baggage and equipment. Two days later, at one o'clock, FitzRoy had the vessel loosed from her moorings and sailed to Barnett Pool beneath Mount Edgcumbe, her holding position prior to departure. Darwin enjoyed the brief voyage, "everything so new and different to what one has ever seen, the Coxswains piping, the manning the yards [*sic*], the men working at the hawsers to the sound of a fife." Most striking was "the rapidity and decision of the orders and the alertness with which they are obeyed."

The enormous quantity of supplies had been stowed so neatly that "not one inch of room" remained—a revelation to one whose "notions on the inside of a ship were about as indefinite as those of some men on the inside of a man, viz a large cavity containing air, water and food mingled in hopeless confusion." Provisions included more than five thousand canisters of Kilner and Moorsom's preserved meat (many would in fact rust and explode on coming into contact with salt water), soup, dried vegetables like peas and beans, and great quantities of pickles, dried apples, and lemon juice—antiscorbutics to guard against scurvy, which had afflicted many on the *Beagle*'s previous voyage. The iron water tanks could accommodate nineteen tons of fresh water. There were also about six tons of coal and wood for the patent stove and about two thousand yards of spare sail canvas. Replacement spars, yard arms, and rigging rope had to be stowed, together with spare anchors of different types and, of course, powder and shot for the ship's guns and the muskets of the eight marines aboard.

On December 3, Darwin slept aboard for the first time in the poop cabin. Getting into a hammock nearly defeated him until he realized the solution was not to insert his long legs first but to sit in the hammock's center and then give his body "a dexterous twist" so that "head and feet come into their respective places." Although the ship rolled a good deal, not feeling ill gave him hope of escaping seasickness, although, as he wrote to Henslow, "I look forward even to sea-sickness with something like satisfaction, anything must be better than this state of anxiety."

On December 7, Darwin noted, "we only wait for the present wind to cease and we shall then sail." Three days later, the moment seemed to have come as the *Beagle* nosed out beyond the breakwater protecting Plymouth Sound. Almost immediately a heavy gale blew up, and Darwin discovered he was not immune to seasickness: "such a night I never passed, on every side nothing but misery; such a whistling of the wind and roar of the sea, the hoarse screams of the officers and shouts of the men made a concert that I shall not soon forget." He was relieved when FitzRoy ordered the *Beagle* back to harbor until the wind changed. Still giddy and nauseous, Darwin tried to clear his head by brisk walks ashore, accompanied by twelve-year-old Charles Musters, a Volunteer First Class and aspiring midshipman with whose mother the young Lord Byron had been infatuated. Musters, a protégé of FitzRoy, like Darwin, suffered badly from seasickness.

To pass the time, Darwin practiced dressing and undressing in the confined space of the poop cabin and held shooting competitions with Sulivan, for which the prize was bottles of wine to be purchased and drunk when the *Beagle* reached Madeira. He also listed in his diary his future occupations: ". . . without method on ship-board I am sure little will be done.—The principal objects are 1st, collecting, observing and reading in all branches of Natural History that I possibly can manage.

Observations in Meteorology.——French and Spanish, Mathematics and a little Classics, perhaps not more than Greek Testament on Sundays."

The delays improved no one's mood. "The ship is full of grumblers and growlers," Darwin noted. ". . . The sailors declare there is somebody on shore keeping a black cat under a tub, which it stands to reason must keep us in harbour." Decades later in his autobiography, Darwin recalled the weeks of waiting as "the most miserable which I ever spent . . . I was out of spirits at the thought of leaving all my family and friends for so long a time . . ." He was "troubled with palpitations and pain about the heart, and . . . convinced that I had heart-disease." However, "resolved to go at all hazards," he did not consult a doctor for fear of being declared unfit for the voyage.

The weather did not improve until December 21 when FitzRoy again attempted to sail, only for the *Beagle* to strike a rock near Drake's Island in Plymouth Sound. The ship stuck fast for thirty minutes until, by dint of the crew running from side to side to set up a rocking motion, she finally floated free, fortunately undamaged. FitzRoy sailed on, but eleven miles off the Lizard Peninsula strong southwesterly winds again drove the *Beagle* back to harbor.

Though FitzRoy continued to be kind and considerate, Darwin was starting to suspect he might not be quite the "beau ideal" he had first thought. He was taken aback when FitzRoy lost his temper with a Plymouth shopkeeper who refused to exchange a piece of crockery purchased for the *Beagle*. Fitzroy asked the price of the most expensive item in the shop, then told the man, "I should have purchased this if you had not been so disobliging." FitzRoy sensed Darwin's dismay, asking him as they walked away whether he had believed what he had told the shopkeeper about intending to buy the expensive china. When Darwin admitted he hadn't, FitzRoy acknowledged, "You are right, and I acted wrongly in my anger at the blackguard."

FitzRoy had also refused Darwin's request to take the brother of a young woman he seems to have been in love with, Fanny Owen—"the prettiest, plumpest, Charming personage that Shropshire possesses"—as a midshipman because the ship's complement had already been agreed.

Darwin and Fanny Owen had been enjoying a flirtation for some time, taking rides together into the woods when he stayed with her family, lying gorging together in the strawberry beds in the kitchen garden and exchanging sexually charged letters conjuring a fantasy world in which he was a postillion, she a housemaid, and her home, Woodhouse—a grand house on an estate—"the Forest."

Fanny Owen wrote her "postillion" several teasing notes while he waited to sail: "You say what changes will happen before you come back—'and you hope I shall not have quite forgotten you'—I doubt not you will find me in *status quo* at the Forest only grown *old and sedate*—but wherever I may be whatever changes may have happen'd *none* there will ever be in my opinion of you—so do not my dear Charles talk of *forgetting!!*"A few weeks later she wrote, "what a steady *old sober body* you will find me when you return from your Savage Islands . . . how I do wish you had not this horrible *Beetle* taste you might have stayed 'asy' [*sic*] with us here I cannot bear to part with you for so long."

Sulivan too was secretly in love—in his case with a vice admiral's daughter, Sophia Young, whose family were staying in Plymouth. He had planned to take a final leave of her at a ball ashore, but, exhausted by hard work, had fallen deeply asleep aboard the *Beagle*, half waking to appear trance-like "in his night-shirt and night cap, shouldering a big duck gun" to drink tea with his fellow officers before marching back to his bunk. He woke next morning to learn of his sleepwalking—and that he had missed the ball and the opportunity to say goodbye.

The weather improved at last on Christmas Day, a time seamen traditionally devoted to getting drunk—"that sole and never failing pleasure to which a sailor always looks forward," Darwin wrote—though they knew they would be severely punished. So inebriated were the *Beagle*'s crew that midshipman King had to stand sentry. Even the next day the ship was still in a "state of anarchy" owing to "the drunkenness and absence of nearly the whole crew." Darwin grumbled that the weather had eased but that the opportunity to sail had been lost.

Finally at 11:00 A.M. on December 27, 1831, with what FitzRoy called "a light 'cat's paw'" rippling the water, the *Beagle* weighed anchor and with some difficulty—the sailors' throbbing heads doubtless had as much to do with it as the conditions—tacked out beyond the Plymouth breakwater. Around 2:00 P.M. a yacht brought out Darwin and Sulivan to join the ship. Darwin blamed "the total absence of sentiment which I experienced on leaving England" on the lavish farewell lunch of mutton chops and champagne they had just consumed. Almost immediately a fresh easterly wind filled the sails, and the *Beagle* scudded forward. A few weeks before, in a letter to his cousin William Darwin Fox, Darwin had conjured a heady vision of "date and cocoa trees, the palms and ferns so lofty and beautiful—everything new, everything sublime." Now the chance to see such wonders had come.

PART TWO

The Voyage of the *Beagle*

CHAPTER THREE

"Like Giving a Blind Man Eyes"

Darwin remembered his first full day at sea, December 28, 1831, for two reasons—his own acute seasickness and the screams of four crewmen bound by their wrists and ankles to upright gratings on deck and brutally flogged. FitzRoy was quick to mete out punishment for the drunken Christmas revelry, sentencing seaman John Bruce to twenty-five lashes for "drunkenness, quarrelling and insolence"; seaman Elias Davis to thirty-one for "neglect of duty"; David Russel, one of the carpenter's team, to thirty-four for "breaking his leave and disobedience of orders"; and seaman James Phipps to forty-four for "breaking his leave, drunkenness and insolence." In addition, FitzRoy disrated five other crew members for similar offenses—a demotion that brought with it loss of pay.

Darwin wrote in his diary that "several have paid the penalty for insolence, by sitting for eight or nine hours in heavy chains." The punishments were "an unfortunate beginning," but he did not doubt that strict discipline was "absolutely necessary . . . amongst such thoughtless beings as Sailors." FitzRoy himself, though he abhorred corporal punishment, believed there were "too many coarse natures which cannot be restrained without it." What he called a "timely stitch" had its effect—he would only feel compelled to order a few further floggings during the voyage.[1]

The following day, as the deeply laden *Beagle* rolled across the notorious Bay of Biscay in a heavy swell, Darwin still felt the misery of seasickness "far far beyond what I ever guessed at" but with "one great difference between my former sea sickness and the present; absence of giddiness; using my eyes is not unpleasant; indeed it is rather amusing . . . to watch the moon or stars performing their small revolutions in their new apparent

---

1. Flogging in the British navy would not end until 1879.

orbits." His only relief was to lie flat, and all he could bear to eat were biscuit and raisins—the latter a remedy suggested by his father—"but of this as I became more exhausted I soon grew tired and then the sovereign remedy is Sago, with wine and spice and made very hot." He would later recall FitzRoy's kindness at this time and how "you came and arranged my hammock with your own hands."

Darwin cheered himself by concentrating on "Humboldt's glowing accounts of tropical scenery." Five of von Humboldt's works were in the *Beagle* library of four hundred volumes of books, journals, pamphlets, encyclopedias, and maps accommodated on forty-six feet of shelving in the cabin Darwin shared with Lort Stokes and midshipman King. They included Darwin's personal collection since FitzRoy had encouraged him to bring whatever he wished. Accounts of voyages and travels—"to me much more interesting than even novels"—and works on natural history accounted for 70 percent of the *Beagle*'s library, with geology the next largest category. The only novel, as such, was Samuel Richardson's seven-volume *The History of Sir Charles Grandison*, published half a century previously. The only poetry was Darwin's copy of John Milton's *Paradise Lost* and two volumes of both poetry and prose by William Shenhurst, published in 1764 and little known today.

Darwin also spent time chatting to FitzRoy, who was relieved the ship's motion was not upsetting his chronometers, hanging snugly in their gimbals, pivoted supports allowing them to rotate about a single axis. Whenever he felt strong enough, Darwin staggered on deck to watch porpoises leaping through the ship's wake and stormy petrels skimming the waves. However, when Madeira appeared on the horizon he was too sick even to get up to view the distant outline of the island FitzRoy had bypassed because of a sudden violent squall.

Darwin was, however, on deck on January 6 as the mountainous northeastern peninsula of the island of Tenerife—a Spanish possession since 1496—appeared, its snowy tips poking through a dense cloud bank. Fired by Humboldt's account of its great volcano, Mount Teide, more than 12,000 feet high, Darwin had of course originally contemplated his own expedition to Tenerife. As the *Beagle* drew nearer, he was anticipating a pleasurable visit when a boat from the island's health office approached, bringing the British vice consul and some local officials, one of whom announced that because of reports of cholera in England, all on board had to complete twelve days' strict quarantine before going ashore. Darwin recalled "a death like stillness in the ship; till the Captain cried 'Up Jib'" and set course southwestward to the Cape Verde islands.

FitzRoy had planned to measure Mount Teide's precise longitude but was not prepared to waste twelve days, even though he realized his decision greatly disappointed his naturalist companion. As the ship sailed on, Darwin found some compensation in the glorious sunsets and the speed with which the sun sank beneath the horizon. For the first time he experienced the magic of the tropical night, which "does its best to smooth our sorrow—the air is still and deliciously warm . . . the sky is so clear and lofty, and stars innumerable shine so bright that like little moons they cast their glitter on the waves." On January 10, the *Beagle* crossed the Tropic of Cancer in weather so warm that all donned lighter clothing, and Darwin exulted that "the miserable wet weather" of Britain was behind them. He experimented with a plankton net that he had had made before the voyage, using it to scoop small marine creatures—his first specimens—from the sea. Their "exquisite" forms and "rich colours" made him marvel "that so much beauty should be apparently created for such little purpose." Less exquisite was the acute stinging pain when he got slime from a Portuguese man-of-war—a relation of the jellyfish with long venomous tentacles—on his fingers, then put them in his mouth without thinking.

By now Darwin was growing accustomed to a daily routine. Provided he was not too seasick, at 8:00 A.M. he breakfasted with FitzRoy in the captain's cabin. The two met again at 1:00 p.m. for what was then known as dinner and the main meal of the day—and at 5:00 p.m. for supper. Meals were brief affairs. As FitzRoy had forewarned, their basic diet was plain—rice, dried peas, ship's biscuit, pickled vegetables and dried apples to ward off scurvy, and sometimes canned meat and cheese. As FitzRoy had also warned, there was no alcohol. Neither waited for the other to finish. "The invariable maxim is to throw away all politeness . . . and bolt off the minute one has done eating" to return to their tasks.

On January 16, FitzRoy anchored the *Beagle* in the broad bay of St. Jago [São Tiago], the largest of the Cape Verde islands colonized by the Portuguese in the fifteenth century and 350 miles off the coast of West Africa. A few hours before, Darwin had collected some of the soft yellow-brown dust he thought wind had carried from Africa. He also found on deck a brightly colored cricket, which, like the dust, might have been blown from the African coast. He would later remember these discoveries when considering how far seeds might travel in the air from the mainland to the islands and from island to island.

Darwin was among the party FitzRoy quickly dispatched ashore to make a courtesy call on the islands' governor in the main town, Porto Praya—a meeting conducted "in a very ludicrous mixture of Portuguese, English and French." Retreating "under a shower of bows," Darwin strolled through the town, "a miserable place" with its ramshackle streets teeming with goats and pigs. He enjoyed the cheap, sweet oranges on sale but was unimpressed by his first taste of a banana, which he had seen in Cambridge but not eaten, finding it "maukish and sweet with little flavor," though he wrote home advising his father to purchase a banana plant for the hothouse at The Mount.[2]

Beyond the town, he came to a deep valley that fulfilled all his expectations: "I first saw the glory of tropical vegetation. Tamarinds, Bananas and Palms were flourishing at my feet.—I expected a good deal, for I had read Humboldt's descriptions and I was afraid of disappointments: how utterly vain such fear is, none can tell but those who have experienced what I today have." The song of birds unknown to him, the sight of "new insects fluttering about still newer flowers" overwhelmed him. In his diary he described "the numberless and confusing associations that rush together on the mind . . . It has been for me a glorious day, like giving a blind man eyes . . ."

---

2. In 1633 Thomas Johnson displayed the first bananas ever seen in England, brought from Bermuda, in his shop window in London, but the fruit remained an expensive luxury until the late nineteenth century.

The next day, Darwin accompanied FitzRoy to nearby Quail Island (Ilhéu de Santa Maria), less than a mile in circumference, where the captain planned to set up a temporary observatory to take chronometric and astronomical measurements. Though the island seemed desolate, Darwin found "corals growing on their native rock." In Edinburgh he had poked about chilly Scottish tidal pools with Robert Grant and examined the minute corals there but never in "the wildest castles in the air" imagined he would one day have a chance to see larger varieties or how exquisite they would be. He also collected his first geological specimens of the voyage. "The first examining of Volcanic rocks must to a Geologist be a memorable epoch," he noted in his diary.

In subsequent days, he explored more of St. Jago, finding the island's geology "preeminently interesting and I believe quite new," though he worried whether he was noting "the right facts and whether they are of sufficient importance to interest others." However, he had already acquired enough knowledge of geology to identify interspersed layers of lava and calciferous rock, all of which appeared to have risen and subsided at various times, and he hoped his specimens might interest other geologists. On St. Jago, and throughout the voyage, he used his geological hammer to chip out rock specimens of the dimensions advised by experts, three inches square and a maximum of three-quarters of an inch thick—"the size of a common flat piece of Windsor soap." Inevitably he sometimes broke the specimen. He would then wrap the pieces in paper before putting them into his geology collecting bag, always carefully marking specimens with location and date of acquisition. Looking back in later years, he claimed that while observing St. Jago's "very striking yet simple" geology, "it . . . first dawned on me that I might perhaps write a book on the geology of the various countries visited, and this . . . was a memorable hour to me."

After lunching on ripe tamarinds and biscuit, Darwin carried his "rich harvest" of geological and marine specimens aboard the *Beagle*.

In subsequent days, as he explored further, Darwin caught himself "thinking of England and its politics" and in particular the Reform Bill, noting, "It is my belief that the word reform has not passed the lips of any man on board since we saw Madeira.——So absorbing is the interest of a new country." Of growing interest to him were the local and largely black human inhabitants. Approaching a village celebrating a feast day, he and some companions overtook a group of twenty young women "dressed in most excellent taste.——their black skins and snow white linen . . . adorned with gay coloured turbans and large shawls." As the *Beagle* men drew closer, the women flung down their shawls and "sung with great energy a wild song: beating time with their hands upon the legs."

On another occasion, while Darwin and ship's surgeon Robert McCormick rested beneath a tamarind tree, "two black men brought us some goats milk." Darwin held out a handful of copper coins in payment, but they took only a farthing. When he insisted they should take at least a penny, "we hardly could prevent them pouring a quart of milk into our very throats." He added, "I never saw anything more intelligent than the Negros, especially the Negro or Mulatto children . . . they examine everything with the liveliest attention, and if you let them the children chattering away, will pull everything out of your pockets to examine it.——My silver pencil case was pulled out and much speculated upon." However, the sight in one village of "black children, perfectly naked and looking very wretched . . . carrying bundles of fire wood half as big as their own bodies" and of badly dressed men and women who "looked much overworked" distressed him.

Also accompanied by McCormick, Darwin took a long walk deep into the interior of St. Jago. Following the course of a broad river they eventually reached a huge baobab tree about which Darwin had read. Measuring its trunk, Darwin found it over thirty-six feet in girth. Its bark was incised with as many dates and initials "as any one in Kensington Gardens." McCormick also noted "the remarkable Baobab Tree" in his diary but did not mention Darwin's presence——a sign perhaps of his growing resentment of the younger man whose role as naturalist overlapped with his own but who enjoyed so many privileges he did not. Four days later Darwin took a "merry and pleasant walk" back to the tree with FitzRoy and Lieutenant Wickham but not McCormick to measure the tree more accurately.

FitzRoy took an angle with his pocket sextant to calculate its height. Then, to be certain, he shinnied up the trunk and let down a string. According to Darwin, "both ways gave the same result, viz forty-five in height."

Darwin's scientific collection was growing rapidly—so much so that he sometimes felt like "an ass between two bundles of hay—so many beautiful animals do I generally bring home with me." He started to worry "that there will be nobody in England who will have the courage to examine some of the less known branches." One intriguing creature he hoped would be unknown to others was a type of octopus that, when stranded by the outgoing tide in rock pools, tried to hide from predators either by injecting the surrounding water with "a dark, chestnut-brown ink" or by "a very extraordinary, chameleon-like power of changing their colour . . . according to the nature of the ground over which they pass; when in deep water, their general shade was brownish purple, but when placed on the land, or in shallow water, this dark tint changed into one of a yellowish green." Darwin wrote to Henslow excitedly describing his new discovery, but later in the voyage a deflating reply informed him that Henslow had seen an octopus behave exactly the same way in a far less exotic location—Weymouth on England's south coast.

Though Darwin had doubted his readiness for his role aboard the *Beagle*, his time in Edinburgh and Cambridge had exposed him to much new thinking, not only about geology but other branches of science, which had given him some reference points and suggested potential lines of research. Both geology and zoology were then emerging disciplines and, because of their implications, not least for religious belief, something of a battle-ground between opposing groups, some of whom still shared the view of James Ussher, the seventeenth-century Anglican archbishop of Armagh in Northern Ireland, who, by totting up all the generations listed in the Bible since Adam and Eve, had concluded that God had completed his seven-day creation of the world in exactly its current state at 9:00 A.M., October 23, 4004 BC, and that no change had since taken place.

In Edinburgh several of his professors had introduced Darwin to the ideas of Jean-Baptiste Lamarck (1744–1829). In Revolutionary Paris in 1793—the year King Louis XVI and Marie Antoinette went to the guillotine—Lamarck had been appointed professor at the newly established Muséum National d'Histoire Naturelle—a republican reorganization of the former royal botanical garden where he had been an assistant botanist. His new professorial remit was to study "inferior animals" like insects and worms. He coined the name *invertebrates* for them. Based on studies of fossils—the so-called fossil record—Lamarck theorized that all life had evolved gradually from earlier forms, with humans emerging from apes as the "final result of the gradual development and improvement of all nature's creatures." He constructed a hierarchy of organic life, with his invertebrates at the base to humans at the apex, and suggested that species did not die out but rather transmuted into other species. His seven-volume history of invertebrates was on the shelves in Darwin's cabin.

Darwin's own grandfather, Erasmus Darwin, developed similar theories, though there is no evidence he read Lamarck's work. A wealthy doctor of sufficient repute to be invited to become physician to King George III—as a strong republican and supporter of the American Revolution, he declined—Erasmus was also a poet, inventor, leading scientific thinker, and member of the Lunar Society, a group of progressive thinkers in Birmingham that included Darwin's other grandfather, the founder of the Wedgwood potteries, Josiah Wedgwood I; James Watt, the steam engine pioneer; and Joseph Priestly, among whose achievements were the discovery of oxygen and the invention of soda water. In addition, Erasmus Darwin was a passionate advocate of women's education and, as were all the group's members, the abolition of slavery.

Erasmus Darwin M.D.

Erasmus Darwin, Charles Darwin's grandfather and author of *Zoonomia*

Smallpox had left Erasmus Darwin badly pockmarked, as it did many at the time. The disease's consequences had been far worse for his friend Josiah Wedgwood—Erasmus had directed the amputation of Wedgwood's infected right leg in a gory operation performed without anesthetics. Erasmus was so corpulent that he had a semicircle cut into his dining table to accommodate his belly. However, his charisma, wit, and intelligence won him a wide circle of friends and admirers that included the French philosopher Jean-Jacques Rousseau. An enthusiastic womanizer, he fathered at least fourteen children—five by his first wife, including Charles Darwin's father, Robert; two by the governess employed after the first Mrs. Darwin's death; and seven by his second wife.

In 1794 he published *Zoonomia*, which suggested, albeit with little supporting data and unknowingly foreshadowing DNA, that all forms of life were related, having gradually evolved from one "living filament . . . whether sphere, cube or cylinder." He postulated that human ancestors were aquatic creatures that millions of years ago had emerged from the sea and learned to breathe. What had caused the "living filament" to develop and diversify were, in Erasmus Darwin's view, such factors as the ever-changing environment and sexual selection, in particular competition to breed: "The final course of this contest among males seems to be, that the strongest and most active animal should propagate the species which should thus be improved."

Erasmus Darwin also presented his ideas in verse in the lengthy and sexually suggestive *The Temple of Nature; or, the Origin of Society*, speaking of "soft embraces," "bowers of pleasure," "seductive sighs," and "wanton play," published posthumously in 1803. In the first canto he asserted that:

> *Organic Life beneath the shoreless waves*
> *Was born and nurs'd in Ocean's pearly caves;*
> *First forms minute, unseen by spheric glass,*
> *Move on the mud, or pierce the watery mass;*
> *These, as successive generations bloom*
> *New powers acquire, and larger limbs assume.*

Lamarck theorized that characteristics acquired in life could be handed on to the next generation—for example, that during their lives anteaters' noses lengthened as a result of constant burrowing into anthills and that they passed this characteristic to their offspring, whose noses grew even longer during their lifetimes.

While they were out walking in Edinburgh, Professor Grant had surprised Charles Darwin by suddenly bursting out, as Darwin later remembered, in "high admiration of Lamarck and his views on evolution." Darwin listened "in silent astonishment, and as far as I can judge, without any affect on my mind. I had previously read the Zoonomia of my grandfather, in which similar views are maintained, but without producing any effect on me." Looking back many years later, he claimed the reason for his indifference was that though *Zoonomia* was bursting with ideas, it contained little underpinning evidence.

Another disciple of Lamarck was the Edinburgh geologist Robert Jameson. Ironically, the professor whose lectures so bored Darwin that he had resolved never to study geology may well have been the first to use the word *evolve* to describe the transmutation of species (i.e., the process of change). In October 1826, a year after Darwin arrived in Edinburgh, an anonymous article titled "Observations on the Nature and Importance of Geology" appeared in the *Edinburgh New Philosophical Journal*, of which Jameson was editor. It suggested that geology was the key not only to understanding the history of Earth but of the life upon it and that "various forms have evolved from a primitive model . . . species have arisen from an original generic form." Style and content both strongly suggest the author was Jameson himself.

Charles Lyell

The geologist who most influenced Darwin both aboard the *Beagle* and later was Charles Lyell. Ten years older than Darwin and the son of a wealthy naturalist, he had studied classics at Oxford University but also attended lectures by William Buckland, a theologian and geologist whose life's work would be struggling to reconcile emerging geological evidence with scriptural accounts in the Bible and, in particular, the Old Testament. Lyell had gone on to study law in London but soon gave it up in favor of what had become his overriding passion—geology, which he wanted "to free . . . from Moses." Another reason for abandoning law was poor eyesight, which sometimes made reading difficult. Shortly before the *Beagle* sailed from Britain, FitzRoy—who, unlike Darwin, had met Lyell and agreed to collect data for him during the voyage—presented Darwin with the first volume of Lyell's *Principles of Geology*, inscribing it "From Capt FitzRoy." Darwin had been reading it as the *Beagle* headed for the Cape Verde islands.

Only published a year earlier and to be followed by two further volumes that would reach Darwin during the voyage, Lyell's central thesis was that the Earth had been—and was still being—shaped by natural forces operating uniformly over immense periods of time to produce relatively small gradual and successive changes—an approach the Cambridge philosopher William Whewell would term *uniformitarianism*. In Lyell's scheme of things, the Earth's crust was constantly rising or subsiding, a process without any direct divine intervention, dictated by the activity of molten rocks beneath. Lyell, though, was no atheist, arguing that after creating the world and life upon it, God had simply given nature free rein. He firmly rejected Lamarckian ideas about a hierarchy of all living organisms, which implied that through transmutation of species, humans and beasts were related.

In developing his theories, Lyell built on the work of the eighteenth-century Edinburgh geologist James Hutton, who believed that the Earth was "alive," continuously changing through geological forces such as volcanic eruptions, earthquakes, and erosion, and that the strata that now compose our continents had been built up on the ocean floor from the debris of continents that no longer existed. Hutton deduced that to allow for the time such processes would take, the Earth must be millions of years old rather than only six thousand, as the church, citing Archbishop Ussher's calculations, still formally maintained. Hutton suggested that "from what has actually been, we have data for concluding with regard to that which is to happen hereafter."

Taking Hutton's emphasis on data to heart, Lyell had traveled extensively through Europe, gathering information on such phenomena as glaciers and the Mount Etna volcano. All he saw convinced him that the ever-dynamic Earth was, just as Hutton thought, far older than many believed and continuously if gradually changing. To illustrate his theory, the frontispiece to the first volume of his *Principles of Geology* was a sketch of three curiously striated forty-foot-high limestone columns from the Temple of Serapis north of Pozzuoli near Naples. The columns were "smooth and uninjured to the height of about twelve feet above their pedestals. Above this is a zone, twelve feet in height, where the marble has been pierced by a species of marine perforating bivalve—Lithodomus," a genus of burrowing clam whose shells were still embedded in the rock. The only logical explanation, Lyell proposed, was that the temple, originally built above sea level, had subsequently been partially submerged. The lowest twelve feet of its columns had been embedded in the sea floor, while the next twelve feet had been in seawater and exposed to the action of marine creatures. The volcanic eruption that had produced Monte Nuovo, "New Mountain," near Pozzuoli in the sixteenth century had raised the temple to its present level. (The present elevation of the encrustations on the columns shows that they have risen a further 3.15 meters since Lyell's day.)

Darwin later paid tribute to Lyell's influence: "The very first place which I examined, namely St. Jago . . . showed me clearly the wonderful superiority of Lyell's manner of treating geology . . ." In particular, Lyell's theories provided Darwin with a rational explanation for why "a perfectly horizontal white band" that he examined on an otherwise dark sea cliff in the Cape Verde islands was studded with fossilized seashells and compressed coral, although forty-five feet above sea level. The white band had once formed part of the seabed but had been gradually elevated by natural forces, just as the pillars of the Temple of Serapis had.

Today, technology (satellite measurements) proves Lyell right in principle, showing that some mountains grow by as much as a quarter of an inch a year and have done so for millennia. We also know, as Lyell could not, the mechanism that promotes this—plate tectonics. The Earth's outer shell, known as the lithosphere, comprises seven large, rigid plates and several smaller ones. They are generally about sixty miles thick and lie on top of a hotter rocky layer beneath, known as the asthenosphere. The plates move relative to one another at different rates, from one to three inches a year, said to be around the speed at which human fingernails grow. This movement results in such geological phenomena as earthquakes and creates mountains, volcanoes, and ocean trenches. Mountains rise where plates push together, and continents split and oceans form where they pull apart.

However, in the 1830s, Lyell's views had many detractors. Although many scientists found it easier to debate the age of the Earth than the origins of humankind and by no means all believed literally in the six days of Creation, many geologists still interpreted the Bible as showing that God had shaped the Earth in six or seven stages and criticized Lyell for apparently taking any direct divine intervention out of geology. Those particularly opposed to Lyell included "catastrophists" like French scientist Georges Cuvier, who maintained that the Earth had been shaped by a few traumatic convulsions such as the biblical flood. Adam Sedgwick, who had taught Darwin how to "geologize" in North Wales, believed in the Noachian flood and was for a while just such a catastrophist, maintaining that "the investigations of geology prove that the accumulations of alluvial matter . . . were preceded by a great catastrophe which has left traces . . . in the diluvial detritus . . . spread out over all the strata of the world." However, his own researches, to his surprise, showed that diluvial deposits of gravel and sand were not uniformly distributed as to be expected if created by a single catastrophic event but occurred in layers of distinct types that could only be produced by millennia of geological activity.

In February 1831, ten months before the *Beagle* sailed, Sedgwick had publicly announced his change of heart to the Geological Society and denounced the diluvial theory. However, somewhat illogically, he continued to believe in the flood and was no disciple of Lyell, whose contention that the gradual changes constantly occurring in the Earth's crust were random and divorced from divine oversight were anathema to Sedgwick's religious beliefs as a clergyman. Henslow, another clergyman, doubted Lyell. Though he had recommended Darwin read *Principles of Geology*, he cautioned him "on no account to accept the views therein advocated."

At Cambridge, as part of his bachelor of arts syllabus, Darwin had studied *Natural Theology; or Evidences of the Existence and Attributes of the Deity*, by the clergyman and philosopher William Paley (1743–1805), who had also attended Christ's College and whose college rooms Darwin had occupied for a while. Paley's central thesis—supported both by Henslow, who praised Paley to Darwin, and Sedgwick—was that entities as complex as living creatures must be the work of an intelligent designer. To make his point, Paley used the metaphor of a watch discovered on the ground, suggesting that it could no more have created itself or placed itself there than a living creature could have placed itself upon the Earth. At Cambridge and for a time afterward, Paley's ideas greatly impressed Darwin. Later in life, reflecting on his own intellectual evolution, he remembered: "I do not think I hardly ever admired a book more . . . I could almost formerly have said it by heart."

After three weeks in the Cape Verde islands, FitzRoy had completed his scientific measurements, including establishing the islands' precise longitudes. He had also assessed St. Jago's possible usefulness to British shipping, concluding that the ample supplies of water, chickens, pigs, turkeys, and abundant fruit at low prices coupled with the obliging local authorities made the port "of more consequence . . . than is usually supposed." He now set course for Brazil, from where Darwin planned to ship his first specimens home. In the meantime he busily examined them, often through his small microscope, wrote up his notes meticulously, elaborating upon the brief jottings he had made ashore in pencil in the small, two-by-five-inch leather-backed pads he always carried with him, and kept his diary up to date—practices he continued throughout the voyage. His notes and his diary and the analysis their detail permitted were key to his future work and theories.

He had an array of tools to help him analyze and describe his geological specimens. These included a hand lens, a magnet to check for iron-bearing rocks, and hydrochloric acid (known to him as muriatic acid) to establish whether carbonate was present in the specimens, as in the case of lime-stone and marble. The standard and simple test for hardness was to see if a mark could be made in ascending order with first a thumbnail, then a copper penny, followed by a shard of glass, and lastly a steel knife. He used a device known as a reflective goniometer to measure accurately the angles of crystal faces to allow them to be judged against reference works. He had a blowpipe to intensify and direct flames onto specimens to help their identification by understanding the effects of heat on them. When he was satisfied that he had assembled as much information as possible, he gave each specimen a unique number.

He also found time to write lengthy letters to friends and family ready to be posted when they next reached land or encountered a homeward-bound vessel. In one of the only two letters he sent directly to his father during the voyage, both early in the expedition (his father wrote to him once confirming "your money accounts are all correct"), he assured him that FitzRoy "continues steadily very kind and does everything in his power to assist me.——We see very little of each other when in harbour, our pursuits lead us in such different tracks.——I never in my life met with a man who could endure nearly so great a share of fatigue.——He works incessantly, and when apparently not employed, he is thinking.——If he does not kill himself he will during this voyage do a wonderful quantity of work." Whether this was simply a lighthearted comment about FitzRoy overworking himself or whether Darwin was alluding to FitzRoy's volatile nature or the suicide of his predecessor Pringle Stokes is difficult to judge.

Darwin also wrote that he himself was "incessantly occupied by new and most interesting animals" and that "as yet everything has answered brilliantly, I like everybody about the ship, and many of them very much. The Captain is as kind as he can be. Wickham is a glorious fine fellow. And what may appear quite paradoxical to you is that I *literally* find a ship (when I am not sick) nearly as comfortable as a house.——It is an excellent place for working and reading . . ." Perhaps aiming to dissipate his father's initial skepticism that the voyage would be a waste of time, he added,

"I am thoroughly convinced, that such a good opportunity of seeing the world might not [come] again for a century.—I think . . . I shall be able to do some original work in Natural History."

Darwin's depiction of the comforts of shipboard life was a little rose-hued, probably designedly so to reassure his family. Given that he shared a cabin, sleeping arrangements were a particular challenge in a small space filled not only with the ship's library but instrument cabinets, a large chart table, small chests, and a wash stand. At night, although John Lort Stokes slept in a separate small cubicle, midshipman King slung his hammock on the starboard side of the cabin while Darwin slung his on the port over the chart table. Being six feet tall, the only way to make enough room for his feet when he lay in his hammock was to remove a drawer from his locker. Some nights were so hot in his hammock he felt he was being "stewed in very warm melted butter." The *Beagle* could also be noisy: "Oh a ship is a true pandemonium and the cawkers who are hammering away above my head veritable devils."

Yet what Darwin said about the camaraderie on board was true. Darwin appears to have shown himself amiable and eager to please and fit in, rather than displaying any outward aloofness or sharply marked or strong-willed character traits. Many years later King would recall how "after being a few days at Sea I found a firm friend in the person of Mr. Charles Darwin to whom my fancy was to relate my experiences in my former voyage." Darwin was also becoming popular with the sailors who nicknamed him "the Flycatcher" for all the insects and spiders among his multiplicity of specimens they often helped him carry on board. FitzRoy labeled him the "Ship's Philosopher"—soon shortened to Philos.

Curiously, though he must have frequently encountered them, at this stage in his diary Darwin made no mention of the three Fuegians—Fuegia Basket, in the potentially difficult situation of being the only female on the ship, and her male companions Jemmy Button and York Minster—nor their accompanying missionary Richard Matthews. Neither did he refer to them in letters home. Yet they must have intrigued him, just as had the inhabitants of the Cape Verde islands. The Fuegians were the least "civilized" people he had met and a living experiment of FitzRoy's, having

Top row: Fuegia Basket (Yokcushlu) (left) and Jemmy Button's wife (right); middle row: Jemmy Button (Orundellico); bottom row: York Minster (Elleparu), sketched by Robert FitzRoy.

been exposed by him to many bewildering experiences during their time in England: "Sometimes I took them with me to see a friend or relation of my own who was anxious to question them." The latter had included his sister, Fanny Trevor-Rice, whom they called "Cappen Sisser" and who gave them generous presents and took them shopping in her carriage. The eldest of the three, York Minster, could be surly, but FitzRoy noted that "the two younger ones had become great favourites wherever they were known."

With the support of the Church Missionary Society, FitzRoy had sent them for eleven months to an infant school in Walthamstow on the northeastern outskirts of London run by an evangelical clergyman, where he hoped they would learn "English, and the plainer truths of Christianity . . . and the use of common tools, a slight acquaintance with husbandry, gardening . . ." A few months before the *Beagle* sailed, King William IV, curious to see them, invited them to St. James's Palace. FitzRoy, who accompanied them, described how, charmed by young

Fuegia Basket, Queen Adelaide left the room for a minute and returned
with one of her own bonnets, which she put upon the girl's head. "Her
Majesty then put one of her rings upon the girl's finger, and gave her a
sum of money to buy an outfit of clothes when she should leave England
to return to her own country."

While being dressed in stiff, starchy clothes and being inspected by
London's great and good must have been a bewildering and unsettling
experience, Fuegia Basket and Jemmy Button seem to have adapted to
some aspects of their new life, Jemmy turning into quite a dandy about
his appearance, according to FitzRoy. He was convinced that their exposure to "civilised society" had done them good—even that their features
had been "much improved by altered habits, and by education." As a
believer in phrenology, FitzRoy had all three Fuegians examined. The
detailed analysis submitted to him stated that, among other things, Fuegia Basket was "strong in attachment . . . a little disposed to cunning,
but not duplicity . . . fond of notice and approbation . . . It would not
be difficult to make her a useful member of society in a short time, as
she would readily receive instruction." Jemmy Button would "have to
struggle against anger, self-will, animal inclinations, and a disposition
to combat and destroy" but was "strongly inclined to benevolence" and
"like the female . . . might be made a useful member of society, but it
would require great care, as self-will would interfere much." York Minster was assessed as having "passions very strong, particularly those of
an animal nature . . . disposed to cunning and caution . . . grateful for
kindness but reserved in showing it" but "will be difficult to instruct,
and will require a great deal of humouring and indulgence to lead him
to do what is required."

St. Paul's Rocks (today known as Saint Peter and Saint Paul Rocks), remote,
uninhabited Brazilian islets fifty miles north of the equator, came into
view on February 15. Darwin was among those who went ashore. Despite
his inexperience, as the first geologist to land on them, he identified the
rocks as not of volcanic origin, correctly suggesting, "Is it not the only
island in the Atlantic which is not Volcanic?" In fact, unknown to him,

these islands are one of the few places where an underwater ridge, pushed up by the action of plate tectonic forces, breaks through the surface of the sea. The non-volcanic rock is peridotite.

The crew also intended to shoot for food some of the seabirds whose guano made the rocks a dazzling snowy white. However, the enormous number of boobies—a species of gannet—and noddies—a kind of tern—were, Darwin noted in his diary, "so unaccustomed to men that they would not move," putting "shooting . . . out of the question." Instead the sailors attacked them with anything that came to hand. Darwin's zest, laying about the birds like an excited schoolboy "with all the force of his own right-arm" with his geological hammer, amused FitzRoy. Soon they had collected piles of birds and filled their hats with eggs. Meanwhile other sailors in the ship's boats attempted to fish, but circling sharks grabbed anything they hooked unless they beat the water with their oars.

In his short time ashore Darwin noted what we would now call the ecological or interdependent relationship between the few living things on the island. Large crabs survived by stealing fish brought by the seabirds to their nests. Flies, ticks, and a species of moth lived as parasites on the birds. A beetle and a louse fed on the guano, and "numerous spiders . . . prey on these small attendants and scavengers of the waterfowl . . . The smallest rock in the tropical seas, by giving a foundation for the growth of innumerable kinds of seaweed and compound animals, supports likewise a large number of fish [including] sharks."

On February 17 the *Beagle* crossed the equator, and FitzRoy ordered the firing of the ship's guns—the traditional salute to the Southern Hemisphere. This was also the signal for all the "shellbacks" or "griffins"—as those who had not crossed before were nicknamed—to be summoned before King Neptune, Darwin included. FitzRoy described the raucous ceremony much as Bligh had done forty years earlier aboard the *Bounty*, as a "disagreeable practice," which many thought "an absurd and dangerous piece of folly." Nevertheless, also like Bligh, FitzRoy understood the value for morale: "Its effects on the minds of those engaged in preparing for its mummeries, who enjoy it at the time, and talk of it long afterwards, cannot easily be judged of without being an eye-witness."

Midshipman King described how "the captain received his godship [Neptune] and Amphitrite, his wife, with becoming solemnity; Neptune was surrounded by a set of the most ultrademoniacal looking beings that could be well imagined, stripped to the waist, their naked arms and legs bedaubed with every conceivable colour which the ship's stores could turn out, the orbits of their eyes exaggerated with broad circles of red and yellow pigments. Those demons danced a sort of nautical war dance exulting on the fate awaiting their victims below. Putting his head down the after companion [way] the captain called out 'Darwin look up here!'"

Darwin was the first blindfolded griffin led up from the lower deck by "four of Neptune's constables." As buckets of water "thundered all around," he found himself "placed on a plank, which could be easily tilted up into a large bath of water. They then lathered my face and mouth with pitch and paint, and scraped some of it off with a piece of roughened iron

HMS *Beagle* crossing the line (equator) celebrations, by Augustus Earle

hoop; a signal being given I was tipped head over heels into the water, where two men received me and ducked me." As King noted, Darwin's "high standing on board as a friend and messmate of the captain" meant he could easily have been excused, but instead he "readily entered into the fun." Yet his status might explain why he got off more lightly than most others who had "dirty mixtures . . . put in their mouths and rubbed on their faces.——The whole ship was a shower bath: and water was flying about in every direction: of course not one person, even the Captain, got clear of being wet through."

Darwin congratulated himself on at last being in the Southern Hemisphere and able to gaze at the Southern Cross. He also reflected on how rapidly things had changed: "In August quietly wandering about Wales, in February in a different hemisphere; nothing ever in this life ought to surprise me."

By the evening of February 19, the *Beagle* was lying off the small, densely forested island of Fernando Noronha, its outline "very grand" in the bright moonlight. Shortly before dark the athletic, sharp-sighted Sulivan had flung a harpoon at a large porpoise with such energy that it passed right through its body. Within minutes, the five-foot-long animal was lying on deck, and "a dozen knives were skinning him for supper." Fernando Noronha was only 230 miles off Brazil of which, primed by Humboldt's descriptions of "the grandeur of the Tropics," with its flowers and hummingbirds, Darwin had such high expectations.

CHAPTER FOUR

"Red-Hot with Spiders"

On the morning of February 28, 1832, the *Beagle* sailed into All Saints Bay, on whose steep northern shores lay the Brazilian port of Bahia (São Salvador de Bahia). "It would be difficult to imagine anything so magnificent," Darwin wrote. "The town is fairly embosomed in a luxuriant wood and situated on a steep bank overlooks the calm waters of the great bay . . . The houses are white and lofty and . . . have a very light and elegant appearance." Going ashore he found the town's attractions "as nothing compared to the vegetation; I believe from what I have seen Humboldt's glorious descriptions are and will for ever be unparalleled: but even he with his dark blue skies and the rare union of poetry with science . . . falls far short of the truth . . . if the eye attempts to follow the flight of a gaudy butter-fly, it is arrested by some strange tree or fruit; if watching an insect one forgets it in the stranger flower it is crawling over . . . The mind is a chaos of delight . . ."

He spent the next few days wandering through the Brazilian forest, trying to take in the "general luxuriance" of elegant grasses, parasitical plants, and gaudily gorgeous flowers—"enough to make a florist go wild." Brazilian vegetation was "nothing more nor less than a view in the Arabian Nights, with the advantage of reality." One day during a long walk with King, he collected numerous small beetles, did a little geologizing, and, he mentioned casually in his diary, some shooting—"King shot some pretty birds and I a most beautiful large lizard.—It is a new and pleasant thing for me to be conscious that naturalizing is my duty, and that if I neglected that duty I should at same time neglect what has for some years given me so much pleasure."

Darwin was becoming as much interested in how creatures behaved as in their appearance. One day, he noticed "many spiders, cockroaches, and other insects, and some lizards, rushing in the greatest agitation across a bare piece of ground." A little behind them, blackening "every stalk and leaf," followed an army of ants. Catching up with their prey, the ants divided and enclosed them— "the efforts . . . the poor little creatures made to extricate themselves from such a death, were wonderful."

William Dampier
(1651–1715)

Darwin was not the first British naturalist on a Royal Naval expedition to explore Bahia's rainforests and marshes. In March 1699, William Dampier, commanding HMS *Roebuck*, had also anchored in Bahia. Dampier was a former buccaneer from the west country of England and the first Briton, with his fellow pirates, to land on the mainland of Australia—then known as New Holland—when, in 1689, they came ashore on the west coast near modern-day Broome. By this time, Dampier was transitioning from pirate to naturalist and recording the natural world around him. On his return to England, his account of his twelve-year circumnavigation became a literary sensation, inspiring writers Daniel Defoe and Jonathan Swift, while his observations—in particular of Australia—caught the Admiralty's attention. They gave him the rank of captain, a ship—the HMS *Roebuck*—and orders to return to Australia to assess its possibilities, while the Royal Society asked Dampier to bring back natural history specimens. However, the *Roebuck*'s officers and crew resented serving under a former buccaneer and suspected Dampier of plotting with confederates to seize the ship. Mid-Atlantic, a fistfight broke out on deck between Dampier and his first lieutenant. Dampier put his lieutenant in chains and headed for Bahia to arrange for him to be shipped home for court-martial.

While the Bahia authorities sorted out the paperwork, Dampier roamed the Brazilian forest, notebook in hand like Darwin. A pioneer of descriptive botany and zoology—the careful, detailed, and objective recording of the world's living things—he cataloged Bahia's crops, from indigo to the cotton trees whose pods burst open, revealing lumps of

cotton "as big as a man's head." Following the Royal Society's instructions, he collected a number of plants, which he "dried between the leaves
of books," and carefully described Bahia's animals and reptiles, from
armadillos to alligators. Yet what struck him most was the vivid, varied
birdlife. The four types of "long-legg'd fowls" wading in the swamps
were, he decided, as "near a-kin to each other, as so many sub-species of the
same kind." This was the first ever use of the term and concept *subspecies*
and—though fifty years had passed since Archbishop Ussher's calculation
that the world and all living things had been created in 4004 BC, distinct
from each other and not changed since—still a dangerously radical and
heretical idea.

Dampier's books influenced later naturalists including the wealthy and
well-connected Joseph Banks, who sailed aboard the *Endeavour* from 1768
to 1771 on the first of Captain Cook's three Pacific voyages and who carried Dampier's descriptive approach to nature to new levels. The *Endeavour*
too put in to Brazil, and the large collection of specimens Banks brought
back from there and elsewhere did much to establish botany and zoology
as sciences, not least because Banks soon became president of the Royal
Society—a position he held for forty-one years—and was instrumental
in the founding of Kew Gardens. He influenced almost every important
British scientific venture or overseas expedition in this period, including
the first fleet to Australia and Bligh's breadfruit voyage to Tahiti. He had
only died in 1820.

Darwin was of course well aware of Banks but also of the earlier
Dampier, whom his idol Humboldt lauded as "the remarkable English
buccaneer" to whose works "the subsequent studies of great European
scholars, naturalists and travellers had added little." Darwin had Dampier's
account of his first circumnavigation with him and referred to him familiarly as "old Dampier."

When not in the forests, Darwin explored the town. The tall houses
and narrow streets down by the wharfs reminded him of Edinburgh's old
town. The air stank, just as it did in Edinburgh. He wrote, "I observe
they have the same need of crying 'gardez l'eau' as in Auld Reekie"—a
Scottish nickname for Edinburgh, meaning literally "Old Smoky," where

to warn of chamber pots being emptied from windows into the streets below, for centuries people had shouted a warning borrowed from French, *gardez l'eau*—"watch out for the water." He also noticed how black slaves did all the hard work. Though warned of the dangers of walking through the streets on the riotous first day of Carnival, marking the start of Lent, Darwin, Wickham, and Sulivan spent "an hours walking the gauntlet," being pelted relentlessly "by wax balls full of water" and soaked through "by large tin squirts." Two days later, Darwin's knee became badly swollen, which he attributed to having "pricked" it. Over following days it became so painful and infected he was confined to his hammock.

He used the time to reflect on slavery in Brazil. In Britain slavery had never been authorized by statute. In 1772, in a historic judgment, Lord Chief Justice Lord Mansfield had ruled slavery illegal in Britain itself under common law and "so odious that nothing can be suffered to support it" and freed a slave who had reached Britain but whose "owner" was trying to reclaim him. However, Britain had, of course, long been active in the slave trade to its American and Caribbean colonies and during the eighteenth century transported more Africans than any other maritime nation.

So powerful were the vested commercial interests that the abolitionist movement in Britain struggled to make progress. Though the politician and social reformer William Wilberforce first brought the subject before Parliament in 1788, trading in slaves was not made illegal in British dominions until 1807—a move reinforced in 1811 when trafficking humans became a capital offense and the British navy began patrolling the Atlantic waters off Africa, South America, and the Caribbean to prevent maritime slave trading by other nations. In 1830, for example, the navy captured twenty-four slave ships and freed more than seven thousand slaves. Freedom for those already enslaved in British possession would not come until 1833, while the *Beagle* was still at sea and the year Wilberforce died. Dom Pedro I of Brazil had accepted the abolition of slave trading in a treaty with Britain in 1828 and thus permitted British anti-slave vessels to use his country's harbors. Nevertheless, Brazil would become the last country in the Americas to abolish the owning of slaves when it did so in 1888. Historians estimate four

million Africans—40 percent of the total brought to the Americas—were
transported as slaves to Brazil from Africa.

Descended from vigorously abolitionist stock on both sides of his
family—his Wedgwood cousins attended anti-slavery rallies; his grandfather, Josiah Wedgwood I, designed a pottery plaque of an African slave
in chains under the legend "Am I Not a Man and a Brother?" and with
Darwin's other grandfather, Erasmus Darwin, helped fund the abolitionist cause—the plight of Bahia's slaves horrified Darwin. In his diary,
he described how Captain Paget of the British warship HMS *Semarang*,
anchored in the harbor of Bahia at the end of a six-month anti-slavery
patrol, paid frequent visits to the *Beagle*. There he

> mentioned in the presence of those who would if they could
> have contradicted him, facts about slavery so revolting, that if
> I had read them in England, I should have placed them to the
> credulous folly of well-meaning people: The extent to which
> the trade is carried on; the ferocity with which it is defended; the
> respectable (!) people who are concerned in it are far from being

Anti-slavery medallion produced by Josiah Wedgwood,
grandfather of both Charles and Emma Darwin, at the
Wedgwood potteries

exaggerated at home. I have no doubt the actual state of by far the greater part of the slave population is far happier than one would be previously inclined to believe. Interest and any good feelings the proprietor may possess would tend to this.——But it is utterly false . . . that any, even the very best treated, do not wish to return to their countries.——"If I could but see my father and my two sisters again, I should be happy. I can never forget them." Such was the expression of one of these people, who are ranked by the polished savages in England as hardly their brethren, even in God's eyes.——From instances I have seen of people so blindly and obstinately prejudiced, who in other points I would credit, on this one I shall never again scruple to disbelieve . . .

Darwin would never waver in his view that humanity was a single species.

Darwin's reference to blindly prejudiced people included FitzRoy, who, he had discovered, "when out of temper . . . was utterly unreasonable" and with whom he had just had a violent argument about slavery. The memory remained sufficiently vivid for him to describe it in detail forty years later in his autobiography:

Early in the voyage at Bahia in Brazil [FitzRoy] defended and praised slavery, which I abominated, and told me that he had just visited a great slave-owner, who had called up many of his slaves and asked them whether they were happy, and whether they wished to be free, and all answered "No." I then asked him, perhaps with a sneer, whether he thought that the answers of slaves in the presence of their master was worth anything. This made him excessively angry, and he said that as I doubted his word, we could not live any longer together.

Darwin was perhaps not being entirely fair to FitzRoy, or FitzRoy failed to explain his views in their entirety. Though he indeed thought that in general the Brazilians treated their slaves "humanely," as he himself later wrote, FitzRoy neither admired nor defended what he saw in Brazil. He believed the presence of so many slaves had made the Brazilians extremely indolent and was stifling economic growth. Furthermore, the ever-growing slave population would surely one day turn on them. The remedy was for the Brazilians to "emancipate the slaves now in their country and decidedly prevent the introduction of more." As for the slave trade itself, FitzRoy denounced it as an "abominable traffic" that should be suppressed.

Having quarreled with his former "beau ideal" of a captain, Darwin feared "I should have been compelled to leave the ship; but as soon as the news spread, which it did quickly, as the captain sent for the first lieutenant to assuage his anger by abusing me, I was deeply gratified by receiving an invitation from all the gun-room officers to mess with them." As frequent victims themselves of FitzRoy's biting tongue, the officers had every sympathy with Darwin. When the junior officers relieved one another on duty, they would ask "whether much hot coffee had been served out," meaning "how was the Captain's temper?"

In the event, Darwin had no need to rely on the officers' hospitality. "After a few hours Fitz-Roy . . . [sent] an officer to me with an apology and a request that I would continue to live with him." FitzRoy clearly did not want to lose the traveling companion, who, he wrote to Beaufort from Bahia, was "a very sensible, hard-working man and a very pleasant messmate. I never saw a 'shore-going fellow' come into the ways of a ship so soon and so thoroughly as Darwin. I cannot give a stronger proof of his good sense and disposition than by saying 'Everyone respects and likes him.'" Darwin had indeed fitted in remarkably quickly with his ship-mates. In later years John Lort Stokes affectionately recalled his "cheery companionship" and his favorite expressions—"by the Lord Harry" and "beyond belief."

Shortly before the *Beagle* was due to leave, Darwin took a final stroll with Philip King. The bright, clear evening was perfect for "fixing in the mind the last and glorious remembrances of Bahia." Yet this Garden of Eden was tainted. "If to what Nature has granted the Brazils, man added his just and proper efforts, of what a country might the inhabitants boast. But where the greater parts are in a state of slavery, and where this system is maintained by an entire stop to education, the mainspring of human actions, what can be expected; but that the whole would be polluted by its part."

The *Beagle* headed next for the small, rocky, uninhabited Abrolhos Archipelago of islands, some 350 miles south of Bahia, where Beaufort had instructed FitzRoy to determine the exact longitude and to take soundings around its shallow shoals. Thousands of gannets, frigate, and tropic birds rose as members of the crew rowed ashore, intent on slaughtering as many as they could for food. Darwin examined the islands' geology—telling FitzRoy they were formed of "gneiss and sandstone in horizontal strata"—and noted the huge numbers of lizards and spiders as well as the rats that had run ashore from visiting ships to flourish on the ready food provided by "nests full of eggs, or young unfledged birds [that] absolutely covered the ground."

With his orders fulfilled, FitzRoy set course for Rio de Janeiro. April 1, 1832—April Fool's Day—was another occasion, like crossing the equator, for shipboard high jinks. Darwin was amused that "at midnight nearly all the watch below was called up in their shirts; Carpenters for a leak: quartermasters that a mast was sprung.—midshipmen to reef top-sails; All turned in to their hammocks again, some growling, some laughing." He was included in the tomfoolery—another sign of his popularity and approachability. Sulivan set the bait, crying out, "'Darwin, did you ever see a Grampus: Bear a hand then.' I accordingly rushed out in a transport of Enthusiasm, and was received by a roar of laughter from the whole watch." No one seems to have been foolhardy enough to attempt to make an April Fool out of FitzRoy.[1]

As the *Beagle* approached Rio, Darwin continued systematically cataloging his growing specimen collection, numbering and recording the identity of each, where he had found it, and—where relevant—what liquid it was preserved in, usually "Spirits of Wine." Despite the cramped conditions, he was still finding living aboard "a most excellent time for

---

1. "Grampus"—a corruption of *gran pisce*—"big fish"—was maritime slang for large sea creatures.

all sorts of study." Interruptions were few unless they were to watch the shipping in the busy seas off Brazil or the many sharks, porpoises, and turtles in the clear, warm waters.

On April 4, the *Beagle* passed beneath the Sugarloaf Mountain, or Pão de Açúcar, into Rio's island-dotted harbor, Guanabara Bay, that extended nineteen miles inland. On January 1, 1502, Portuguese navigators had mistaken the harbor entrance for the mouth of a great river, which they named Rio de Janeiro, or "River of January." Sixty years later a town— São Sebastião do Rio de Janeiro, named for St. Sebastian—was founded on Guanabara Bay's western shore. The beautiful situation—broad white sand, blue seas in front, the domed granite and gneiss Serra do Mar mountains behind—moved a visiting Portuguese Jesuit, Fernão Cardim, to comment that it "appears to have been devised by the supreme painter and architect of the world, Our Lord God." Rio grew rich on gold and diamond mining and in 1763 became the capital of Portugal's Brazilian empire. In 1822, just ten years before the *Beagle* arrival, Brazil had gained independence from Portugal, with Rio as the national capital.

Darwin stared at a city "gaudy with its towers and Cathedrals" and a bay "studded with men of war, the flags of which bespeak every nation." He recorded with some pride how "in most glorious style," the *Beagle* lowered her sails beside the seventy-four-gun third-rate British flagship HMS *Warspite*, a veteran of the Napoleonic Wars. The Royal Navy maintained a squadron of a dozen ships in Rio—headquarters of its South American station—ready both to protect British interests in "those disturbances almost usual in South America, especially in Brazil," as FitzRoy wrote, but also to undertake anti-slavery patrols. Enthused by the *Beagle* crew's "beautiful order and discipline," Darwin joined in. King recalled how "Mr Darwin was told to hold to a main royal sheet in each hand and a top mast studding sail tack in his teeth. At the order 'Shorten Sail' he was to let go and clap on to any rope he saw was short-handed—this he did and enjoyed the fun of it, afterwards remarking 'the feat could not have been performed without him.'" Though his comrades were doubtless humoring him, Darwin clearly saw himself as a sailor, writing to his sister Caroline that he was "becoming one, ie. knowing ropes and how to put the ship about etc. . . ."

Four months had passed since the *Beagle* ploughed her way past the Plymouth breakwater. Like everyone on board, Darwin anticipated "the ecstasies of opening letters" as mail was brought aboard. But naval discipline intervened—"Send them below," thundered Wickham, "every fool is looking at them and neglecting his duty." Darwin had to wait an hour. When he did receive his post he found a letter from Fanny Owen informing him of an "awful and important event"—her marriage to Myddleton Biddulph, owner of a grand estate, Chirk Castle. She had become engaged within days of the *Beagle*'s departure, and by the time her letter reached Darwin she was already married. Anxious and affectionate letters from his sisters also broke the news. Catherine Darwin wrote, "I hope it won't be a great grief to you, dearest Charley." He was indeed wounded, writing to Caroline Darwin, "if Fanny was not perhaps at this time Mrs Biddulph I would say poor dear Fanny till I fell to sleep . . . I am at a loss what to think or say; whilst really melting with tenderness I cry [out] my dearest Fanny . . . I find my thoughts and feelings and sentences are in such a maze, that between crying and laughing I wish you all good night."

Next day Darwin went ashore with the ship's artist, Augustus Earle, who knew Rio, having lived there awhile—to look for lodgings for a few weeks while FitzRoy concentrated on his survey work. They found "a most delightful house" with a garden "overwhelmed by flowers" on Botafogo Bay about five miles from the center of the city and where many European merchants had their fine mansions. This was a much more difficult journey then than now when tunnels have been dug through the hills that previously had to be circumvented on horseback. Conscious of what he was costing his father, Darwin calculated his board and lodging would be only twenty-two shillings a week. Midshipman King was to share the house with them. He and Darwin swiftly set to collecting butterflies and catching beetles by "hanging an umbrella upside down in the bushes and shaking the branches."

Darwin soon began preparing for a trip 150 miles on horseback to a large estate to the north on the Rio Macaé that belonged to an Irish merchant, Patrick Lennon, who had lived in Rio for twenty years and made a fortune selling spectacles and thermometers. Lennon had agreed to let

Darwin join the small group he was conducting there. Brazil could seldom have seen such an "extraordinary and quixotic set of adventurers," Darwin thought on meeting his six companions, who included Lennon's nephew, "a sharp youngster" already "making money," and a "clever Scotsman, *selfish unprincipled* . . . by trade partly Slave-Merchant partly Swindler."

The journey required a permit to travel from the Brazilian authorities. "It is never very pleasant to submit to the insolence of men in office . . . But the prospect of wild forests tenanted by beautiful birds, Monkeys and Sloths and Lakes by Cavies and Alligators, will make any naturalist lick the dust even from the foot of a Brazilian," Darwin consoled himself. And his submission was indeed worth it. Riding through the Brazilian forest, Darwin noted the "large and brilliant butterflies, which lazily fluttered about," graceful tree ferns and cabbage palms, thick twisting creepers, and "enormous conical ants' nests . . . nearly twelve feet high." In places the vegetation was so thick and matted that "we were obliged to have a black man to clear the way with a sword." They passed a steep granite hill on top of which were the remains of grass huts built, Darwin learned, by runaway slaves who had "contrived to eke out a subsistence" there until soldiers were sent to recapture them. One woman had jumped to her death rather than be taken prisoner. "In a Roman matron this would have been called the noble love of freedom: in a poor negress it is [called] mere brutal obstinacy," Darwin reflected.

At night the party halted in ramshackle hostelries called *vendas*, built of "thick upright posts, with boughs interwoven, which are afterwards plastered" with a veranda at the front. This accommodation—all that was available—was cheap, but guests had to sleep on thin straw mats. Darwin thought most of the owners disagreeable and "often filthily dirty," just as their houses were. Getting something to eat could be problematic—"it not infrequently happens that the guest is obliged to kill with stones the poultry, for his own dinner." Even then there was often no cutlery to eat it with, as Darwin, used to comfortable, well-run, well-equipped houses, observed with mingled surprise and disdain. Arriving at one venda near dusk, he felt so "miserably faint and exhausted" that he feared he might fall from his horse. "All night felt very unwell: it did not require much imagination to paint the horrors of illness in a foreign country, without being able to speak one word or obtain any medical aid." He cured himself with a mixture of cinnamon and port wine.

As the journey continued, the abundant crops cultivated on land "cut out of the almost boundless forest"—coffee, the most profitable, but also cassava used to produce tapioca, beans, sugarcane, and, in swampier areas, rice—impressed Darwin as he gathered specimens and made notes. Cattle, goats, sheep, and horses grazed on cleared pastureland, and he admired the efficiency of a sawmill, where felled trees were cut into thick planks, ready to be floated downriver. "If many were to imitate the example [of clearing land] . . . what a difference a few years would produce in the Brazils," Darwin wrote. Later, elsewhere in South America, he would perceive the dangers of deforestation, but here the forests seemed infinite.

They teemed with game. Dining at the house of a wealthy friend of Lennon's proved a very different experience from the sparse provender of the grubby vendas: ". . . if the tables do not groan, the guests surely do.—Each person is expected to eat of every dish; . . . having, as I thought, nicely calculated so that nothing should go away untasted, to my utter dismay a roast turkey and a pig arrived in all their substantial reality." He also noted "it was the employment of a man to drive out sundry old hounds and dozens of black children which together at every opportunity crawled in.—As long as the idea of slavery could be banished, there was something fascinating in this simple and patriarchal style of living.—It was such a perfect retirement and independence of the rest of the world.—As soon as any stranger is seen arriving, a large bell is set tolling and generally some small cannon are fired."

When the party finally reached Lennon's estate, Darwin's vision of peaceful, patriarchal communities evaporated as "a most violent and disagreeable quarrel"—Darwin never identified the cause—erupted between Lennon and his estate manager, during which Lennon "threatened to sell at the public auction an illegitimate mulatto child" to whom the agent was "much attached"—perhaps his son. Lennon also "nearly put into execution taking all the women and children from their husbands and selling them separately at the market at Rio. Can two more horrible and flagrant instances be imagined? And yet I will pledge myself that in humanity and good feeling Mr. Lennon is above the common run of men. How strange and inexplicable is the effect of habit and interest! Against such facts how weak are the arguments of those who maintain that slavery is a tolerable evil!"

Soon after, he again saw the evil firsthand. While crossing a river on a ferry manned by a slave, he gestured with his hands to show the man where he wished to go: "Instantly, with a frightened look and half-shut eyes, he dropped his hands. I shall never forget my feelings of surprise, disgust, and shame, at seeing a great, powerful man afraid even to ward off a blow, directed, as he thought, at his face. This man has been trained to a degradation lower than the slavery of the most helpless animal."

By late April, after a return journey through forest where Darwin admired "the extreme elegance of the leaves of the ferns and mimosae" shaded by lofty trees, he was back aboard the *Beagle* with his specimens, including "many insects and reptiles." The following day, as he moved his possessions by boat from the ship to his lodgings in Botafogo Bay, Darwin experienced "the horrors of shipwreck.—Two or three heavy seas swamped the boat, and before my affrighted eyes were floating books, instruments and gun cases and everything which was most useful to me." Luckily nothing was lost or damaged beyond repair.

Towering over Botafogo from behind his lodgings was the 2,329-foot steep granite mountain named Corcovado, "hunchback," for its shape, atop which Christ the Redeemer now stands. Darwin climbed it twice with companions from the *Beagle*—no easy task since on one side it was "so precipitous, that it might be plumbed with a lead." The astonishing view from the summit made him "suppose that the view from a Balloon would be exceedingly striking."[2]

---

2. The first manned flight by hot air balloon was half a century earlier in 1783. Designed by the French Montgolfier brothers and powered by a wood fire, the balloon flew some five miles. Among those watching was Benjamin Franklin, who, when a French officer derided the balloon as a mere toy, amusing but useless, replied, "Of what use is a new-born baby?"

Darwin had heard that the Corcovado was known for the runaway slaves hiding out there. On his first climb, with Augustus Earle and the *Beagle*'s first mate Alexander Derbyshire, they encountered three slave hunters—"most villainous-looking ruffians, armed up to the teeth" who received "so much for every man dead or alive whom they may take." In the former case, all they had to do to be paid was to produce the dead slave's ears. A disgusted Darwin wrote in his diary: "Amongst other things which the anti-abolitionists say, it is asserted that the freed slave would not work," but "I repeatedly hear of run-away ones having the boldness of working for wages in the neighbourhood of their masters. If they will thus work when there is danger, surely they likewise would when that was removed . . . What will not interest or blind prejudice assert, when defending its unjust power or opinion?"

Darwin explored the large Royal Botanic Garden beneath the Corcovado, laid out in 1819. Most plants seemed to him to be grown for their utility rather than their interest, among them acres of tea trees—"an insignificant little bush with white flowers" whose leaves, when infused, "scarcely possessed the proper tea flavour." Yet the leaves of camphor, cinnamon, clove, and pepper trees "had a delightful aromatick taste and smell." He also identified two handsome species of fruit tree that he had admired around Bahia without knowing their names: jackfruit and mango. Of the dense-leaved mango he wrote, "I had no idea any tree could cast so black a shadow."

In a long letter to his sister Caroline, Darwin told her he was sending a packet containing his diary but warned, "I have taken a fit of disgust with it and want to get it out of my sight, any of you that like, may read it.—a great deal is absolutely childish: Remember . . . that it is written solely to make me remember this voyage and . . . is not a record of facts but of my thoughts.—and in excuse recollect how tired I generally am when writing it." He wrote he was giving both letter and diary to the ship's surgeon, Robert McCormick, who was "returning to England, being invalided i.e. being disagreeable to the Captain and Wickham." "He is no loss," he added of the man to whose goodwill he was entrusting his diary and whom, even before the voyage began—throughout his life more waspish in writing than in person—he had described to Henslow as "an ass."

McCormick had wasted no time after arriving in Rio in trying to transfer to another naval ship. When that failed, he asked to be invalided home, and his request was granted. As well as any disagreement with FitzRoy and Wickham, his desire to leave the *Beagle* reflected his uneasy relationship with Darwin. Even if he did not show it publicly, Darwin privately thought McCormick's methods of working antiquated, writing to Henslow that this "philosopher of rather an antient [*sic*] date . . . at St. Jago by his own account . . . made *general* remarks during the first fortnight and collected particular facts during the last." McCormick, in turn, resented Darwin as the cuckoo who had flung him out of his rightful nest—that of ship's naturalist. In his memoirs, written many years later, he claimed: "Having found myself in a false position on board a small and very uncomfortable vessel, and very much disappointed in my expectations of carrying out my natural history pursuits, every obstacle having been placed in the way of my getting on shore and making collections, I got permission from the admiral in command of the station here [Rio] to be superseded and allowed a passage home . . ." He regretted "so much time, health, and energies utterly wasted." The *Beagle*'s assistant surgeon, the equable twenty-eight-year-old Benjamin Bynoe, filled his place.

If Darwin was dismissive of McCormick, he wrote enthusiastically to his family of other *Beagle* companions—the fourteen-year-old King was "the most perfect, pleasant boy I ever met with and is my chief companion," while Wickham was "a fine fellow—and we are very good friends." However, following their recent quarrel, what he now wrote about FitzRoy was more nuanced, indeed more critical, than previously. FitzRoy was no longer his "beau ideal" although still

a very extraordinary person.—I never before came across a man whom I could fancy being a Napoleon or a Nelson.—I should not call him clever, yet I feel nothing is too great or too high for him.—His ascendancy over everybody is quite curious: the extent to which every officer and man feels the slightest rebuke or praise would have been, before seeing him, incomprehensible. It is very amusing to see all hands hauling at a rope they not supposing him on deck, and then observe the effect, when he utters a syllable: it is like a string of dray horses, when the waggoner gives one of his aweful smacks.——His candor and sincerity are to me unparalleled: and using his own words his "vanity and petulance" are nearly so.——I have felt the effects of the latter: but the bringing into play the former ones so forcibly makes one hardly regret them.——His greatest fault as a companion is his austere silence: produced from excessive thinking: his many good qualities are great and numerous: altogether he is the strongest marked character I ever fell in with.

Darwin was not, of course, alone in suffering from FitzRoy's brooding personality. Yet FitzRoy was in many ways not an inconsiderate captain. He issued standing orders reflecting his view that "attention to duty . . . must never be relaxed but while duty is properly done I shall always be anxious to increase as much as possible the few comforts that a small vessel can possess." While the *Beagle* was in port, the crewmen should be allowed three-quarters of an hour for breakfast and supper and an hour and a half for dinner. While at sea, they should have half an hour for breakfast and supper and an hour for dinner. The off-duty watch should not "be disturbed without *absolute necessity* or on the order of the commanding officer." Any men away from the vessel without provisions——on survey work in a ship's boat, for example——could expect hot food on their return. The crew could sit, read, and smoke on any part of the *Beagle*'s upper deck, except the quarterdeck, and, while off-duty, wear what they wanted. Conscious that frock coats and cocked hats were uncomfortable and impractical on a small vessel, FitzRoy allowed his junior officers "to avoid the use of them as much as possible."

Darwin reassured his sister that whatever its vagaries, "I like this sort of life very much: I can laugh at the miseries of even Brazilian travelling," except for one morning "when I did not get my breakfast till one o'clock having ridden many miles over glaring sand." But he also assured her——doubtless aware his letter would be read by or read aloud to their father——that "although I like this knocking about, I find I steadily have a distant prospect of a very quiet parsonage, and I can see it even through a grove of Palms." Perhaps he really was still contemplating a tranquil future as a country clergyman. He was a regular attender at shipboard divine service and even sought out Anglican churches when ashore. In his autobiography he wrote, "Whilst on board the *Beagle* I was quite orthodox, and I remember being heartily laughed at by several officers (though themselves orthodox) for quoting the Bible as an unanswerable authority on some point of morality." At this stage in his life, he shared both his religious views and an enthusiasm for science with FitzRoy. What divided them was politics.

On May 3, he and FitzRoy observed the inspection of the *Warspite* by Admiral Sir Thomas Baker, commander in chief of the South American station and——while the *Beagle* was in the region——FitzRoy's commanding officer. It involved the staging of a simulated attack. Four hundred seamen manned the *Warspite*'s yardarms——"from the regularity of their movements and from their white dresses, the men really looked more like a flock of wild-fowl than anything else." Darwin thought it an impressive sight: "Everything is done precisely the same as if she was engaged with an enemy . . . One almost wished for an enemy, when the aweful words were shouted in the great batteries below——'Clear for Action . . .' the most glorious thing was when the Bugle gave the signal for the Boarders; the very ship trembled at so dense a body rushing a long [the decks] with their drawn cutlasses."

On May 10, the *Beagle* weighed anchor to return north to Bahia because, to his consternation, FitzRoy had spotted "a difference, exceeding four miles of longitude . . . between the meridian distance from Bahia to Rio" on some French charts compared to his own measurements. Checking and rechecking, he could find no mistake or oversight on his part and decided his only option was to take fresh measurements. He would prove himself right and the French charts wrong and, despite acting without formal Admiralty authority, be commended for his initiative.

Instead of sailing with the *Beagle*, Darwin remained in Rio to build his collection of specimens. Shortly before leaving, FitzRoy told him that three of a small party who had taken the *Beagle*'s cutter up the Rio Macacu to go snipe shooting had contracted high fevers. Darwin congratulated himself that he had not joined them but was soon unwell again himself— this time with an inflamed arm. "Any small prick is very apt to become in this country a painful boil," he wrote. Earle was suffering agonies of rheumatism, and they were joined by another invalid, Beazeley, the *Beagle*'s sergeant of marines. FitzRoy was also leaving behind in Rio, in the care of an English family, "Miss Fuegia Basket, who daily increases in every direction except height," Darwin wrote—his first reference to her in his diary. A quick learner, she soon increased her knowledge of Portuguese.

Once recovered, Darwin began collecting again—"The naturalist in England enjoys in his walks a great advantage . . . in frequently meeting with something worthy of attention; here he suffers a pleasant nuisance in not being able to walk a hundred yards without being fairly tied to the spot by some new and wondrous creature." He noticed how, just as in England, the odor of fungi attracted beetles, musing, "We here see in two distant countries a similar relation between plants and insects of the same families, though the species of both are different." A butterfly "that uses its legs for running" astonished him, as did a fight between a wasp and a large spider.

Darwin wrote to Henslow, "I am at present red-hot with Spiders, they are very interesting, and if I am not mistaken, I have already taken some new genera.—I shall have a large box to send very soon to Cambridge." He added, in an indication that he and FitzRoy sometimes discussed politics, "The Captain does everything in his power to assist me . . . but I thank my better fortune he has not made me a renegade to Whig principles: I would not be a Tory, if it was merely on account of their cold hearts about that scandal to Christian Nations, Slavery." He asked Henslow to remember him to Sedgwick, who "does not know how much I am indebted to him for the Welch expedition—it has given me an interest in geology which I would not give up for any consideration." To his friend and cousin William Darwin Fox he wrote, "My mind has been since leaving England in a perfect *hurricane* of delight and astonishment . . . Geology carries the day; it is like the pleasure of gambling, speculating on arriving what the rocks may be; I often mentally cry out 3 to one Tertiary against primitive . . ."

When the *Beagle* returned to Rio in early June, she brought "most calamitous news"—all three men taken ill on the snipe shooting expedition were dead. They included twelve-year-old volunteer first class Charles Musters, with whom Darwin had taken walks. "My poor little friend," wrote FitzRoy of the boy who had learned of his mother's death just two days before falling ill himself. FitzRoy noted that the snipe hunters had camped in a place notorious for "pestilential malaria"—"As far as I am aware, the risk, in cases such as these, is chiefly encountered by sleeping on shore, exposed to the air on or near the low banks of rivers, in woody or marshy places subject to great solar heat. Those who sleep in boats, or under tents, suffer less . . ." He speculated whether the cause of disease was "a vapour, or gas, formed at night or a failure to sweat properly."[3]

As the *Beagle* prepared to depart, FitzRoy reorganized the stores— some tins of preserved meat had been damaged during rough weather and were jettisoned—and rearranged the guns to accommodate two extra brass nine-pounders, newly purchased at his own expense, bringing the total number of cannons onboard to the nine that he, unlike the Admiralty, thought necessary. However, he hoped he would never have to fire them for fear of disrupting the delicate functioning of his twenty-two chronometers. He also recruited several crewmen from other British naval vessels stationed in Rio to replace losses. Going aboard, Darwin found the ship in the "same inextricable confusion which she was in in Plymouth." He was dismayed "to see so many new faces on the deck."

The knowledge that the *Beagle*—"a floating prison" he now called it, albeit affectionately—would soon sail made him feel like a schoolboy whose holidays were ending. He made farewell visits to his favorite spots—a lake to watch "for the last time its waters stained purple by the last rays of twilight" and a spot where he lay to "watch the setting sun gild the bare side of the Sugar Loaf." He also reflected on Brazilian

---

3. The link between mosquitoes and malaria would only be shown in 1897 by Sir Ronald Ross, a doctor in the British Indian Medical Service, for which he received the Nobel Prize for medicine in 1902, becoming the first British Nobel Laureate.

society—"Everybody can here be bribed.—A man may become a sailor or a physician or any profession, if he can afford to pay sufficiently.—It has been gravely asserted by Brazilians that the only fault they found with the English laws was that they could not perceive rich respectable people had any advantage over the miserable and the poor."

He characterized Brazilians as "ignorant, cowardly, and indolent in the extreme; hospitable and good-natured as long as it gives them no trouble; temperate, revengeful, but not quarrelsome . . . Their very appearance bespeaks their little elevation of character.—figures short, they soon become corpulent." To Darwin, raised as a low church Protestant, the Catholic monks were the worst, their faces clearly revealing "persevering, cunning, sensuality and pride." While Brazilian matrons, "surrounded by slaves . . . become habituated to the harsh tones of command and the sneer of reproach . . . they are born women but die more like fiends." Some kept thumb screws in the house to punish their slaves—"Mr Earle has seen the stump of the joint . . . wrenched off . . ."

In contrast to their short, fat masters, the black slaves had impressive athletic figures. Darwin thought "their intellects have been much underrated.—they are the efficient workmen in all the necessary trades . . ." Their leading characteristics were "wonderful spirits and cheerfulness, good nature and a 'stout heart' mingled with a good deal of obstinacy.—I hope the day will come when they will assert their own rights and forget to avenge their wrongs."

By June 28, Darwin was installed "once again in the intricacy of my own corner . . . It is something quite cheering to me to hear the old noises.—the men foreward singing; the sentinel pacing above my head and the little creaking of the furniture in the cabin . . ." Several days later, he attended divine service on the *Warspite* and admired the sight of 650 men simultaneously removing their hats to sing "God save the King." Chauvinistic Briton of his time, he wrote, "Seeing, when amongst foreigners, the strength and power of one's own Nation, gives a feeling of exultation which is not felt at home.—This ship would be in exactly the same state, if she was going to fight another battle of Trafalgar . . . Can one wonder at pride in the Captain, when he knows that all and everything bends to his will." Dining in the *Warspite*'s wardroom, the "comforts and luxuries," compared to the *Beagle*, made him "a little envious." The rendering by the ship's band of the overture from Rossini's *The Barber of Seville*, which premiered in Rome in 1816, made him realize he was missing music.

In a hastily scribbled letter to his sister Catherine dispatched just before sailing, Darwin predicted he would probably suffer badly from seasickness on the next leg and asked for books, including his copy of Humboldt's *Tableaux de la Nature*, to be sent out to him—he had told family and friends that for the next eighteen months Montevideo would be the best place to write to him—adding, "You cannot imagine what a miser-like value is attached to books, when incapable of procuring them." He never went on a shore expedition without a book and, when only able to carry one, invariably chose his well-thumbed copy of *Paradise Lost*. John Milton, a fellow alumnus of Christ's College, Cambridge, wrote the epic poem, first published in 1667, to "justify the ways of God to men." It depicted the rebellion of Lucifer, most beautiful of the angels, and how, in the new guise of Satan, he subverted the first humans created by God—Adam and Eve. Milton's description of Creation—

> . . . The Earth obey'd, and strait
> Op'ning her fertil Woomb, teem'd at a Birth
> Innumerous living Creatures, perfet formes
> Limb'd and full grown . . .

—of course differed greatly from the theory Darwin would adopt. However, Milton's view of science was modern for his time; the only scientist he mentioned in the poem is the astronomer Galileo, persecuted for his views only three decades earlier. But Milton's republican, anti-tyrannical beliefs, which led him to support Cromwell in England's civil war, differed little from those of Darwin's grandfather Erasmus, and his austere Puritan faith matched that of the Wedgwoods. Nevertheless, for Darwin to make *Paradise Lost* his favorite work must have relied much more on his appreciation of the story of its tragic, romantic anti-hero Satan and the pace and majestic rhyme of the poetry.

Darwin also again asked his sister for news about the Reform Bill, since "we are all very anxious about reform." However, uppermost in his mind was not parliamentary reforms and a widening of the franchise but what lay ahead. As he told Catherine, "I long to put my foot, where man has never trod before . . ."

On July 5, 1832, three months after first arriving in Rio de Janeiro and with a gentle breeze rippling the water, FitzRoy set course for Montevideo on the River Plate in Uruguay. Sailors high in the rigging of HMS *Warspite*, riding at anchor, cheered the *Beagle* while the *Warspite*'s band played "To Glory You Steer." FitzRoy was gratified: "Strict etiquette might have been offended at such a compliment to a little ten-gun brig . . . unless she were going out to meet an enemy, or were returning into port victorious: but although not about to encounter a foe, our lonely vessel was going to undertake a task laborious, and often dangerous . . ."

CHAPTER FIVE

"Gigantic Land Animals"

"The weather has been most provoking," wrote Darwin as almost immediately his predictions of seasickness came true. For a week he could barely stir to view a passing whale. However, by July 13, the sea was smooth, the sun bright, and his enthusiasm revived, particularly as he anticipated seeing the remote southern regions of Patagonia, where the *Beagle* was bound after calling at Montevideo.

He was not alone: "Everybody is full of expectation and interest about the undescribed coast of Patagonia," he recorded in his diary. "Endless plans are forming for catching Ostriches, Guanaco, Foxes . . . I believe the unexplored course of the Rio Negro will be investigated.——What can be imagined more exciting than following a great river through a totally unknown country?——Every thing shows we are steering for barbarous regions, all the officers have stowed away their razors, and intend allowing their beards to grow in a truly patriarchal fashion."[1] He wrote to his sister Susan that, "my face at present looks of about the same tint as a half-washed chimney sweeper.——With my pistols in my belt and geological hammer in hand, shall I not look like a grand barbarian? . . . I expect grand things in Natural History . . . the only thing unpropitious is the ferocity of The Indians .——But I would sooner go [upriver] with the Captain with 10 men than with anybody else with 20.——He is so very prudent and watchful . . . and so resolutely brave when pushed to it."[2]

---

1. Throughout the voyage in his diary and notes Darwin referred to the large flightless birds he saw in South America as "ostriches." They were actually rheas, which are related to ostriches and emus. "Ostrich" has been retained when quoting Darwin, but "rhea" is used elsewhere.
2. Darwin, and others, described the indigenous peoples he encountered in South America as "Indians." I have retained the term but in quotation marks.

Patagonia would indeed be quite different from anywhere Darwin had yet seen—a huge plateau, part steppe, part desert, covering some 260,000 square miles of southern Argentina and Chile down to the Strait of Magellan. The name Patagonia appears to derive from the name "Patagones," given by the Portuguese explorer Ferdinand Magellan to the indigenous Tehuelche people he encountered when, in 1520, he landed in a bay eleven hundred miles south of Buenos Aires. Some suggest Patagones derives from the Portuguese *pata grau*, which means "big feet," and that Magellan intended the word to mean "giant animal feet" for the huge fur boots the nomadic hunters there wore; others believed that the tall, well-built Tehuelches reminded him of a giant, Patagon, in a then-popular romance. Whatever Magellan actually meant, the idea of a race of giants was so fancifully appealing that it endured in the popular imagination until well into the late eighteenth century, when British naval explorers like John "Foulweather Jack" Byron, the poet's grandfather, reported that, though local people could indeed be very tall—six foot six inches in some cases—they were not giants.

Early Spanish attempts to colonize the far south of Patagonia failed mainly because of the harsh climate and inadequate food supplies. The Spanish crown largely ignored the region until the eighteenth century when, with other nations showing an interest, it established coastal settlements as a mark of ownership. Argentina and Chile—which had recently claimed independence from Spain—were now beginning an aggressive expansion into the interior in what was fast becoming a war of extermination as they ruthlessly and systematically "cleansed" the indigenous people from the Patagonian plains to make way for the *estancias* of powerful cattle ranchers.

By July 20, 1832, the Southern Hemisphere winter temperature was already dropping, and chill winds alternating with thick fogs reminded Darwin of an English autumn. As the *Beagle* entered the 120-mile-wide estuary of the River Plate—notorious, as FitzRoy well knew, for its very dangerous shoals, strong and irregular currents, and sudden storms— jackass penguins and seals sporting around the ship made such strange noises that the ship's master, Edward Chaffers, reported to Lieutenant

Wickham that he could hear cattle lowing on shore. One night Darwin watched as vivid lightning illuminated the dark sky. "The tops of our masts and higher yards ends shone with the Electric fluid playing about them . . . To complete these natural fireworks.——the sea was so highly luminous that the Penguins might be tracked by the stream of light in their wake." St. Elmo's fire had been first identified as an electrical phenomenon by Benjamin Franklin in 1749 and was once thought by sailors to be the cavorting of witches or hobgoblins gleefully warning of impending doom. As intrigued as Darwin, FitzRoy recalled, "[I] was curious enough to go out to a yard-arm and put my hand on a luminous spot; but, of course, could feel nothing, and when I moved my hand the spot reappeared."

As the *Beagle* beat up the wide Plate estuary against a powerful current, strong winds blew because, as FitzRoy noted, "The land on each side of the Plata is so low, and those extraordinary plains called pampas, hundreds of miles in extent, are so perfectly free from a single obstacle which might offer any check to the storm, that a pampero [pampas wind] sweeps over land and water with the weight of a rushing hurricane." Though the temperature was a mild fifty degrees Fahrenheit and he was "loaded with clothes," Darwin was surprised how much colder he felt than others, surmising "my constitution in a shorter time becomes habituated to a warm climate.——and therefore on leaving it more strongly feels the contrary extreme."

On July 26, the Beagle entered the harbor of Montevideo, Uruguay's capital since independence in 1828, and founded a century earlier by the Spanish to deter Portuguese rivals from encroaching southward from Brazil. In 1806–1807 during the Napoleonic Wars, when Spain was an ally of France, the British had seized both Montevideo, on the eastern banks of the Plate estuary, and Buenos Aires, on the western shore, but had quickly been driven out again. So heavy were the British losses that FitzRoy's uncle, Lord Castlereagh, then Secretary for War, had concluded that attempting conquests in the region was a hopeless task. Henceforth, he decreed, the Royal Navy should only send men ashore when British lives and property were threatened.

As fate would have it, the *Beagle* had arrived in Montevideo at just such a moment. As she was about to anchor, HMS *Druid*, a British frigate stationed there to protect British interests, signaled the *Beagle* to "clear for action" and "prepare to cover our boats." Minutes later as the *Beagle* ran out her guns, six boats from the *Druid* approached packed with heavily armed sailors and marines ready for action. The *Druid*'s captain, Gawan Hamilton, boarded the *Beagle* to alert FitzRoy that the present Uruguayan regime was "a military usurpation" that had seized four hundred horses belonging to a British subject. His little flotilla was on its way "to convince the inhabitants they must not plunder British property."

British property in Montevideo was considerable. Though Britain's attempts at colonization in South America had failed, the government had remained keenly interested both in exporting British manufactured goods to the vast market presented by newly independent South American countries and in accessing the region's rich natural resources such as hides and wool. Uruguay's river system offered good routes into the interior, while Montevideo itself was one of the region's best ports. So many British merchants had settled there and so blatant were British attempts to promote an Anglophile Uruguayan ruling elite that in 1826, a disgruntled US envoy called Uruguay a British colony in disguise.

That resentment of British influence might be a partial cause of the current instability did not occur to Darwin, who wrote, "The revolutions in these countries are quite laughable; some few years ago in Buenos Ayres, they had 14 revolutions in 12 months.—things go as quietly as possible; both parties dislike the sight of blood; and so that the one which appears the strongest gains the day.—The disturbances do not much affect the inhabitants of the town, for both parties find it best to protect private property." His greatest fear was that the disturbances would prevent him from landing to explore. However, the dispute about the four hundred horses was quickly resolved, and he was able to climb the 450-foot-high, fortress-topped hill above Montevideo that had given the city its name.

The view from the summit of an undulating green plain grazed by great herds of cattle with not a single tree, house, or sign of cultivation to "give cheerfulness to the scene" was to Darwin "one of the most uninteresting I ever beheld"—exactly like Cambridgeshire if every tree there was uprooted and arable land turned to pasture. Neither did Montevideo itself impress when Darwin toured it with FitzRoy: "it is of no great size, possesses no architectural beauties, and the streets are irregular and filthily dirty." However, the inhabitants were "a much finer set" than in Rio. The men had "handsome expressive faces and athletic figures; either of which it is very rare to meet with amongst the Portuguese."

Darwin was pleased with how his work was progressing, writing to his sister Susan that "Natural History goes on very well, and I certainly have taken many animals . . . which would be interesting to Naturalists.— Independent of this satisfaction, I have begun so many branches, previously new to me . . ." Conscious as ever about the cost to his father and as usual using his sisters as intermediaries, he informed her he intended to draw twenty-five pounds—Dr. Darwin had instructed his bank, Robarts, Curtis & Co. of Lombardy Street in London, to honor every bill that arrived in his son's name from an overseas bank or agent. This would make eighty pounds in total drawn since leaving England—some of which had been spent on scientific equipment—but he reassured her, as proxy for his father, that he was about to sail for such remote regions that "even with my ingenuity, I do not think I shall be able to spend a penny." He also told her that before heading south for Patagonia, FitzRoy intended to sail farther up the River Plate to Buenos Aires to consult some old Spanish charts of the Patagonian coast that he had heard were stored in archives there. "I am glad of it, the more places the merrier."

On July 31, after taking aboard Lieutenant Robert Hamond from HMS *Druid*, who had sailed previously with FitzRoy and was being loaned for the voyage south, the Beagle sailed for Buenos Aires. Two centuries older than Montevideo, the city was founded in 1536 by Spanish settlers, who named it Ciudad del Nuestra Señora Santa María del Buen Ayre (City of Our Lady Saint Mary of the Fair Winds). Fierce attacks by the indigenous people led to the settlement's abandonment five years later, but in 1580 it was refounded and subsequently flourished, exporting grain, hides, and dried beef. When the United Provinces of the River

Plate—which would later become the Republic of Argentina—declared independence from Spain in 1816, Buenos Aires became its capital. By the time the *Beagle* arrived, its population was close to fifty thousand people.[3]

Once again, the *Beagle* had arrived at a tense moment—"Peace flies before our steps," wrote Darwin after a Buenos Aires guard ship fired a blank at the *Beagle* as she sailed past. Moments later came a further explosion, this time "accompanied by the whistling of a shot over our rigging"—sounds Darwin had never heard before though he did not doubt their meaning—"We Philosophers do not bargain for this sort of work and I hope there will be no more," he wrote to a friend. Before the guard ship could fire again, FitzRoy took the *Beagle* out of range. After dropping anchor, FitzRoy dispatched Wickham ashore to inform the senior British official there "of the insult offered to the British flag." Darwin accompanied him, but as they approached the shore a quarantine boat intercepted them. The local officials on board rebuked them for the *Beagle* not heeding the guard vessel's warning and ordered them to "return on board, to have our bill of health inspected, from fears of the Cholera.—Nothing which we could say about being a man of war, having left England 7 months . . . had any effect."

They had no option but to obey, the boat crew rowing against blustering winds and a strong tide nearly three miles back to the *Beagle*. FitzRoy, meanwhile, had dispatched fierce messages to the governor of Buenos Aires and the commander of the guard ship that had fired on the *Beagle*. According to Darwin, their import was that FitzRoy "was sorry he was not aware he was entering an uncivilized port, or he would have had his broadside ready for answering his shot."

Abandoning hopes of consulting the Patagonian charts himself and deciding instead to send "a capable person" to Buenos Aires to copy them for him, FitzRoy gave the order to run out the guns, weigh anchor, and retreat to Montevideo. Darwin recorded how, with guns loaded and trained, the *Beagle* "ran down close along the guard-ship. Hailed her and said that when we again entered the port, we would be prepared as at

3. Tierra Argentina, or "Land of Silver," was the name given by the conquistadores.

present and if she dared to fire a shot we would send our whole broadside into her rotten hulk." By the evening of August 3, the *Beagle* was back in Montevideo, and FitzRoy, still fuming, boarded HMS *Druid* to report being fired on at Buenos Aires. He returned with news that the *Druid* would sail the next morning to Buenos Aires to demand an apology. "Oh I hope the Guard-ship will fire . . . . at the Frigate; if she does, it will be her last day above water," Darwin wrote.

Two days later, after the larger HMS *Druid* had departed for Buenos Aires, the chief of the Montevideo police boarded the *Beagle* and "begged for assistance against a serious insurrection of some black troops." The British consul-general Thomas Hood also implored FitzRoy "to afford the British residents any protection" in his power. FitzRoy went ashore to judge for himself whether there was indeed a risk to British lives and interests. Deciding there was, he signaled to the *Beagle* from the mole "to hoist out and man our boats." In minutes they were in the water with fifty-two men "heavily armed with Muskets, Cutlasses and Pistols" aboard, Darwin among them. Once ashore they marched to the government headquarters, the fortress of St. Lucia in the center of the town. FitzRoy ordered them only to fire if threatened. It was a tense walk, "like treading on cracked ice," FitzRoy recalled. He had been involved in confrontations before, but to Darwin, "It was something new . . . to walk with pistols and cutlass through the streets of a town."

The insurgents had broken open the prison, armed the prisoners, positioned artillery in the streets, and occupied the citadel close to the harbor where the munitions were stored. However, with neither side seemingly yet ready or willing to fight, a standoff ensued. "We remained at our station and amused ourselves by cooking beefsteaks in the Courtyard," Darwin wrote. Toward sunset, FitzRoy dispatched a boat back to the *Beagle* to fetch warm clothing for the men who would have to bivouac in the fort that night. Complaining of a bad headache—something that often afflicted him later in times of stress—Darwin returned with the boat. Back aboard the *Beagle* the arrangements made by the ship's master, Edward Chaffers, whom FitzRoy had left in command, in case of attack impressed him: "They have triced up the Boarding netting, loaded and pointed the guns,——and cleared for action.——We are now at night in a high state of preparation so as to make the best defence possible . . ." However, the night passed without incident.

The next day, August 6, with a growing number of government troops and armed citizens surrounding the rebels in the citadel, FitzRoy decided danger to British residents was over and, not wishing to become further embroiled in local politics, withdrew his men without firing a shot. FitzRoy's own account of the incident is muted, but Darwin's not so: "There certainly is a great deal of pleasure in the excitement of this sort of work.——quite sufficient to explain the reckless gayety with which sailors undertake even the most hazardous attacks." Three days later, FitzRoy went back into Montevideo, returning with news of fresh skirmishes and of yet another party contending for political power, caus-ing even the liberal Darwin to wonder "whether despotism is not better than such uncontrolled anarchy."

The crackle of musketry from the city the following night, August 10, suggested "anarchy" had indeed broken out, though the reality, as FitzRoy and Darwin discovered, was that "not even one has been wounded.——in fact both parties are afraid of coming within reach of musket range of each other." Reflecting on what seemed to him an ongoing farce, Darwin wrote that though "one is shocked at the bloody revolutions in Europe," having witnessed the "imbecile" and interminable political disturbances in South America, "it is hard to determine which of the two is most to be dreaded."

The situation had calmed sufficiently by August 13 for Darwin to accompany several officers ashore to go shooting. He hoped to see "ostriches" and spotted one from a distance. At first it looked like "a very large deer running like a race-horse.——as the distance increased it looked more like a large hawk skimming over the ground.——the rapidity of its movements were astonishing." He was also busily collecting—"Under stones were several scorpions about 2 inches long; when pressed by a stick to the ground, they struck it with their stings with such force as very distinctly to be heard."

By now, Darwin was beginning to realize the scale of his task, some-times returning to the ship with so many "animals of all sorts" that he

Cartoon attributed to Augustus Earle titled "Quarter Deck of a Man of War on Diskivery [sic] or Interesting Scenes on an Interesting Voyage" and believed to be dated 1832. The tall figure in the top hat is Charles Darwin.

felt like a human Noah's ark. His companions had grown used to him bringing strange things on board. Wickham, "a very tidy man who liked to keep the decks so that you could eat your dinner off them," would tease him, saying, "If I had my way, all your d . . . d mess would be chucked overboard, and you after it old Flycatcher."

Darwin had recently acquired an informal assistant—Syms Covington, listed in the ship's muster as "Fiddler and Boy to the Poop Cabin"—to help him gather specimens, shoot, skin, and stuff birds and animals, and pack them up for sending home. Darwin was also increasingly aware how much he owed John Henslow for agreeing to take charge of the boxes of specimens, the first of which he was dispatching from Montevideo, having failed to do so from Rio. He was also indebted to FitzRoy for authorizing them to be sent as official Admiralty cargo, thus saving him from paying the cost himself. As well as quantities of plants, spiders, and beetles, Darwin warned his mentor to expect "a good many geological specimens . . . I have endeavoured to get specimens of every variety of rock."

After receiving and inspecting this consignment, Henslow replied the following January with advice about how best to package specimens—while the insects had traveled quite well, two mice had gone "rather mouldy," several birds were unlabeled, and the tail feathers of one were crumpled. As for Darwin's plant collection, Henslow told him: "Avoid sending *scraps*. Make the specimens as perfect as you can, *root, flowers* and *leaves* and you can't do wrong. In large ferns and leaves fold them back upon themselves on *one* side of the specimen and they will get into a proper sized paper." Henslow even included a sketch of a correctly folded leaf. He praised the collection of lichens but asked, "For goodness sake what is No. 223 it looks like the remains of an electric explosion, a mere mass of soot . . ."

By August 19, 1832, FitzRoy was ready to sail south to survey the Rio Negro and surrounding coast—"our real wild work," Darwin called it. Once underway, he was soon seasick again: "I have never seen so much spray break over the *Beagle* and I have not often felt a more disagreeable sensation in my stomach." Every day when he felt well enough, he noted the vegetation along the shoreline, larks, flycatchers, doves, and butcher-birds resting in the *Beagle*'s rigging, and the increasing clarity of the water as they approached the mouth of the Plate estuary. Leaning over the side, he cast his net to add to his collection of sea creatures.

He also observed FitzRoy's skill both as an interpreter of weather patterns and reader of instruments—talents on which the captain would build his later career. On August 26, a day of such torrential rain and thick mist that the *Beagle* could only lie at anchor, Darwin wrote: "We had today a beautiful illustration how useful the barometer is at sea.—During the last three or four fine days it has been slowly falling.—The Captain felt so sure that shortly after it began to rise we should have the wind from the opposite quarter, the South, that when he went to bed he left orders to be called when the Barometer turned. Accordingly at one o'clock it began to rise, and the Captain immediately ordered all hands to be piped up to weigh anchor.—In the course of an hour from being . . . calm it blew a gale right on shore, so that we were glad enough to beat off.——By the morning we were well out at sea . . . If we had not a Barometer we probably . . . should have been in a most dangerous situation."

Three days later, Darwin realized exactly a year had passed since his return from geologizing in North Wales with Sedgwick to learn of the *Beagle* voyage: ". . . it is amusing to imagine my surprise, if anybody on the mountains of Wales had whispered to me, this day next year you will be beating off the coast of Patagonia:——And yet how common and natural an occurrence it now appears to me."

On September 6, the *Beagle* entered an inlet near the fortified settlement of Baia [Bahia] Blanca to encounter a maze of mud banks and treacherous shoals—"an unexpected dilemma," FitzRoy called it. Fortunately an Argentine schooner appeared, and an Englishman on board, James Harris, who lived on the Rio Negro, offered to guide the *Beagle* to safe anchorage. From Harris, Darwin learned that Baia Blanca had been founded only six years earlier as "a frontier fort against the Indians." The land had been purchased before independence "in the time of the old Spaniards" from "the native chief of the place," but relations between the ranchers intent on establishing vast *estancias* and the indigenous peoples had since degenerated into "a barbarous and cruel warfare."

The next day, FitzRoy, Rowlett (the *Beagle*'s purser), Harris, and a curious Darwin set off for the settlement in one of the ship's boats. For a while they became lost amid mud banks and rocky creeks so narrow that the sailors' oars scraped the sides, but as night fell they finally approached a creek only four miles from the settlement. Some "wild Gaucho cavalry" watched their arrival, Darwin wrote, ". . . by far the most savage picturesque group I ever beheld.——I should have fancied myself in the middle of Turkey by their dresses.——Round their waists they had bright colored shawls forming a petticoat, beneath which were fringed drawers. Their boots were very singular, they are made from the hide of the hock joint of horses hind legs, so that it is a tube with a bend in it; this they put on fresh, and thus drying on their legs is never again removed." They had enormous spurs and "all wore the Poncho, which is [a] large shawl with a hole in the middle for the head.——Thus equipped with sabres and short muskets they were mounted on powerful horses." Even more remarkable to Darwin than their exotic, quixotic dress were the men themselves: "the greater number were half Spaniard and Indian.—some of each pure blood and some black.—The Indians, whilst gnawing bones of beef, . . . half recalled wild beasts.—No painter ever imagined so wild a set of expressions."

Many, though not all, gauchos were indeed mixed race. Though the term *gaucho*—of uncertain origin—had been in common usage for less than twenty years, gauchos had emerged as an identifiable group in the previous century as highly skilled and mobile horsemen and cattle drovers—the cowboys of the South American pampas, with their own culture and traditions. A favorite gaucho saying was that a man without a horse was like a man without legs. Some were ferocious fighters, playing an active part in the war of extermination against the indigenous peoples, as Darwin would later discover.

With night fast closing in, leaving the boat crew to bivouac on the shore, FitzRoy accepted a horse from the gauchos and took up the purser behind him, while Darwin and Harris each mounted up behind a gaucho, and the group set off at a gallop across a flat plain to the settlement. Its commandant and his second in command, an old major, gave the *Beagle* men a dusty welcome. FitzRoy was amused that the major, "poor old soul, thought we were very suspicious characters." The major could not understand why an armed British naval vessel, even one as small as the *Beagle*, had arrived in Baia Blanca and asked "endless questions" about the size of the crew. Darwin decided that he was already imagining British marines seizing his fort. He seemed particularly wary of Darwin, whom Harris introduced as "Un naturalista"—a term, FitzRoy recalled, "unheard of by any person in the settlement." Harris's explanation that it meant "a man that knows everything" only seemed to make matters worse.

The next morning, when FitzRoy requested horses to return to the *Beagle*'s boat, "trifling excuses were made about the want of horses and fear of Indians arriving." When he insisted that if the commandant would not provide them, he and his companions would walk, the commandant—reluctant to detain them by force—gave in but insisted on sending an escort with them. By midday, all were safely back on the *Beagle*, where Darwin had much to ponder, including further information he had gleaned about the fighting between the Spanish and the indigenous people. "The War is carried on in the most barbarous manner. The Indians torture all their prisoners and the Spaniards shoot theirs."

The commandant had posted detachments of gauchos on rising land around the bay to watch the *Beagle*. However, when the crew went ashore, the gauchos were helpful rather than hostile, showing them where to find fresh water and even hunting for them. FitzRoy was startled to be offered a live puma "in hopes I should offer a good price, and embark it alive." However, "having no wish for so troublesome a companion on our little crowded vessel," he instead bargained only for its skin and watched the gauchos kill it and make a hearty meal of the flesh.

Darwin was interested to see "these hardy people fully equipped for an expedition.—They sleep on the bare ground at all times and as they travel get their food; already they had killed a Puma or Lion . . . also an Ostrich, these they catch by two heavy balls, fastened to the ends of a long thong." These balls, or "bolas," consisted of two round stones covered with leather linked by an eight-foot-long plaited thong. Darwin watched how they held one ball while whirling the other round and round "and then with great force send them both revolving in the air towards any object.—Of course the instant it strikes an animal's legs it fairly ties them together." The gauchos gave the *Beagle* men rhea eggs, showing them nests containing as many as two dozen. "It is an undoubted fact that many female Ostriches lay in the same spot."

FitzRoy decided to hire two schooners from Harris to help survey the intricate coast between Baia Blanca and the Rio Negro to the south. Lieutenant Wickham, assisted by Harris as pilot, would command the larger, fifteen-ton *La Paz*, while John Lort Stokes would command the nine-ton *Liebre*, helped by Harris's friend, Roberts, long settled in the region and equally knowledgeable about the vagaries of its currents and tides. Not only were these small boats ideally suited to surveying in shallow waters, but hiring them would allow this survey work to continue while the *Beagle* was otherwise occupied, whether resupplying at Montevideo, taking the Fuegians home, surveying in the far south, or calling at the Falkland Islands.

The only serious difficulty, FitzRoy acknowledged, was that he was not "authorized to hire or purchase assistance on account of the Government." Characteristically, he went ahead anyway, paying Harris the substantial sum of £1680, "which I could so ill spare," himself and trusting in the Admiralty to approve his actions in retrospect and reimburse him. In the circumstances it seemed reasonable enough. As FitzRoy knew, during the first South American surveying expedition, despite having two vessels—the *Adventure* and the *Beagle*—its commander Philip Parker King had purchased an additional vessel, the schooner *Adelaide*, and been reimbursed, though admittedly he had taken the precaution of obtaining Admiralty approval first.

Harris left at once to fetch the schooners, which, currently used for sealing, were in the Rio Negro. Darwin meanwhile went shooting with Wickham. The thought struck him that "I am spending September in Patagonia, much in the same manner as I should in England, viz in shooting" though with "the extra satisfaction of knowing that one gives fresh provisions to the ship's company.—Today I shot another deer and an Agouti or Cavy.—The latter weighs more than 20 pounds; and affords the very best meat I ever tasted." He realized their diet would have been considered "very odd" in England, but as befitted a member of the Cambridge Glutton Club who had dined on squirrel and hawk, he enjoyed rhea dumplings, finding the meat similar to beef, while armadillo cooked without its shell both looked and tasted "like a duck."

In mid-September, some gauchos invited Darwin hunting and lent him a horse. The stirrups were so narrow that even without shoes he could only insert his first two toes—the gauchos used their own big toes like talons to maintain a grip, a practice that over time so enlarged and deformed their toes that some could scarcely walk. His companions were "9 men and one woman; the greater part of the former were pure Indians, the others most ambiguous; but all alike were most wild in their appearance and attire." The woman "dressed and rode like a man, and till dinner I did not guess she was otherwise." At night, they fed on rhea eggs and armadillos roasted in their shells over a fire. Again Darwin witnessed hunting with bolos as, whirling them above their heads, his companions hurtled after and brought down a fine rhea. A good horseman himself, he admired the speed and precision with which they turned their mounts. Sitting around the fire with them at night, Darwin pondered the self-sufficiency of his "half-savage hosts" who did not much care where they lived: "Like to snails, all their property is on their backs and their food around them."

On September 22, Darwin joined FitzRoy, Sulivan, and others for a cruise around the bay during which they arrived at Punta Alta, a twenty-foot-high bank of stratified gravel and reddish mud studded with shells. Assisted by Syms Covington, Darwin began digging into it and before long made a discovery that was one of the most important of the voyage—the fossilized remains of "gigantic land-animals." FitzRoy described the moment: "My friend's attention was soon attracted to some low cliffs . . . where he found . . . huge fossil bones . . . and notwithstanding our smiles at the cargoes of apparent rubbish which he frequently brought on board, he and his servant used their pick-axes in earnest . . ."

The following day Darwin was back at Punta Alta where, to his "great joy," he unearthed "the head of some large animal, imbedded in a soft rock.——It took me nearly 3 hours to get it out: As far as I am able to judge, it is allied to the Rhinoceros." Though he did not yet grasp the implications, his belief that the creature he had found was related to a species still living would later be key in developing his ideas on the transmutation of species. Darkness had fallen by the time he got his gargantuan prize back to the *Beagle*. A few days later, his further digging uncovered "a jaw bone which contained a tooth: by this I found that it belongs to the great ante-diluvial animal, the Megatherium." In 1796, Georges Cuvier had given the name Megatherium—Latin for "large beast"—to a skeleton the size of a baby elephant discovered in South America and, as Darwin knew, subsequently sent to Madrid. Cuvier had classified it as extinct but noted its apparent similarity to sloths living in the rain forests of South America. Nearby Darwin found some "osseous"—bony—plates, which

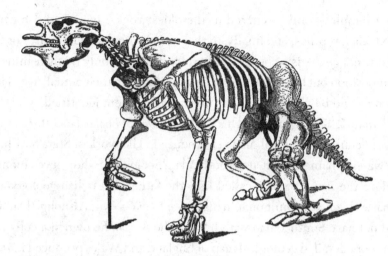

Drawing of a Megatherium skeleton, 1911

correctly made him wonder whether the creature had possessed some kind of armored hide like the armadillo.[4]

Darwin was careful to record not only the location but the stratum of rock in which he found his fossils, as Adam Sedgwick advised. In the previous century, German mineralogist Johann Lehmann had suggested that successive layers of rock recorded successive geological events. Lehmann called the earliest rocks *primary* to distinguish them from subsequent *secondary* rocks formed from debris washed from the primary group. Many shared his view that rock strata were a record of how natural forces had shaped the Earth. Early in the nineteenth century Briton William Smith provided a means of assessing the relative ages of rock strata.

Smith, a land surveyor and self-taught geologist, noticed that rocks in the west country of England were arranged in a distinct order, with stratum succeeding stratum in a discernible sequence, with the oldest logically at the bottom. He observed that the fossils found in each stratum were distinct from the fossils found in others. Invertebrate species—trilobites,

---

4. Today the site of Darwin's fossil discoveries lies beneath Argentina's Puerto Belgrano naval base. Modern techniques have dated his fossil findings in South America to between ten thousand and five hundred thousand years old.

for example—only occurred in the oldest rocks, fossilized fish came next, then reptiles, and finally, in the youngest rocks, mammals. For the first time, geologists could classify rock strata not only by their mineral composition but by their relative, if not at this time their actual, age. Thus strata could be mapped and any movement in them identified.

In 1815, Smith published his *Geological Map of England and Wales*—the most detailed geological map yet produced. The work of Stata Smith, as he was nicknamed, influenced both Adam Sedgwick, who, eleven months before the *Beagle* sailed, called him the father of English geology and Charles Lyell who built on it in his *Principles of Geology*. Although Darwin did not have Smith's map with him on the *Beagle*, he owned a copy of a yet more detailed geological map of England and Wales produced in 1820 by George Greenough, who drew on Smith's work and whose geological essays were in the *Beagle* library.

By mid-October, a group of the *Beagle*'s men, including the sailmaker, armorer, and cooper, had under Wickham's supervision refitted the two schooners. When they had arrived from the Rio Negro three weeks before, FitzRoy was unimpressed. The *Paz* was "as ugly and ill-built a craft as I ever saw, covered with dirt, and soaked with rancid oil," and the *Liebre* little better. But now, with "spars altered, and improved rigging, well-cut sails, fresh paint, and thorough cleanliness," FitzRoy congratulated himself that they "had been transformed." Nevertheless, Sulivan realized conditions on board would be challenging: "The cabin in Stokes' craft is seven feet long, seven wide, and thirty inches high. In this three of them stow their hammocks, which in the daytime form seats and serve for a table. In a little space forward, not so large, are stowed five men. The larger boat carries the instruments. Her cabin is the same size, but is four feet high, and has a table and seats." Darwin would miss the sloops' commanders Wickham and Stokes, and his friend young King who was to sail with them—"our society on board can ill afford to lose such very essential members." He predicted they would face "many privations."

On October 18, cheered by those on the *Beagle*, the fifteen men aboard the schooners sailed south to begin their survey work. The following day,

FitzRoy set course north back to the River Plate to take longitudinal measurements of Montevideo and ready the *Beagle* for her voyage to Tierra del Fuego. One pitch-dark night Darwin recorded a strange sight: "The sea from its extreme luminousness presented a wonderful . . . appearance; every part of the water, which by day is seen as foam, glowed with a pale light. The vessel drove before her bows two billows of liquid phosphorus, and in her wake was a milky train.—As far as the eye reached, the crest of every wave was bright, and from the reflected light the sky just above the horizon was not so utterly dark as the rest of the Heavens . . ." The sight reminded him of Milton's descriptions "of the regions of Chaos and Anarchy" from his favorite book, *Paradise Lost*. Presumably it was St. Elmo's fire again.

In Montevideo, letters from family and friends—written four months earlier—awaited. Darwin wrote, "No half-famished wretch ever swallowed food more eagerly than I do letters." They brought firm news about the Reform Bill—a matter of keen interest aboard. The night before receiving his mail, Darwin had begun a long letter to his sister Caroline: "We are all very curious about politics . . . whether there is a King or a republic according to the Captain, remains to be proved."

The Bill had finally passed into law, but only after dramatic political upheavals. In April, the hereditary House of Lords had thrown the Bill out in defiance of the elected House of Commons, only for the Whig Prime Minister, Lord Grey, to demand King William create fifty new lords to enable it to pass. He had refused, and Grey had resigned. The leader of the opposition, the Tory Duke of Wellington, victor at Waterloo, had then tried but failed to form a government. Grey duly returned to office and, with unrest sweeping the country, Wellington agreed to abstain with enough colleagues in the lords to allow the Commons Bill to pass into law on June 4, 1832. It had received royal assent from a reluctant king three days later.

A letter from his cousin William Darwin Fox related the country's mood during the passage of the Bill: "For some days we certainly were on the very verge of revolution. The excitement in the country was quite extraordinary . . . We have however now I trust safely passed this grand corner upon which so much hung." Nevertheless, to some the pace of reform seemed disappointingly sluggish. The new Act had broadened the property qualifications to vote to include small landowners, tenant farmers, shopkeepers, and some householders, but the majority of working men were still denied the vote. The Act also formally disenfranchised all women. Before 1832 there were occasional, albeit rare, instances of propertied women having the vote, but the new Act for the first time specifically confined the vote to "male persons." Darwin's brother Erasmus complained that although many wished to see progress on "vote by ballot, abolition and commutation of tithes, abolition of slavery . . . Now that we have got the Reform Bill people seem disinclined to make any use of it . . ." He added, entirely mistakenly, "I have written to you all this politics tho' I suppose you are too far from England to care much about it. Politics won't travel."

Also waiting for Darwin was the second volume of Lyell's *Principles of Geology*—most timely given his recent fossil discoveries. In it Lyell discussed the extinction of species, suggesting evidence from fossils showed that some species disappeared, perhaps due to local conditions, and fresh species appeared to replace them. He did not suggest by what mechanism the latter occurred.

A few days later, as the *Beagle* made for Buenos Aires, Darwin found the ship's ropes "coated and fringed with gossamer web"—the work of tiny, dusky red "aeronaut spiders," which, he calculated, must have been blown some sixty miles from the shore. This time the *Beagle* sailed unchallenged by the guard ship into Buenos Aires. Darwin was among the first ashore: "after being for some months in a ship, the mere prospect of living on dry land is very pleasant, and we were all accordingly in high spirits.—It is from this cause, I suppose, that most foreigners believe that English sailors are all more or less mad."

Darwin thought Buenos Aires, unlike Montevideo, well laid out and handsome. He went to the theater, where, though he could not understand a word of the play, he admired the brio with which the actors delivered their lines in Spanish. Visiting several of the ornate churches during mass, he was struck by "the fervor which appears to reign during the Catholic service as compared with the Protestant" and by the apparent social equality, at least before God—"The Spanish lady with her brilliant shawls kneels by the side of her black servant in the open aisle." Walking through the streets, often with Hamond, the young former mate of the *Druid* with whom he had become friendly, Darwin positively ogled the passing women. "In the hair (which is beautifully arranged) they wear an *enormous* comb; from this a large silk shawl folds round the upper part of the body. Their walk is most graceful, and although often disappointed, we never saw one of their charming backs without crying out, 'how beautiful she must be.'" He described them to his sister Caroline as "angels," adding "how ugly Miss sounds after Signorita; I am sorry for you all; it would do the whole tribe of you a great deal of good to come to Buenos Aires."

However, he encountered one woman who was certainly no angel when FitzRoy took him to call on Mrs. Clarke, or Donna Clara, as she was known. This wrinkled old lady had, Darwin learned, as a young woman been transported from Britain "for some atrocious crime." Aboard the convict ship, she had become the captain's mistress. As the vessel approached Buenos Aires, the female convicts revolted, and she killed her lover with her own hands. Helped by mutinous sailors, they brought the ship into Buenos Aires. There she settled down, married a wealthy man, and during Britain's failed attempt to occupy the city during the Napoleonic Wars, won fame for nursing wounded British soldiers. Scrutinizing her, Darwin thought her face decidedly masculine and that, despite her great age, she retained "a most ferocious mind"—her favorite expressions were "I would hang them all Sir" and "I would cut their fingers off." Darwin thought "the worthy old lady looks as if she would rather do it, than say so."[5]

---

5. Mary Clarke was transported from Britain aboard the *Lady Shore* bound for Australia's Botany Bay and involved in the mutiny off Brazil. Her "atrocious crime" is uncertain since two female prisoners on the *Lady Shore* were named Mary Clarke. However, Darwin's Donna Clara was probably a London linen draper convicted of theft.

Sometimes Darwin and Hamond rode into the hinterland. One day, they came across the huge public corral where cattle were slaughtered. Darwin watched horsemen first lasso the beasts by the horns, then drag them bellowing before the matador, who "with great caution cuts the hamstrings and then being disabled sticks them . . . a horrible sight: the ground is made of bones, and men, horses and mud are stained by blood." Before leaving Buenos Aires, despite the money he had withdrawn in Montevideo, Darwin drew a further twenty pounds to keep him in funds, as dutifully reported to his sister Caroline. He also told her that FitzRoy had asked him to pay a year in advance for his food on board. He had agreed to pay the fifty pounds requested, "for I could not, although, perhaps I ought, refuse a person who is so systematically munificent to everyone who approaches him." The reason for FitzRoy's request was perhaps shortage of funds after acquiring and fitting out the schooners.

On November 10, the *Beagle* returned to Montevideo, where Darwin found further letters from family and friends waiting. They were brimful of gossip, including details of the not entirely felicitous married life of his kittenish "housemaid," Fanny Owen. They told him that the cholera epidemic afflicting Britain—the reason the Buenos Aires authorities had sought to quarantine the *Beagle*—was easing.

The letters also brought welcome news that the part of his diary he had sent home had reached The Mount safely. His sisters asked whether they might show it to their Wedgwood cousins at Maer. Darwin replied, "I leave that entirely in your hands. I suspect the first part is abominably childish, if so to not send it to Maer. Also, do not send it by the coach, (it may appear *ridiculous* to you) but I would as soon lose a piece of my memory as it." He assured Caroline he had "become quite devoted to Nat: History—you cannot imagine what a fine miserlike pleasure I enjoy, when examining an animal differing widely from any known genus" and that "I have been wonderfully lucky with fossil bones—some of the animals must have been of great dimensions: I am almost sure that many of them are quite new; this is always pleasant, but with the antediluvian animals it is doubly so." Darwin's belief that some of the fossils he unearthed in South America were new would later be confirmed.

As the *Beagle*'s departure approached, Darwin attended a grand ball, admiring the "splendidly dressed" guests, stately music—"in very slow time"—and graceful dancing. It was held in the theater and Darwin was surprised that "every part not actually occupied by the dancers was entirely open to the lowest classes of Society . . . And nobody ever seemed to imagine the possibility of disorderly conduct on their parts. How different are the habits of Englishmen, on such Jubilee nights!" He returned the next night to watch Rossini's opera, *Cenerentola*.

He also filled further casks with specimens, including the giant fossil bones, to send Henslow. In a letter from which Henslow would have extracts published by the Cambridge Philosophical Society, Darwin told him, "I have been very lucky with fossil bones" and to expect "fragments of at least 6 distinct animals; as many of them are teeth I trust . . . they will be recognised. I have paid *all the attention* I am *capable* of to their geological site . . . If it interests you sufficiently to unpack them, I shall be *very curious* to hear something about them." He was also sending fossilized shells, dried plants, fish preserved in alcohol, seeds, and pillboxes filled with beetles, which he asked Henslow to be sure to open "as they are apt to become mouldy." He assured Henslow everything was going well—"the only drawback is the fearful length of time between this and day of our return."

By late November, the *Beagle* was ready for her long voyage south from Montevideo to Tierra del Fuego, with food for the next eight months stored everywhere, even in the officers' cabins, and supplies including extra iron and coal for the ship's forge "in case of any serious accident." Augustus Earle, whose health had still not improved, was remaining in Montevideo to recuperate, but Darwin was more than ready to sail— "Anything must be better, than this detestable Rio Plata.—I would much sooner live in a coal-barge in the Cam [river in Cambridge]." As for Fitz-Roy, he wrote to his sister, "I am again quitting the demi-civilised world and am returning to the barbarous regions of the south."

CHAPTER SIX

Land of Fire

In early December 1832, near San Blas Bay on the Patagonian coast, five hundred nautical miles south of the River Plate, the *Beagle* crew were reunited with their shipmates on the two small schooners, *La Paz* and *Liebre*, who had been surveying between Baia Blanca and the Rio Negro. Constant exposure to sun and wind had burnished the latter's faces. Fitz-Roy thought that if Wickham, the expedition's commander, had "been half-roasted his . . . appearance could hardly have been more changed. Notwithstanding the protection of a huge beard, every part of his face was so scorched and blistered by the sun that he could hardly speak, much less join in the irresistible laugh at his . . . expense."

Though Wickham dismissed the coastline he had surveyed as "even more uninteresting than that of Baia Blanca," FitzRoy was pleased with his work and issued him fresh instructions to survey, time permitting, as far south as Port Desire before rendezvousing with the *Beagle* in March on its return from Tierra del Fuego and the Falkland Islands. FitzRoy had the schooners reprovisioned but was somewhat alarmed by the size of Roberts, the *Liebre*'s pilot—"one of the largest of men . . . his little vessel looked, by comparison, no bigger than a coffin." Wickham told FitzRoy the portly Roberts had broken the *Liebre*'s mast while attempting to climb it but assured him "his moveable weight answered admirably in trimming the craft; and . . . when she got a-ground, Mr Roberts stepped overboard and heaved her afloat."

On December 4, propelled by "a rattling breeze," the *Beagle* scudded southward to begin the voyage to Christmas Sound, on the Pacific side of Cape Horn, where FitzRoy intended to return York Minster and Fuegia Basket home. Ten miles from San Blas, FitzRoy noticed the horizon "strangely distorted by refraction" and anticipated a violent change in the weather. Instead, "myriads of white butterflies" suddenly enveloped the ship, so that "the men exclaimed, 'it is snowing butterflies.'" He estimated the cloud, numerous "as flakes of snow in the thickest shower," at two hundred yards high, a mile wide, and several miles long. Scooping up some of the insects, Darwin discovered several species of butterfly but also several types of moth and "a fine beetle." He decided they had not been blown all the way from the shore but had voluntarily taken flight to be caught by the strong breeze.

The next days' sailing was so smooth Darwin called it "gliding." He was busy using his microscope to examine some small crustaceans he could not identify from reference works. They were, he decided, "not only of new genera, but very extraordinary." However, by December 11—exactly one year since the *Beagle*'s first abortive attempt to sail from Plymouth—the barometer heralded one of the heaviest squalls FitzRoy had ever seen, but he had the *Beagle* ready, with all sails furled "and the ship put before the wind." As black clouds piled the sky and the wind whistled through the rigging, for once Darwin felt exhilarated rather than seasick—"it is always interesting to watch the progress of a squall . . . the line of white breakers, which steadily approaches till the ship heels over . . ." With the temperature between 45° and 50°F, the air had "the bracing *feel* of an English winter day."

On December 15, as thick fog slowed the *Beagle*'s progress, Darwin complained, "Everything conspires to make our passage long." However, by evening the crew could make out land south of the Strait of Magellan, through which Ferdinand Magellan's ships had sailed in 1520—the first European vessels to cross from the Atlantic into the Pacific—and which separated the large, triangular main island of Tierra del Fuego from the South American mainland. The next day the *Beagle* anchored near Cape Santa Inez on Tierra del Fuego's eastern shore, 130 miles south of the strait. The *Beagle* had not visited this part of Tierra del Fuego on her first voyage, but smoke rising from the shore showed it was inhabited: "By the aid of glasses we could see a group and some scattered Indians evidently watching the ship with interest," Darwin wrote. They were tall, nearly naked, and accompanied by several large dogs. He wondered

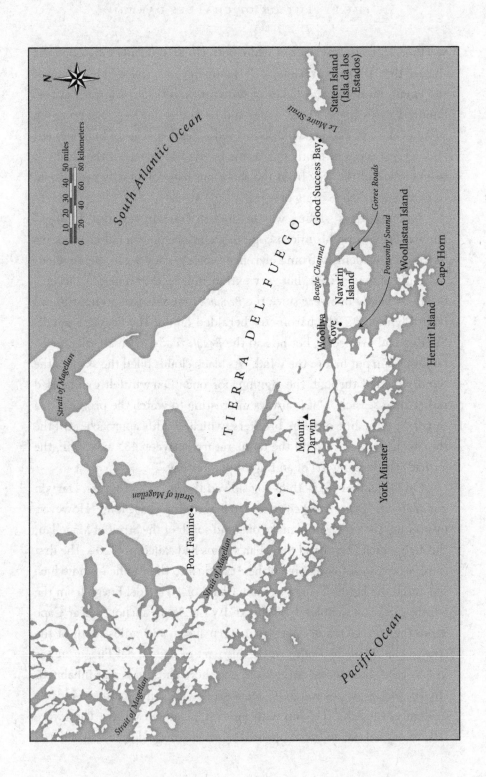

whether they had lit fires to signal the *Beagle*'s arrival to others or possibly to attract the ship's attention. Struck by all the smoke he had seen curling skyward, Magellan had named the region Tierra del Humo, the Land of Smoke, but Charles V of Spain, his sponsor, changed it to Tierra del Fuego—the Land of Fire.

The three Fuegians aboard the *Beagle* became increasingly elated as they neared home, Jemmy Button never tiring of assuring FitzRoy "how excellent his land was—how glad his friends would be to see him—and how well they would treat us in return for our kindness to him." However, the Fuegians on the shore unsettled Jemmy and York Minster, who recognized them as members of the Ona tribe—also known as the Selk'nam—who inhabited the main island. The Ona were often in conflict with York's and Fuegia's people, the Alakaluf of the western isles, and Jemmy's people, the canoe-going Yamana of the Beagle Channel—the passage at the extreme southern tip of South America that FitzRoy had partially explored on the *Beagle*'s first voyage and later named for his ship. Jemmy implored FitzRoy to fire on them, "saying that they were 'Oens-men—very bad men.'"

However, with a sudden heavy swell threatening to drive the *Beagle* ashore, FitzRoy ordered the anchor to be weighed. As a freshening breeze carried the *Beagle* out of danger, Darwin noticed "horizontal strata of some modern rock" along the shoreline, which in places formed "abrupt cliffs facing the sea." Despite the lowering skies, the green thickets and trees gave the land "a cheerful appearance." To the far south, the summits of "a chain of lofty mountains" glittered with snow.

The next day under low drifting cloud, the *Beagle* entered the Le Maire Strait between Staten Island and the eastern extremity of Tierra del Fuego and anchored in Good Success Bay, where Cook had called on his first and second expeditions.[1] Spotting the ship, a group of Fuegians "perched on a wild peak overhanging the sea," waved their animal-skin

---

1. In 1616, the Dutch mariners Jacob Le Maire and Willem Shouten discovered the Le Maire Strait while seeking a more southerly route than the Strait of Magellan between the Atlantic and Pacific—a voyage during which they also discovered and named Cape Horn.

cloaks, and gave "a loud sonorous shout." As darkness fell, the *Beagle* crew heard more cries and saw their fires begin to burn. "Being at anchor in so wild a country as Tierra del F" was everything Darwin could have imagined as a schoolboy reading *Wonders of the World*: "The very name of the harbour we are now in, recalls the idea of a voyage of discovery; more especially as it is memorable from being the first place Capt. Cook anchored in on this coast; and from the accidents which happened to Mr. Banks and Dr. Solander." (While out climbing, Banks, Solander, and their party had been caught in a blizzard, and two had died.)

Even more excitingly, Darwin was seeing real "savages" for the first time, "as savage as the most curious person would desire." He clearly regarded Jemmy, York, and Fuegia, after their "civilizing" stay in England, entirely differently, believing "3 years has been sufficient to change savages, into, as far as habits go, complete and voluntary Europeans." The following day, December 18, he accompanied FitzRoy, Hamond, the young missionary Matthews, Jemmy Button, and others ashore to try to communicate with the Fuegians. As soon as the ship's boat came within hailing distance, one of a small group of men on the shore began shouting and pointing to a good place to land. All the women and children had vanished, and as the *Beagle* party stepped ashore, the Fuegian men looked wary, though they continued talking and gesticulating vigorously: "Without exception the most curious and interesting spectacle I ever beheld.——I would not have believed how entire the difference between savage and civilized man is.——It is greater than between a wild and domesticated animal, in as much as in man there is greater power of improvement . . ." However, Darwin would never doubt that the single species "man" embraced both the "civilized" and the "savage"——"in the naked barbarian, with his body coated with paint, whose very gestures, whether they may be peacible [*sic*] or hostile, are unintelligible, with difficulty we see a fellow-creature."

Looking around, he pondered the Fuegians' harsh existence, subsisting on mussels and limpets, supplemented by seal blubber, birds, and the occasional guanaco, with few possessions except their hunting weapons—bows, arrows, and spears. The dense vegetation, extending down almost to the high-water mark and too thick to penetrate, allowed scarcely any shelter from the rain and wind other than that from a few scrubby bushes and overhanging ledges of rock. By comparison, "the Southsea Islanders are civilized compared to them, and the Esquimaux, in subterranean huts may enjoy some of the comforts of life."

This was FitzRoy's first close encounter with Fuegians since the previous voyage. He described their rich reddish-brown skin color as somewhere between "rusty iron and clean copper" and compared it to that of "the Dev-onshire breed of cattle." Like Dar-win, a believer in man as a single

A Fuegian, by C. Martens

species, he later wrote, "Disagreeable, indeed painful, as is even the mental contemplation of a savage, and unwilling as we may be to con-sider ourselves even remotely descended from human beings in such a state, the reflection that Caesar found the Britons painted and clothed in skins, like these Fuegians, cannot fail to augment an interest excited by their childish ignorance of matters familiar to civilized man, and by their healthy, independent state of existence."

FitzRoy watched how his companions who had never seen "man in such a totally savage state" reacted—"I can never forget Mr. Hamond's earnest expression" and his exclamation of, "What a pity such fine fel-lows should be left in such a barbarous state!" Hamond's reaction seemed to justify his own actions in removing the young Fuegians from their homeland—"It told me that a desire to benefit these ignorant, though by no means contemptible human beings, was a natural emotion, and not the effect of individual caprice or erroneous enthusiasm; and that his feelings were exactly in unison with those I had experienced . . . which had led to my undertaking the heavy charge of those Fuegians whom I brought to England." Matthews, who for the first time was seeing the people he would shortly be ministering to, did not appear unduly discouraged, telling FitzRoy "they were no worse than he had supposed them to be."

FitzRoy also watched with interest how Jemmy Button interacted with the Fuegians. Despite his European clothing, they clearly recognized him as one of their own. However, Jemmy "would not acknowledge them as countrymen, but laughed at and mocked them." He also pretended not to know their language, though he clearly "understood much" of it.

The leader of the Fuegians on the shore was an old man, but his companions were young and powerfully built—not exactly Patagonian giants but about six feet tall. They reminded Darwin of "representations of Devils on the Stage," like those in von Weber's opera *Der Freischutz*, *The Marksman*, premiered a decade earlier and which Darwin had seen performed in Edinburgh:

> The old man had a white feather cap; from under which, black long hair hung round his face.—The skin is dirty copper colour. Reaching from ear to ear and including the upper lip, there was a broad red coloured band of paint.—and parallel and above this . . . a white one; so that the eyebrows and eyelids were even thus coloured; the only garment was a large guanaco skin . . . thrown over their shoulders, one arm and leg being bare; for any exercise they must be absolutely naked.—Their very attitudes were abject, and the expression distrustful, surprised and startled.

They seemed afraid of guns "and nothing would tempt them to take one in their hands."

The *Beagle* men gave the Fuegians pieces of red cloth, which they hung around their necks, and so, Darwin wrote, "we became good friends." The elderly man patted the visitors' chests while making a noise like people made "when feeding chickens." As he walked with the old man "this demonstration was repeated between us several times; at last he gave me three hard slaps on the breast and back at the same time, and making most curious noises." He invited Darwin to return the compliment, which he did.

Captain Cook had likened the people's language to a man clearing his throat. Darwin thought that an understatement—"certainly no European ever cleared his throat with so many hoarse, guttural, and clicking sounds." The Fuegians seemed chiefly interested in obtaining knives, gesturing that they needed them to cut animal blubber. They were skilled mimics. When some officers squinted and made monkey-like faces, one young man responded with "still more hideous grimaces." The Fuegians were amazed when their visitors began to sing, "and with equal delight they viewed our dancing and immediately began themselves to waltz with one of the officers"—surely one of the strangest sights ever seen in Tierra del Fuego.

Later that day, Darwin accompanied FitzRoy and some officers on a second visit ashore. This time they brought York Minster with them. The Fuegians seemed less nervous—their women and children were with them—and as with Jemmy, they were keenly interested in York Minster, telling him "he ought to shave, and yet he has not 20 hairs on his face." They also examined the color of York's skin, before inspecting that of the Beagle men—"An arm being bared, they expressed the liveliest surprise and admiration." Darwin was convinced that despite their bushy beards several of the shorter officers were "taken for ladies." The Fuegians and the sailors parted "very good friends"—a good thing, Darwin reflected, since the dancing and "skylarking," which had included comparing their respective heights, "had occasionally bordered on a trial of strength."

Eager to explore the rounded slate mountains cloaked in evergreen forests behind Good Success Bay and to collect alpine plants and insects, the next day, Darwin followed a stream into hills, through vegetation so dense he sometimes had to crawl, eventually halting his progress. Nevertheless, he found the wildness of decaying and fallen trees and tumbled rocks exhilarating: "To have made the scene perfect, there ought to have been a group of Banditti." Being armed and roughly dressed he felt "in tolerable unison with the surrounding savage Magnificence." Best of all was the idea that "this part of the forest had never before been traversed

Guanaco in Patagonia, February 2010

by man." However, when Darwin later described his climb, FitzRoy suggested he had probably been on the lower slopes of the very mountain Banks and Solander had climbed.

The following day, now feeling that he "was treading on ground, which to me was classic," Darwin followed the same watercourse, determined to reach the summit. Finding himself so hemmed in by "dead and living trunks" he could not find his way, he climbed a tree to take a compass bearing on the summit. Battling on, he startled two guanaco, which fled. Finally he forced his way through low Antarctic beech trees "as thick as Box in the border of a flower garden" on to the lichen-covered rocky summit to look down on Good Success Bay and "the little *Beagle*." He followed a ridge to an even higher snow-topped peak from where the views eastward toward the Le Maire Strait were magnificent. However, remembering the vicious snowstorm that had overtaken Banks, Solander, and their party at a similar time of year—mid-January—despite Good Success Bay being "in the latitude of Durham!" and despite mid-January being midsummer in the Southern Hemisphere—he did not linger.

Early on December 21, driven by a fine easterly wind, which, Darwin wrote with the knowledge he was fast acquiring of the weather in these regions, was "as lucky and rare . . . as getting a prize ticket in a lottery," the *Beagle* continued on its mission to return York Minster and Fuegia Basket to their homeland. By the following evening they had rounded "the old weather-beaten Cape Horn," but within hours seas around the Horn were living up to their reputation as a ship's graveyard as a terrific westerly gale began to push the *Beagle* back. On Christmas Eve, Darwin "saw this notorious point [the Horn] in its proper form, veiled in a mist and its dim outline surrounded by a storm of wind and water; Great black clouds were rolling across the sky and squalls of rain and hail swept by us with very great violence."

FitzRoy decided to take the *Beagle* into nearby Wigwam Cove, a sheltered bay on Hermit Island, west and a little north of Cape Horn. As Darwin knew, British explorer James Weddell had named Wigwam Cove a decade earlier during a voyage that took him south over the Antarctic Circle to latitude 74°15′ S—exceeding Cook's record south by over three degrees. Darwin was glad to be anchored "in quite smooth water" where "the only thing which reminds us of the gale . . . is the heavy puffs or Whyllywaws [sudden, violent squalls], which every 5 minutes come over the mountains, as if they would blow us out of the water."

The following day was Christmas. Wishing it to be kept "merrily," FitzRoy suspended all duties. Covington wrote, "The Captain indulged the ships company in everything he possible could . . . we could dance, sing, joke or in a word DO anything to make one another happy . . ." After breakfast, despite the cold, squally weather, Darwin landed with Hamond and Sulivan to climb Katers Peak, a steep 1,700-foot-high conical mountain overlooking the cove. With some of the slopes again clothed in thick beech, the three young men found it exhausting but enjoyed themselves taking pot shots at wild fowl and yelling to set off echoes. Sulivan hurled boulders down the mountainside while Darwin hammered "impetuously" away at the rocks with his geological hammer. Looking down at the small islets below, Darwin wondered whether they were "the termination of the chain of the Andes; the mountain tops only being raised above the ocean." He was correct—the southern and western portions of Tierra del Fuego's main island and the rest of the archipelago are indeed a continuation of the Andes and have peaks rising to more than 7,000 feet.

The miserable weather continued over the next few days, with day-time temperatures around 45°F and rain and hail falling almost constantly. "Considering this is the middle of the summer and that the latitude is nearly the same as Edinburgh, the climate is singularly uncongenial," Darwin grumbled. One day he accompanied FitzRoy on a reconnaissance of some surrounding bays and islets. The weather was "so bleak and raw as to render boating rather disagreeable," but he enjoyed examining some Fuegian wigwams. The shape was "like a cock of hay, about 4 feet high and circular; it can only be the work of an hour, being merely formed of a few branches and imperfectly thatched with grass, rushes etc." The reason the dwellings were "so very miserable," he concluded, was that the Fuegians were almost constantly on the move looking for food. However, they usually built their wigwams on mounds of shells and bones, suggesting they frequently returned to the same spot.

The birdlife caught his attention, especially the large goose-like steamer ducks that, though unable to fly, used their wings "to flapper along the water" and the penguins that spent so much time under it that "their habits are like fish" and that had only short feathers on their rudimentary wings. He noted he had seen three types of birds that had adapted their wings "for more purposes than flying; the Steamer as paddles, the penguins as fins, and the Ostrich spreads its plumes like sails to the breeze." Years later in his *On the Origin of Species* he would cite these species as examples of creatures adapted to their environment.

On New Year's Eve, with a slight improvement in the weather, FitzRoy took the *Beagle* back out to sea to continue west toward Christmas Sound. However, further ferocious gales blew up, allowing little progress to be made. With the ship pitching heavily "and the miseries of constant wet and cold," Darwin was "scarcely for an hour . . . quite free from sea-sickness: How long the bad weather may last, I know not: but my spirits, temper and stomach . . . will not hold out much longer." Even so, he noticed "how the Albatross with its widely expanded wings, glided right up the wind."

On January 13 the wind reached 11 on Beaufort's wind force scale, denoting a "violent storm" and only one below the top of the scale, force 12, "hurricane." In the early hours, the *Beagle* "lurched so deeply, and the main-mast bent and quivered so much" that FitzRoy was forced to take in all the sail he could. Later that morning, just as Darwin was remarking with apparent bravado to FitzRoy that "a gale of wind was nothing so very bad in a good sea-boat," to which FitzRoy replied "to wait until we shipped a sea," the seas rose "to a great height" and three huge rollers came surging toward the *Beagle*. Their "size and steepness" at once warned FitzRoy "that our sea-boat, good as she was, would be sorely tried." The *Beagle* rode the first wave, but it took off her headway. The second turned her from the wind. The third almost engulfed her: "taking her right a-beam, [it] turned her so far over, that all the lee bulwark . . . was two or three feet under water. For a moment, our position was critical; but, like a cask, she rolled back again, though with some feet of water over the whole deck. Had another sea then struck her, the little ship might have been numbered among the many of her class ['coffin-brigs'] which have disappeared: but the crisis was past . . ." Like a dog shaking water from its coat, "she shook the sea off her through the ports, and was none the worse . . ." FitzRoy told Darwin it was the worst storm he had ever encountered.

In his narrative, FitzRoy did not refer to the crucial role played by Sulivan and the ship's carpenter, Jonathan May, during the storm. According to Sulivan's son's memoir of his father, FitzRoy had issued a standing order that the ship's gunports on the open main deck should be kept securely shut, which Sulivan himself thought dangerous. He had therefore told the carpenter "always to have a handspike handy for eventualities." At the height of the storm, Sulivan "found the carpenter up to his waist in water, standing on the bulwark, driving a handspike against the port, which he eventually burst open. This probably saved the ship . . .," enabling foaming seawater to drain away more quickly.

Though the ship's rigging was largely intact and no lives had been lost, one of the *Beagle*'s new whaleboats had been washed into the sea, a chronometer had been damaged, and seawater had destroyed some of Darwin's collection. In his "List of Specimens not in Spirits" he recorded, "All specimens from 888 to 900 much injured by the gale . . ." It was "an irreparable loss . . . my drying papers and plants being wetted with salt-water.—Nothing resists the force of an heavy sea; it forces open doors and sky lights, and spreads universal damage.—None but those who have tried it, know the miseries of a really heavy gale of wind.—May Providence keep the *Beagle* out of them." Furthermore, it was "a disheartening reflection that it is now 24 days since doubling [rounding] Cape Horn . . . and we are now not much above 20 miles from it."

With the wind moderating and the seas dropping, FitzRoy tried to sail on, but squalls continued to push the *Beagle* back east around the Horn. On January 15, he found safe anchorage in Goree Sound, north of Cape Horn and near the eastern (Atlantic) entrance to the Beagle Channel. This 150-mile-long, three- to eight-mile-wide passage ran east–west, bounded to the north by Tierra del Fuego and to the south by a string of islands. They were not far from the homeland of Jemmy Button's people, the Yamana.[2] This proximity was perhaps why York Minster suddenly told FitzRoy "he would rather live with Jemmy Button" and his people than return to his own, which was "a complete change in his ideas."

FitzRoy was relieved no longer to have to try to battle to the Pacific side of Cape Horn for the present at least and anyway thought it better that York, Jemmy, and Fuegia Basket should live together. He set out by boat, accompanied by Darwin, to identify a suitable spot nearby for a new settlement to house them and Matthews, but it proved a fruitless quest. Apparently promising flat land "turned out to be a dreary morass only tenanted by wild geese and a few Guanaco," and further exploration convinced them the whole area was merely "a swamp." FitzRoy decided instead to take the Fuegians and Matthews to Jemmy Button's homeland some hundred miles west along the Beagle Channel in Ponsonby Sound before journeying on to chart the channel's western reaches.

---

2. The name *Yamana* that the canoe-going people of the channel used to describe them-selves means simply "people" in their language.

CHAPTER SEVEN

"Truly Savage Inhabitants"

FitzRoy prepared with customary care. Given the region's unpredictable weather, he decided to leave the *Beagle* at a secure anchorage in Goree Sound with a skeleton crew while he set out along the Beagle Channel with the ship's yawl and the three surviving whaleboats. With him would go Bynoe, McCormick's replacement as ship's surgeon, Hamond, several other officers, Darwin, Covington, twenty-four seamen and marines, and, of course, the three Fuegians and Matthews. Rather than show-ing signs of hesitation or reluctance, the young missionary appeared to FitzRoy to be positively eager "to begin the trial to which he had been so long looking forward."

FitzRoy ordered the carpenter Jonathan May to fit a temporary deck to the yawl to allow it to accommodate "a large cargo"—so large that even with her sails set, the other boats would have to tow her. This cargo included some of "the stock of useful things which had been given to [the Fuegians] in England," as FitzRoy in his published narrative called the assortment of items considered indispensable for daily life in any part of the globe by their genteel, well-meaning, if naïve, donors. Darwin was more to the point if less polite in his diary. Items such as decanters, tea trays, soup tureens, and chamber pots were the result of "the most culpable folly and negligence." The money spent could, he felt, instead have purchased "an immense stock" of really serviceable and practical articles. The expedition was also to take five rabbits, given to FitzRoy in England by a well-wisher for the mission, which he had tended carefully aboard the *Beagle* "in spite of their gnawing through every machine which could be contrived for their safety."

By the morning of January 19, the small convoy was ready. As it entered the Beagle Channel, Darwin found the scenery "most curious and interesting." "Indented with numberless coves and inlets, and as the water is always calm, the trees actually stretch their boughs over the salt water," it reminded him of Loch Ness in the Scottish Highlands. That first night they camped in a quiet cove where they built fires so huge that FitzRoy joked they could roast entire elephants over them. However, as one seaman was chopping firewood, his axe slipped, and he almost severed two fingers, leading the party to christen the spot Cutfinger Cove.

As they sailed on the next day, Fuegians watched from the shore. "Nothing could exceed their astonishment at the apparition of our four boats," Darwin thought. "Fires were lighted on every point to attract our attention and spread the news.——Many of the men ran for some miles along the shore . . . Four or five . . . suddenly appeared on a cliff near to us . . . absolutely naked and with long streaming hair . . . they sent forth most hideous yells. Their appearance was so strange, that it was scarcely like that of earthly inhabitants." FitzRoy observed the reactions of his Fuegian charges: York Minster, laughing derisively, called them "large monkeys," "Jemmy assured us they were not at all like his people, who were very good and very clean," while Fuegia Basket, "shocked and ashamed . . . hid herself, and would not look at them a second time." One cannot know what was really going through the minds of the three returnees. Having been shown a new world and now being returned to their old one must have raised difficult thoughts and mixed emotions, not least making them self-conscious with so many eyes, both of the crew and of their fellow Fuegians, on them.

That afternoon the convoy put in to the shore so the stiff, hungry occupants could prepare food. Fuegians who gathered around them were at first not friendly, holding their slings and stones in readiness. "Trifling presents such as tying red tape round the forehead" were well received but encouraged demands for further gifts, Darwin wrote. "The last and first word is sure to be 'Yammerschooner' which means 'give me.'"[1] Hamond thought the Fuegians "miserably thin in the arms and legs with large bodies" and was surprised how finicky they were about unfamiliar

---

1. *Yammerschooner* is actually the Yamana word *yamask-una*, meaning "do be liberal to me."

food. They ate ship's biscuit readily enough but reacted with disgust to a tin of preserved meat.

Hoping to find an uninhabited cove in which to overnight, FitzRoy pushed on "further westward than was at all agreeable, considering the labour required to make way against a breeze and a tide of a mile an hour." In the end he had to bivouac near a group of Fuegians who, Darwin wrote, "were very inoffensive so long as they were few in numbers." The night passed peacefully, but by early the next morning the mood had changed. More Fuegians arrived with first light, many wielding slingshots menacingly and some picking up stones from the shore.

Darwin was convinced an attack was imminent, but it did not materialize: "I was very much afraid we should have had . . . to have fired on such naked miserable creatures.——Yet their stones and slings are so destructive that it would have been absolutely necessary." The mere sight of the crew's muskets was no deterrent since "savages like these . . . have not the least idea of the power of fire-arms." A man armed with a musket "appears . . . far inferior to a man armed with a bow and arrow, a spear, or even a sling." The only way to demonstrate the muskets' superiority was "by striking a fatal blow. Like wild beasts they do not appear to compare numbers; for each individual if attacked, instead of retiring, will endeavour to dash your brains out with a stone, as certainly as a tiger . . . would tear you."

Another reason for being ready to take extreme measures, however regrettable, against the Fuegians was, in Darwin's view, that they were "bold cannibals." He must have discussed this with FitzRoy, who was convinced from his conversations with Jemmy, York, Fuegia, and others that "when excited by revenge or extremely pressed by hunger" their people indeed ate human flesh. In his narrative, FitzRoy would state that the Fuegians commonly killed and ate their prisoners of war——the men devouring the legs, and the women the arms and breast. In times of famine, the oldest woman in the group would be grabbed and held over a fire until the thick smoke choked her, whereupon the rest of the group would eat her every morsel. Jemmy Button, "in telling this horrible story" had seemed "much ashamed" of his countrymen's practices, assuring FitzRoy "he would rather eat his own hands" than do such a thing. When asked why the Fuegians did not kill and eat their dogs, he replied, "Dog catch iappo [otter]."

Despite Darwin and FitzRoy's categoric statements, anthropologists now doubt whether the Fuegians practiced cannibalism often, or indeed at all. The widely believed idea that they did persisted for many years and may have derived from exaggerated sealers' stories or from mutual incomprehension. Perhaps outsiders' credulity resulted from their pre-conceptions and prejudices, generalizing to all "savages" the experiences of Cook and others in the Pacific.

As the voyage continued, Fuegia Basket caught her dress on a hook in the boat she was traveling in. Hamond seemed to find her a nuisance, noting in his diary, "We took the precaution to shift her over upon each tack, for she was more like a bundle of dirty clothes, and much more in the way." Everyone noticed that York was extremely jealous of any atten-tion paid "to his intended wife, Fuegia."[2] As FitzRoy wrote in his account of that day, "He had long shewn himself attached to her . . . If any one spoke to her, he watched every word; if he was not sitting by her side, he grumbled sulkily; but if he was accidentally separated, and obliged to go in a different boat, his behaviour became sullen and morose."

The scenery grew ever more magnificent. Straining for a good view, despite "the lowness of the point of view in a boat," Darwin looked up at an unbroken sweep of three-thousand-foot-high jagged-peaked mountains rising from the water's edge. "The dusky-coloured forest" covering their lower slopes to a uniform height of about fifteen hundred feet created a "truly horizontal" tree line, reminding him of "the high-water mark of drift-weed on a sea-beach."

On the evening of January 22, their craft reached a small bay close to the entrance to Ponsonby Sound, where, FitzRoy wrote, "from a small party of . . . natives, Jemmy's friends, whom we found there, he heard of his mother and brothers, but found that his father was dead." During

---

2. Missionaries and westerners always used the term *wives*, although no formal marriages existed.

the voyage Jemmy had had a premonition that his father had died, telling Bynoe how one night a man had come to his hammock and whispered the news in his ear. Jemmy still appeared to have forgotten much of his own tongue, or to be pretending he had, and York Minster, although from another tribe, did much of the interpreting.

Darwin thought the people "quiet and inoffensive" and noted how, as they joined the seamen around a blazing fire, though naked, "they streamed with perspiration at sitting so near to a fire which we found only comfortable." He had watched a woman in a canoe suckle her newborn baby "whilst the sleet fell and thawed on her naked bosom, and on the skin of her naked child." Their ability to survive extreme cold with little or no clothing struck him forcibly as it had earlier European visitors like James Cook and Joseph Banks. Animal skins—otter, fox, seal, sea lion, and guanaco—worn loose over their shoulders and reaching nearly to their knees or, in the case of women, tied apron-like around them—provided their only clothing. Sometimes they wound strips of raw seal hide around their feet to protect them. Very often they went completely naked.

The Fuegians had dispatched a canoe to spread word of the new arrivals, and early the next morning canoes appeared from every direction until nearly forty had assembled. Other Fuegians came by foot, running so fast, according to Darwin, "that their noses were bleeding," and "they talked with such rapidity that their mouths frothed." With their naked bodies painted red, white, and black they looked "like so many demoniacs." They were good-naturedly ebullient but very curious about the visitors and their possessions—one man managed to steal Hamond's axe but returned it without argument when challenged.

Followed by a string of canoes and guided by Jemmy Button, Fitz-Roy brought his flotilla down Ponsonby Sound to a quiet cove named Woollya on Navarin Island, where Jemmy and his family had lived. The usually restrained FitzRoy described the moment almost lyrically to his sister—"snow-capped mountains glittered in the sun on one side while on the other they threw a deep darkness over the icy smooth dark blue water . . . From the fires in each canoe, small columns of blue smoke ascending added to the novelty and picturesqueness of the scene. It was

Button Island near Woollya, by C. Martens

not what one would expect in Tierra del Fuego . . ." Snowy mountains apart, it was "a scene of the South Sea Islands." As for the curling smoke, FitzRoy was constantly impressed by the Fuegians' ability in their cold, wet, and windy climate to create an effective fire quickly compared with the two or so hours it took his men. Although the Fuegians tried to keep their fires alight, whether in their canoes, wigwams, or even holding burning brands, "they are at no loss to rekindle it, should any accident happen. With two stones (usually iron pyrites) they procure a spark, which received among tinder, and then whisked round in the air, soon kindles into a flame." Their favored tinder was birds' down, fine moss, or dry fungus.[3]

FitzRoy ordered his men to land, set up camp, and dig a boundary ditch around which he posted armed sentries. Meanwhile a Fuegian family whom the new arrivals had found living at Woollya sent a canoe

---

3. The Fuegians laid fires in their canoes on a bed of stones, clay, and coarse sand to prevent the canoes catching light.

to Jemmy's relations with news of his return. The next day "the Fuegians began to pour in," including Jemmy Button's mother, four brothers, two sisters, and an uncle. However, Jemmy's family seemed suspicious rather than joyful about their reunion. FitzRoy told his sister "strange dogs meeting in a street shew more anxiety and more animation than was manifested at this inhuman meeting of a lost child and his afflicted mother and relatives."

Darwin too noticed "no demonstration of affection" but was surprised by "the astonishing distance" at which Jemmy recognized one brother's voice: ". . . their voices are wonderfully powerful.—I really believe they could make themselves heard at treble the distance of an Englishmen [*sic*]." Although Darwin had excellent eyesight himself—few *Beagle* crew could see further, and his brother Erasmus teased him about the "telescopes you call eyes"—during the voyage he had noted the Fuegians' superb vision: "All the organs of sense are highly perfected; sailors are well known for their good eyesight, and yet the Fuegians were as superior as another almost would be with a glass.—When Jemmy quarrelled with any of the officers, he would say, 'me see ship, me no tell.'" At this stage, Darwin simply concluded that the Fuegians' exceptional hearing and vision had developed as survival aids in a challenging environment.

With its sheltered position and fertile soil, Darwin thought Woollya "far better suited for our purposes, than any place we had hitherto seen." FitzRoy described how, "rising gently from the water-side, there are considerable spaces of clear pasture land, well watered by brooks, and backed by hills of moderate height . . . Rich grass and some beautiful flowers . . . augured well for the growth of our garden seeds." Deciding this would be an excellent site for Matthews's mission, he set his men to laying out and planting two gardens with vegetables, including potatoes, turnips, beans, carrots, and onions, and building three large huts with sapling frames and thatched with grass and twigs, like Fuegian wigwams, but cocooned in sailcloth and stout rope to strengthen them against the elements. Matthews's hut had an attic made from ship's boards for his stores and a secret space beneath the floorboards in which to hide his most precious belongings.

For several days all was tranquil. The local people—Darwin estimated they numbered about 120—watched quietly as the work progressed. What seemed to interest them most was the Europeans' white skins revealed as they washed in streams. Darwin himself felt confident enough to take long walks in the surrounding woods. However, by January 27, with the Fuegians stealing whatever they could and communication problems—Jemmy was still not speaking the Yamana language with any fluency—tensions were rising. Then, all of a sudden, most of the men left in their canoes. Shortly afterward the *Beagle*'s sailors realized they were now watching them from a neighboring hill. Some thought the Fuegians had been frightened away by the sight and sound of the crew cleaning and discharging their muskets the previous evening, which FitzRoy had ordered as a deterrent to the thieving. Others suspected their withdrawal was connected with an argument between an old man and a sentry guarding the camp perimeter—"the old man being told not to come so close spat in the seaman's face and then . . . made motions, which . . . could mean nothing but skinning and cutting up a man."

FitzRoy himself suspected the sudden exodus was the prelude to an attack and "thought it advisable not to sleep another night there." However, Jemmy, York, and Fuegia seemed content to remain, as did Matthews. The missionary "behaved with his usual quiet resolution," Darwin wrote, though adding, "he is of an eccentric character and does not appear (which is strange) to possess much energy."

That evening, leaving Matthews and his three charges "to pass rather an aweful night," as Darwin put it, FitzRoy moved the rest of his party to a neighboring cove. Returning next morning to Woollya, he was relieved to find most of the Fuegians were back and quietly spearing fish in the cove. Satisfied all was well but promising to return in a few days, FitzRoy dispatched the yawl and one whaleboat back to the *Beagle* and with the two remaining boats set off with Darwin, Hamond, and others to survey the western reaches of the Beagle Channel.

Darwin found this part of the channel "very striking . . . So narrow and straight a channel and in length nearly 120 miles, must be a rare phenomenon.—We were reminded that it was an arm of the sea, by the number of Whales . . ." At night, they took turns to stand sentry, Darwin included. He found "something very solemn in such scenes; the consciousness rushes on the mind in how remote a corner of the globe you are then in . . . the quiet of the night is only interrupted by the heavy breathing of the men and the cry of the night birds.—the occasional distant bark of a dog reminds one that the Fuegians may be prowling, close to the tents, ready for a fatal rush."

On January 29, disaster nearly overwhelmed the group after they landed at the base of some mountains. As they were admiring some magnificent "beryl blue" glaciers running down to the channel only a couple of hundred yards away, a large mass of ice fell from them "roaring into the water," creating a mini tsunami. Seeing the water surge and great waves rolling toward the boats, Darwin and several others dashed to them, seized the mooring ropes, and hauled them further up the shore in the nick of time before "they were tossed along the beach like empty calabashes." In his diary Darwin did not mention his own role in saving the boats, their equipment, and supplies, hence preventing the marooning of the expedition in a hostile environment. The next day a grateful FitzRoy named a westward continuation of the Beagle Channel, down which they sailed for some way, "Darwin Sound" for "my messmate who so willingly encountered the discomfort and risk of a long cruise in a small loaded boat."

The grandeur of the scenery continued to awe Darwin, as did the realization that "we were sailing parallel . . . to the backbone of Tierra del. [sic]; the central granite ridge which has determined the form of all the lesser ones." In the bone-chillingly cold, rainy weather typical of Tierra del Fuego, the party reached their farthest point west—Stuart Island at the entrance to the Pacific and some 150 miles from the *Beagle*—before turning back for Woollya Cove to check on Matthews's progress. Reentering Ponsonby Sound, they encountered a group of Fuegians in a canoe. Ominously, one woman was wearing material belonging to Fuegia Basket. "The sight of this piece of linen, several bits of ribbon, and some scraps of red cloth" made FitzRoy anxious about Matthews and his party.

Making as much as haste they could back to Woollya, they met other Fuegians also sporting items that could only have come from there. As they landed guns at the ready, Fuegians rushed whooping toward them, but then, FitzRoy wrote, "to my extreme relief, Matthews appeared, dressed and looking as usual. After him came Jemmy and York, also dressed and looking well; Fuegia, they said, was in a wigwam." FitzRoy took Matthews into his boat so they could talk in private while Fuegians watched intently from the shore, reminding FitzRoy "of a pack of hounds waiting for a fox to be unearthed."

Despite his outwardly calm appearance, the missionary was traumatized.

Darwin wrote in his diary, "From the moment of our leaving, a regular system of plunder commenced, in which not only Matthews, but York and Jemmy suffered." He did not mention Fuegia Basket, whose clothes had also been seen to have been stolen, but added, "Matthews had nearly lost all his things; and the constant watching was most harassing and entirely prevented him from doing anything to obtain food etc. Night and day large parties of natives surrounded his house . . . They showed by signs they would strip him and pluck all the hairs out of his face and body [with mussel-shell tweezers].——I think we returned just in time to save his life." Matthews could not wait to escape from Woollya and pleaded with FitzRoy to take him with him to New Zealand to join his brother, a missionary there.

FitzRoy agreed, but the mission's failure was a disappointment to him. He had hoped it might prove a catalyst for improving the lot of the Fuegians. He was unsure whether the Fuegians had any "distinct belief in a future life," although his time with his three Fuegian passengers had convinced him "their ideas were not limited by the visible world. If anything was said or done that was wrong, in their opinion it was certain to cause bad weather. Even shooting young birds, before they were able to fly, was thought to be a heinous offence." He recalled York Minster chiding Bynoe for shooting some very young ducks: "Oh, Mr. Bynoe, very bad to shoot little duck—come wind—come rain—blow—very much blow." Jemmy Button was similarly "very superstitious and a great believer in omens and dreams." This belief that evil spirits tormented the Fuegians in this world—sending storms, hail and snow to punish them for misdemeanors—seemed to exclude any belief in "future retribution" and in an immortal soul. Nevertheless, though he had never witnessed "any act of a decidedly religious nature," FitzRoy believed that "ideas of a spiritual existence—of beneficent and evil powers—they certainly have." However, Matthews, "rather too young, and less experienced than might have been wished," as FitzRoy described him, was clearly not the person to establish a mission in challenging conditions.

Through subterfuge and careful stage management, FitzRoy and his men rescued what remained of Matthews's belongings from his underground hideaway. Jemmy and York, however, were determined to remain. Darwin found it "quite melancholy leaving our Fuegians amongst their barbarous countrymen," especially Jemmy, "a universal favourite" who had consoled him on the *Beagle* during bouts of seasickness by murmuring "poor, poor fellow!" He reassured himself that they "appeared to have no personal fears" and that York, despite being "a full grown man and with a strong violent mind" when brought to England, would "I am certain in every respect live as far as his means go, like an Englishman." He was more concerned about Jemmy, whose own brother had been stealing from him and who "looked rather disconsolate" and in Darwin's view "certainly would have liked to have returned with us."

Darwin did not discuss Fuegia Basket's feelings—perhaps believing that either her sex or age meant she would automatically follow the lead of the other two—but predicted pessimistically for all three that "whatever other ends their excursion to England produces, it will not be conducive to their happiness.—They have far too much sense not to see the vast superiority of civilized over uncivilized habits; and yet I am afraid to the latter they must return." Darwin had noticed how the Fuegians had divided their plunder equally among them, concluding in an apparent argument for capitalism that "the perfect equality" among the Fuegian tribes "must for a long time retard their civilization." Unless a leader emerged, allowing the development and imposition of a hierarchical structure and benefits such as promoting agriculture or domesticating animals, nothing would improve. Yet how could such a leader emerge "till there is property of some sort by [the acquisition of] which he might manifest and still increase his authority?"

FitzRoy promised to return in a few days to check that all was well, and true to his word, a week later took a single whaleboat back to Wool-lya. He found that "very few of the things belonging to Jemmy, York and Fuegia had been stolen and the conduct of the natives was quite peacible." Jemmy was hollowing out a tree trunk to make a canoe of the kind he had seen Brazilians use in Rio de Janeiro. Though the gardens had been trampled, a few hardy vegetables were sprouting leaves. To his sister FitzRoy wrote, "Things might have turned out worse, as well as better." Darwin, backtracking a little from his previous comments, reflected that "this little settlement may be yet the means of producing great good and altering the habits of the truly savage inhabitants."

CHAPTER EIGHT

Res Nullius

On February 26, FitzRoy set course for the Falkland Islands, three hundred miles northeast of the tip of South America. The remote archipelago had a complex, at times bloody, recent past. John Davis, an English navigator aboard the *Desire*, was, in 1592, probably the first to sight the uninhabited islands. John Strong, another Englishman, on the *Welfare*, made the first recorded landing in 1690. He named the channel between the two main islands after Viscount Falkland, Treasurer of the Royal Navy. The British subsequently extended the name to the entire archipelago, prosaically calling the two largest islands East Falkland and West Falkland.

Half a century later, Commodore George Anson passed by the islands during his circumnavigation and later recommended them to the British government as a possible base near Cape Horn. The French government also realized the islands' potential for refuge and supply and was slightly quicker off the mark. In 1764, the French explorer Louis de Bougainville established the first settlement, Port Louis, on East Falkland and named the islands Les Iles Malovines (Islas Malvinas in Spanish) after St. Malo in northern France. The following year, unaware of de Bougainville's arrival, Commodore John "Foulweather Jack" Byron claimed the islands for King George III and founded the first British settlement—Port Egmont on Saunders Island off West Falkland.

The almost simultaneous arrival of the French and the British alarmed Spain, who considered the islands part of its South American empire. Spain and France were close allies, and in response to Spanish pressure, the French government ordered de Bougainville to sell his interests in Port Louis to Spain. The settlement was renamed Puerto de la Soledad, placed under the colonial government in Buenos Aires, and five Spanish frigates were dispatched to the islands with 1,400 marines. Their arrival

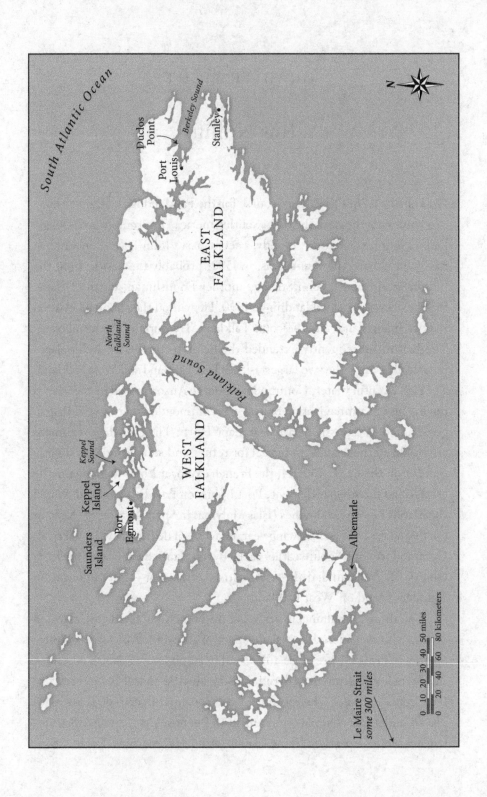

in 1770 convinced Port Egmont's small British garrison to depart, but only after leaving a plaque asserting Britain's claim. Rather than risk war, Spain restored the settlement to Britain the following year. Finding its far-flung colony "expensive and almost useless," as FitzRoy put it—a view some later British governments would share—Britain soon abandoned it but did not renounce sovereignty.

Nothing significant happened until the 1820s when Louis Vernet, a naturalized Buenos Aires citizen, with British agreement, reestablished a settlement at Puerto de la Soledad, which again became Port Louis, as a private venture. Now independent from Spain, in 1829 the Buenos Aires government claimed the islands by right of inheritance from the former colonial power and appointed Vernet as governor, whereupon Britain lodged a formal protest.

In July 1831, Vernet and his business partner Matthew Brisbane, a Scottish Antarctic explorer and sealer, seized several American sealing vessels for illegally operating in the islands' waters. Vernet took some of the arrested crewmen to Argentina for trial. During his absence, Captain Silas Duncan arrived aboard the corvette USS *Lexington* to avenge the insult. He sacked Port Louis, arrested seven senior settlers, including Brisbane, as "pirates" and declared the islands *res nullius*, "belonging to no one."

Still determined to enforce its claim, in 1832 the Buenos Aires government appointed Major Esteban Mestivier the new governor, with orders to establish a penal colony, and dispatched him to Port Louis with a tiny garrison and a few convicts. In response, the British government ordered the commander in chief of the South American naval station to raise the British flag again in the islands. Captain James Onslow, commanding HMS *Clio*, arrived in late December 1832, soon to be joined by HMS *Tyne*.

Onslow quickly discovered Mestivier's men had mutinied and murdered him. He ordered the officer who had taken charge after suppressing the mutiny to depart with his men and put William Dickson, Vernet's storekeeper, in charge of the settlement. Onslow instructed him to fly the British flag every Sunday—and whenever a ship arrived—and helpfully gave him a flagpole to erect. On January 7, 1833, task completed, the *Clio* and the *Tyne* departed.

Settlement at Port Louis, Falkland Islands, by C. Martens

Two months later, on March 1, after days of blustery winds and heaving seas—"the history of this climate is a history of its gales," Darwin complained—the *Beagle* approached East Falkland and anchored in Berkeley Sound, a wide inlet on whose western shore lay Port Louis.[1] Those aboard were, of course, unaware of recent events: "The first news we received was to our astonishment, that England had taken possession of the Falklands Islands and that the [Union Jack] Flag was now flying," Darwin wrote. His informant was the chief mate of *Le Magellan*, a French whaling vessel driven ashore in the great storm of January 13, during which the *Beagle* had lost a whaleboat. According to him, the present inhabitants were only Dickson, who had "charge of the British Flag, 20 Spaniards and three women . . ."

Le Magellan's crew were living in tents made from the sails of their ship on the northern shore of Berkeley Sound. FitzRoy visited their encampment that afternoon. On inspecting the wreck, he decided that with a little energy and enterprise the whalers could have refloated her but that he had "too many urgent duties" himself to help those "who would not help themselves." However, he promised to carry as many of the whalers

---

1. Berkeley Sound is north of modern-day Port Stanley.

with their stores as he could to Montevideo and arranged to buy timber from the wreck for firewood. Meanwhile he had dispatched Edward Chaffers to check the situation in Port Louis. Chaffers reported that "there was no constituted authority whatever resident on the islands" but that Onslow had indeed left the flag in charge of Dickson. "What a strange solitary life his must be," Darwin reflected.

Early next morning, FitzRoy took the *Beagle* across the sound to anchor closer to *Le Magellan* crew's camp and to begin survey work. Beaufort had told him he need not undertake a complete survey but, since the islands were a frequent refuge for sealing and whaling vessels, to ensure ships had reliable information about safe anchorage and available resources, like timber for repairs and food for their crews.

Darwin was quickly ashore, comparing the plants and insects of the Falkland Islands with those of the South American mainland and wondering without reaching any definitive conclusion whether similarities suggested "closer connection than migration." After the jagged, snow-crowned peaks of Tierra del Fuego, the landscape was desolate and dreary, being "low and undulating with stony peaks and bare ridges" with few trees. Covington, tramping across the peat beds covered in coarse, wiry brown grass, found bushes laden with red berries that were "very good eating." He also noted the plentiful "wild oxen"—descendants of cattle brought by French settlers in the 1760s—and the many horses, pigs, and rabbits also introduced into the islands. Geese, ducks, and snipe were prolific.

On March 3, a merchant schooner arrived from Buenos Aires, the crew of which "hastened to make themselves drunk" as reward for a difficult voyage. Among the passengers was Matthew Brisbane, Vernet's business partner, arrested by the commander of USS *Lexington* for piracy but now released. Brisbane immediately boarded the *Beagle* to explain to FitzRoy that he had returned to safeguard what remained of the long absent Vernet's property.

The next day Brisbane took FitzRoy on a tour of Port Louis to show him the extent of the damage inflicted by the *Lexington*, which FitzRoy deplored as crass vandalism. On his return to the *Beagle* tragic news awaited FitzRoy. The seminaked body of his clerk, Edward Hellyer, had been found entangled in thick ribbons of kelp in the sea. Mindful of FitzRoy's standing order that no one should go far unaccompanied, he had gone shooting with one of *Le Magellan*'s crew. However, after about a mile, the whaler had returned. Hellyer had gone on alone, telling the man he wanted to shoot a type of duck he had not seen before. An islander had found his clothes, watch, and gun on the shore and raised the alarm. Darwin and Bynoe were among a search party that discovered the corpse. Darwin believed Hellyer had "shot a bird and whilst swimming for it, the long stalks of the seaweed had caught his legs and thus caused his death."

The following day, March 5, a melancholy cortege preceded by a Union Jack carried at half-mast conveyed Hellyer's coffin for burial at Duclos Point, a "lonely and dreary headland" in sight of the anchored *Beagle*. A distressed FitzRoy wrote: "Mr Hellyer had been much with me, both as my clerk and because I liked his company." He feared Hellyer had plunged into the sea to retrieve the unusual species of duck from "a desire to get it for my collection."

Darwin continued to explore, shooting snipe and collecting rocks and "the few living productions which this Island has to boast of." Walking along the beach, he noticed algae encrusted rocks and, just as in Edinburgh, took specimens of algae. Contemporary opinion was divided over whether these primitive organisms were plants or animals. Peering at their structure through his microscope, he decided they were plants.[2]

Another of his discoveries was a large white Doris sea slug. It produced elegant ribbon-like egg masses, similar to strings of beads, which adhered to rocks and which, Darwin calculated, contained "the enormous number of six hundred thousand eggs." Nature's logic for such "wonderful . . . fecundity" featured in his later evolutionary thinking about how seeming overproduction was designed to compensate for heavy loss of offspring before or immediately after hatching.

---

2. Today algae are classed as being neither animal nor plant but as sharing characteristics of both.

With the ships' boats continuing survey work and sailors busy loading coals and timber from *Le Magellan*, taking on fresh water, making repairs and scrubbing hammocks, Darwin, unusually for him, became bored and decided to walk the few miles to Port Louis. He found the semiruined settlement depressing—"The place bespeaks what it has been, viz a bone of contention between different nations." Yet during that walk "the whole aspect of the Falkland Islands were . . . changed to my eyes . . . for I found a rock abounding with shells and these of the most interesting geological era." These fossil shells, embedded in bands of reddish-yellow sandstone, seemed similar to those he saw in Wales. He told Henslow, "As this is so remote a locality from Europe, I think the comparison . . . with the oldest fossiliferous rocks of Europe will be pre-eminently interesting." He wondered whether such abundant shells in the now so bleak and chilly Falkland Islands might "indicate a climate previously warmer."[3]

The arrival on March 26 of "a notorious and singular man who has frequented these seas for many years and been the terror to all small vessels" also relieved Darwin's boredom. The man in question was Scotsman William Low, the captain of a sealing vessel, the *Unicorn*. FitzRoy described him as returning from a disastrous six-month voyage "a ruined man, with an empty ship."

"It is commonly said," Darwin wrote, "that a Sealer, Slaver and Pirate are all of a trade; they all certainly require bold energetic men and amongst Sealers there are frequently engagements [fights] for the best [seal] 'rookeries' and in these affrays Capt. Low has gained his celebrity.—In their manners, habits etc I should think these men strikingly resembled the old Buccaneers." The *Unicorn* had picked up survivors of yet another ship wrecked during the January 13 storm. On hearing of the loss, Darwin congratulated himself that he had survived "the very worst weather, in one of the most notorious places in the world . . . in [the *Beagle*,] a class of vessel . . . generally thought unfit to double the Horn . . . Few vessels would have weathered it better than our little 'diving duck.'"

---

3. Modern radiometric techniques date Darwin's fossil shells to between 350 and 400 million years old.

Low's arrival with an apparently sound ship and in obvious need of money struck FitzRoy as an opportunity. FitzRoy "had often anxiously longed for a consort, adapted for carrying cargoes, rigged so as to be easily worked with few hands, and able to keep company with the *Beagle*." Recently he had become worried that the "tedious, although not less useful, details of coast surveying" might consume so much time that he would not be able to complete "the most interesting part" of the voyage—the global chain of longitudinal measurements. Hoping Low's roomy, oak-built 170-ton *Unicorn* might be the answer, he ordered carpenter Jonathan May to inspect her. When May pronounced her in good condition, FitzRoy's "wish to purchase her was unconquerable," and he struck a deal with Low. Darwin thoroughly approved. The *Unicorn* was "an excellent sea-boat," he wrote—"this day will be an important one in the history of the *Beagle*." He hoped "it may shorten our cruise" and at least reduce the time needed for the surveying work and thought it "always pleasant to be sailing in company . . . to break the monotonous horizon of the ocean."

FitzRoy hired a few of Low's men to help crew his new acquisition, renamed her *Adventure*, "to keep up old associations"—the *Adventure* had been the *Beagle*'s consort on the first voyage—and placed Edward Chaffers, the ship's master, in command until he could recall Wickham from the *Liebre* to take over permanently. On April 4, after making what improvements he could with the limited resources available, FitzRoy dispatched the *Adventure* to the Rio Negro on the Patagonian mainland, where he intended to rejoin her and oversee her further refitting.

Of course, FitzRoy could not know whether the Admiralty would pay for all this. Meanwhile, as when hiring the smaller *Paz* and *Liebre*, he funded the cost himself—nearly £1,300—to be paid to Low's partners in Montevideo, and several hundred pounds more to refit her. To his London banker, Hugh Hoare of Hoare's Bank, he wrote praising his purchase—"She is a Schooner, English built—and of the very best materials," of his hope that "the wise heads at the Admiralty may give me some assistance—and lighten the heavy burden which now lies entirely upon me," but seeking an advance of £1,060. In a postscript he added, "I hope that my relations will not know what expense I have incurred . . . Their anxiety would be excited unnecessarily. Those who are married and have children do not think as an unmarried sailor whose castle is the hollow oak, whose home is on the sea. Such a voyage, as I am now making is not likely to occur twice in my life and I am quite resolved to sacrifice future comforts—and views of settling on shore—to the effective execution of the service in which I am engaged. If I can afford it, I shall take my consort—(she is dumb—obedient—and does not eat) round the world with me."

To the Admiralty, Fitzroy wrote, "I believe that their Lordships will approve of what I have done, but if I am wrong . . . I alone am responsible . . . and am willing and able to pay the stipulated sum." Naturally, he very much preferred not to pay, explaining to Beaufort why he had bought the Unicorn, asking for additional men to crew her and exhorting his friend, "Now pray fight my battle . . ." Possibly in an attempt at persuasive flattery, he praised the Beaufort scale and how well the crew had adapted to it: "You would have smiled at hearing some of my Shipmates saying . . . 'if Captain Beaufort were here now he'd call this fifteen . . . we certainly had a large share of eleven at the least. The numbers and letters are as familiarly used now as could be desired, and no one would willingly return to the old plan."

Meanwhile, FitzRoy strove to maintain order in the islands. He worried what would happen after the Beagle left, given the frequent arrival of often drunken and violent whalers and sealers: "the men of several American vessels all armed with rifles; the English sealers with their clubs . . . the crews of several French whalers—who could not or would not see why they had not as good a right to the islands as Englishmen." Many months "might elapse without the presence of a [British] man-of-war, or the semblance of any regular authority."

The "cut-throat looking gauchos," brought by Vernet from Argentina to work at Port Louis, who gambled, quarreled, and fought with their long knives, also concerned him. Yet he knew they were essential to the settlement as "the only useful labourers . . . in fact, the only people on whom any dependence could be placed for a regular supply of fresh beef."

Though most of the dozen or so gauchos wished to leave the islands, he "induced" seven to remain, presumably by offering them money. Three days before the *Beagle*'s departure, with their lassos and bolas, they brought down several wild cattle "bellowing in impotent rage" so that 690 pounds of beef could be brought aboard.

Darwin kept busy in the kelp "with the Zoology of the Sea; the treasures of the deep to a naturalist are indeed inexhaustible." Unfortunately in the process he broke his watch, reminding himself in his notebook to "send watch to be mended." He did not regret leaving islands about which he shared FitzRoy's view that "there is very little either remarkable or interesting" and "no lack of the elements of discord."

On April 6, FitzRoy set course for the Rio Negro, intending to rendezvous with the *Adventure* but also with Wickham and Stokes on the two schooners. Driven by gales, the *Beagle* maintained a cracking pace, but Darwin found working on his specimens difficult—"every place dark wet and the very picture of discomfort." A week later, the *Beagle* arrived off the Rio Negro "in baffling winds" and a heavy swell that prevented FitzRoy from approaching the shore. With no sign of the *Adventure*, *Paz*, or *Liebre*, he ordered the firing of a gun to signal the *Beagle*'s arrival and spent the rest of the day cruising back and forth, which Darwin found "exceedingly annoying."

Around noon the following day, a sail sighted to the southwest proved to be the *Adventure*, which FitzRoy now ordered Chaffers to take to Maldonado on the northern bank of the Plate estuary and there await the *Beagle*. The next day, the *Beagle* encountered a small vessel leaving the estuary of the Rio Negro carrying letters for FitzRoy from Lieutenant Wickham on the *Liebre*, reporting the sad news that a marine, Corporal Williams, had drowned in the Rio Negro. Wickham also wrote that the two schooners were heading for the Gulf of San Matias, about a hundred miles to the south, and FitzRoy decided to try to locate them there.

After the rough passage from the Falklands, Darwin enjoyed the voyage south beneath "celestial" blue skies, with soft, warm breezes and smooth seas. In the unbroken line of cliffs, the rock strata ran for miles "exactly parallel to the surface of the sea"—an "El Dorado," which he was certain was studded with fossils. Porpoises played around the ship, and a female was harpooned. FitzRoy sketched it while Darwin measured its length and preserved the head to send home—later to be identified as belonging to a new species and named *Delfinus fitzroyi* after the captain.

By April 19, with northwesterly winds rising, FitzRoy abandoned the search for the schooners and headed north again. Five days later the *Beagle* was off the Plate estuary, with lightning streaking the sky and St. Elmo's fire illuminating one of the mastheads. Finding the current had pushed the ship several miles off course, FitzRoy made first for Montevideo, west of Maldonado, where Darwin found things had "been going on pretty quietly, with the exception of a few revolutions." He went ashore to visit Augustus Earle, whose health had still not sufficiently improved for him to rejoin the *Beagle*. FitzRoy would replace him in July with Conrad Martens, who, hearing the *Beagle* needed an official artist, offered FitzRoy his services.

Waiting for Darwin were several letters from his sisters. They reported the death after a short illness of his twenty-six-year-old Maer cousin Fanny Wedgwood, whom they had once teased him about as a possible wife, and the subsiding of the cholera outbreak in Shrewsbury. On a lighter note, they cataloged marriages, hunt balls, and the travails of the erstwhile Fanny Owen with her "horrid" new mother-in-law. They also told him Dr. Darwin had followed his advice and ordered a banana tree for which a deep hole had already been dug in the hothouse at The Mount, and that the British press had reported the *Beagle* crew's armed intervention during the insurrection in Montevideo.

The *Beagle* finally reached Maldonado on the Plate estuary on April 28 to find the *Adventure* anchored there "all safe and snug." FitzRoy hired a small boat to take Alexander Usborne—assistant to Chaffers, the ship's master—to the Rio Negro to replace Wickham and to bring the latter back to command the *Adventure*. He then took the *Beagle* straight back to Montevideo to purchase materials to refit the *Adventure*.

Darwin remained with Syms Covington in Maldonado, where he found somewhat unsatisfactory lodgings in the house of an elderly lady. The rooms were large with high ceilings but "almost destitute of furniture," with tiny windows. "The very existence of what an Englishman calls comfort never passed through the builder's mind," Darwin grumbled. For the first few days torrential rain prevented him exploring the hinterland—"a bridge is an invention scarcely known in these parts." He spent long hours in his gloomy quarters writing letters. He told a Cambridge friend, "It does ones heart good to hear how things are going on in England.—Hurrah for the honest Whigs.—I trust they will soon attack that monstrous stain on our boasted liberty, Colonial Slavery." Though he did not yet know it, his Whig uncle Jos Wedgwood had been elected Member of Parliament for Stoke on Trent in the recent general election—the first since the passing of the Reform Bill, which had swept the Whigs under Lord Grey back into power.

By May 9, the storms had finally eased, and Darwin set out on a short excursion seventy miles northward with Covington, two guides, and, because they were so cheap, a dozen spare horses so they could always have fresh mounts. Since the local practice was "to ask for a night's lodging at the first convenient house" and his guides had "plenty of friends and relations in the country [they] were just the people for my purpose." They were also very well armed. Initially Darwin wondered whether their "pistols and sabres" were strictly necessary, "but the first piece of news we heard . . . that, the day before, a traveller from M. Video had been found dead on the road, with his throat cut," quelled any doubts.

At pulperias—drinking shops—or in the private houses where he overnighted, Darwin discovered he could cause a sensation by demonstrating how he used his pocket compass and map. It "excited the liveliest admiration, that I a perfect stranger should know the road . . . to places where I had never been." His "promethians" also caused excitement. These were a type of match patented in London five years before by Samuel Jones and consisting of a glass bead containing a single drop of vitriol, wrapped in paper coated with a mixture of chlorate of potash, powdered sugar, and gum arabic. When the tip of the glass bubble was nipped with pliers or bitten off—as Darwin did—the vitriol reacted with the primed paper to create a flame—"it was thought so wonderful that a man should strike fire with his teeth."

The apparent ignorance of even wealthy ranch owners surprised Darwin: "I was asked whether the earth or sun moved; whether it was hotter or colder to the north; where Spain was . . ." Some believed "England was a large town in London!" He attributed this to few foreigners passing through the region and realized his novelty was an exploitable asset.

The general astonishment at the compass and other things was . . . advantageous, as with that and the long stories my guides told of my breaking stones, knowing venomous from harmless snakes, collecting insects etc I paid them for their hospitality.——Being able to talk very little Spanish, I was looked at with much pity, wonder and a great deal of kindness . . . time after time they pile heaps of meat on your plate; having eat[en] a great deal too much and having skilfully arranged what is left so as to make as little show as possible, a charming Signorita will perhaps present you with [a] choice piece from her own plate with her own fork; this you must eat . . . Oh the difficulty of smiling sweet thanks, with the horrid and vast mouthful in view.

Journeying through seemingly endless green hills, Darwin found the gauchos drinking spirits and smoking cigars in the pulperias far more interesting. Tall, handsome, with "proud, dissolute" expressions, curling mustachios, long black hair, bright clothing, great clanking spurs, and sharp-bladed knives at their waists, they never drank "without expecting you to taste it; but, as they make their exceedingly good bow, they seem quite ready . . . to cut your throat at the same time." Darwin watched a gaucho disarm a drunken elderly man who, having taken offense, had drawn his knife from beneath his poncho. "After this, to frighten the old gentleman the others in jest pretended to stab him.——the method with which they dashed across the room, struck him upon the heart and then sprang out of the door, showed it to be the result of practice . . ." Darwin kept his own large knife in a scabbard on a string round his neck "in the manner of sailors"——something the gauchos thought strange.

As he circled back toward Maldonado, Darwin climbed a nine-hundred-foot-high hill—the Sierra de las Animas—to discover on the summit piles of stones, which his guide called "the work of the Indians in the old times." They reminded Darwin of cairns on the Welsh mountains: "How universal is the desire of Man to show he has ascended the highest points."

Back in Maldonado, Darwin wrote to his sister Catherine: "I most devoutly trust that next summer (your winter) will be the last on this side of the Horn: for I am becoming thoroughly tired of these countries: a live Megatherium would hardly support my patience . . ." He asked for "those most valuable of all valuable things," more scientific books, four pairs of "very strong walking shoes," and another box of "promethians," adding, "When you read this I am afraid you will think that I am like the Midshipman in 'Persuasian' who never wrote home excepting when he wanted to beg . . ."

He assured her, "I have worked very hard (at least for me) at Nat History and have collected many animals and observed many geological phenomena: and I think it would be a pity having gone so far, not to go on and do all in my power in this my favourite pursuit: and which I am sure, will remain so for the rest of my life." He was not exaggerating. In the past year he had cataloged more than fifteen hundred specimens.

Possibly some of these comments were intended to smooth the way for what followed: "The following business piece is to my father: having a servant of my own would be a really great addition to my comfort." Though from the start of the voyage FitzRoy had allowed him to share the services of his steward Henry Fuller's young assistant, Darwin wanted someone to assist him full time both at sea and ashore and had identified a man "willing to be my servant and <u>all</u> the expences would be under sixty £ per annum.—I have taught him to shoot and skin birds . . ." Since it would take such a long time to receive Dr. Darwin's reply, and because he had kept his expenses since leaving England to around two hundred pounds per annum, he told Catherine he was certain his father "would allow me this." In other words he was going ahead. A year passed before he received a letter from his sister Susan telling him Dr. Darwin was "*exceedingly* glad you have engaged a Servant as he is sure it will conduce very much to your comfort and only regrets you had not one sooner." She added, "Pray tell us next time what countryman he is? if he had been a Negro, you wd have said so I am sure."

Even at Cambridge, Darwin had employed helpers in his natural history work. The person he had in mind on the *Beagle* was Syms Covingon, already his informal assistant, though he did not name him in his letter and had as yet not obtained FitzRoy's agreement. About three years Darwin's junior, Covington, a religious teetotaler, was capable, resourceful, and keen to better himself. When commenting on the *Beagle*'s ports of call in his journal—a small brown notebook with a waterproof cover in which he covered some fifty pages in a neat hand—he often noted the opportunities they offered. Of Montevideo he wrote, "Here if a man is willing to work, he can save money." Even before Darwin formally employed him, but probably under his influence, he had begun making observations about wildlife and geological features. He was also a competent artist—which Darwin was not—and produced several sketches and watercolors during the voyage. Curiously for a man whose role included being the ship's fiddler, he was deaf in one ear.

When the *Beagle* returned from Montevideo, FitzRoy made no objection to Darwin's plan and generously kept Covington on the ship's books for his rations, saving Darwin thirty pounds a year. However, in return Covington still had to perform ship's tasks, remaining one of the water party regularly sent ashore with shovels and empty casks to find drinking water to fill the *Beagle*'s huge iron tanks.

While the *Beagle* had been in Montevideo, Robert Hamond, who had only joined the ship the previous July but been Darwin's frequent companion ashore, had quit the navy. This was no surprise to Darwin, who had known Hamond wanted to leave because of his "stammering and disliking the Service." However, another friend, Wickham, soon arrived to take command of the *Adventure*, bringing midshipman King, who confessed they were "heartily tired" of living in confined conditions, struggling to keep clothes and bedding dry and being almost constantly seasick and were delighted to be back on the *Beagle*.

While FitzRoy concentrated on "calculations and chart-work" and refitting the *Adventure*, Darwin, still living ashore, continued "one day, shooting and picking up my mouse traps, the next preserving the animals which I take." His collection from around Maldonado—eighty kinds of bird, some twenty animals including a water-dwelling capybara, "the largest gnawing animal in the world" with broad teeth and strong jaws well suited to grinding the aquatic plants it fed on, and large numbers of reptiles—was "becoming very perfect." For small payments, he had "enlisted all the boys in the town . . . and few days pass, in which they do not bring me some curious creature."

On June 29, Darwin moved back to the *Beagle* with his "menagerie," having become such a landlubber that "I knock my head against the decks." When FitzRoy told him he planned to round the Horn the coming summer (winter in the Northern Hemisphere), the gloom he had recently complained of to his sister Catherine lifted, and he wrote, "My heart exults whenever I think of all the glorious prospects of the future."

A few nights later, shortly before midnight, a shout of "reeve the running rigging and bend sails" startled Darwin. Soon after the anchor was weighed. The reason the *Beagle* was departing at such an hour and so abruptly was FitzRoy's discovery that one of Low's crew he had taken on to help man the *Adventure* had formerly sailed on the *President*—a piratical vessel notorious for seizing a packet ship and murdering everyone on board. *The (London) Times* had reported that many were forced to walk the plank. The *President* had also attacked a British naval ship, HMS *Black Joke*, engaged in anti-slaving patrols off West Africa. FitzRoy had immediately decided to sail to Montevideo to hand the man over to the British consul to face justice.

In Montevideo, Fitzroy found letters from Beaufort written on March 9 and April 3 waiting for him. They contained disappointing news. Although he had supported FitzRoy's case to the Admiralty, Beaufort advised caution over hiring the *Paz* and *Liebre*. FitzRoy replied, regretting "my outstepping would have caused any uneasiness to yourself . . . Injury to *myself alone* was all I anticipated . . ." Although he had suffered a serious financial blow, FitzRoy added philosophically, "The work is done, the vessels *are paid for* . . ." He also told Beaufort that, had he received his letters before purchasing the *Adventure*—about which he had subsequently written to him—he would not have involved him in "another fracas. I am now upon thorns to know the result [of his request to the Admiralty to fund the purchase of the *Adventure*]."

Finally, late on July 24, "a wild looking night to go to sea," with lightning again flashing, the *Beagle* and her new consort the *Adventure* headed back to the Rio Negro, three hundred miles south. Before departing Darwin had dispatched half his latest specimens aboard a packet ship bound for Rio on the first leg of their long journey home. In an accompanying letter he told Henslow there should be four barrels. One, packed with skins and plants, should be opened immediately on arrival; two contained geological specimens that could be left, unless Henslow wanted to examine them; and the fourth and smallest contained fish preserved in spirits. He added, "I have now got a servant of my own, whom I have taught to skin birds etc, so that for the future I trust, there will be rather a larger proportion of showy specimens."

Arriving at the Rio Negro, FitzRoy located the *Liebre*. After hearing that all was well with the two schooners, FitzRoy took the *Beagle* to survey some of the outer banks, while Darwin boarded the *Liebre*, which was to explore around the Rio Negro. When the *Liebre* rejoined the *Paz*, he heard how the crews had survived the vicious mid-January storm during which the *Beagle* had lost her whaleboat. Darwin wondered whether the schooners' small size had saved them, "for the sea instead of striking them sends them before it.—I never could understand the success of the small craft of the early navigators."

Darwin found the Rio Negro geologically interesting. Sandstone cliffs south of the river mouth contained a "remarkable" layer "composed of a firmly-cemented conglomerate of pumice pebbles, which must have travelled more than four hundred miles, from the Andes." He rode with Stokes eighteen miles upriver to the small town of Patagones [El Carmen]—established fifty years before and "still the most southern position (lat. 41°) on this eastern coast . . . inhabited by civilized man." On their way, they passed ruined *estancias*, which, a man told Darwin,

Arucanian people from southern Chile had attacked a few years before. Removing their fur cloaks they ran naked into the fight armed with *chusas*—long, sharp bamboo spears decorated with rhea feathers—"My infomer seemed to remember with the greatest horror, the quivering of these Chusas as they approached."

Patagones lay beneath a sloping cliff where the Rio Negro flowed past, deep, rapid, and "about four times as wide as the Severn at Shrewsbury," surging around islets covered in willow trees. Some houses were carved cavelike into the sandstone cliffs. Of Patagones's few hundred inhabitants, many were "Indians of pure blood" who subsisted by making horse blankets and items of riding equipment and ate the flesh of old, worn horses. To Darwin, "these Spanish colonies do not, like our British ones, carry within themselves the elements of growth."

One nearby source of prosperity was, however, "the great Salina"—a shallow, briny lake that dried up in summer, leaving "a large field of snow white salt." At that time, the entire population of Patagones camped on the riverbank, and lines of bullock wagons rolled continuously back and forth between the glistening salt pan and the riverbank where the salt was piled in great heaps. Given the great amount of salt produced here and elsewhere in the area—"a quantity sufficient to supply the world"—Darwin wondered why Montevideo's inhabitants instead used English salt to mix with their butter. He was reminded of Buenos Aires, where, despite the huge amounts of wheat grown in the region, the citizens imported flour from North America. He attributed the latter to laziness—"Killing an animal and flaying it does not give much trouble, and hides in consequence are nearly the only produce . . . these indolent people care about." He later discovered the supposed reason English salt was used in Montevideo was that Patagonian salt's "excessive purity" made it a less effective food preservative.

He noticed large crystals of gypsum and sulfate of soda embedded in the black, fetid mud around the lake's edge. Poking about in the mud, Darwin found a number of worm casts. Surprised any creature could survive such a briny environment, he wondered what became of the worms during the long summer when the entire surface of the lake hardened into a solid layer of salt. Seeing flamingos wading about, he concluded they were probably searching for the worms to feed upon—"Thus we have a little living world within itself, adapted to these inland lakes of brine . . . Well may we affirm that every part of the world is habitable!"

From Patagones, Darwin decided to ride across the pampas to Baia Blanca, some 170 miles to the north, where he knew that FitzRoy intended to call with the *Beagle*. If that went well he would continue over land all the way to Buenos Aires. Though "the wandering tribes of horse Indians . . . have always occupied the greater part of this country" he hoped the journey would be tolerably safe since the Buenos Aires government had dispatched an army "under the command of General Rosas to extermi-nate the Indians." This extermination was part of the attritional conflict Darwin and FitzRoy had heard about during their stay in Baia Blanca.

James Harris, who had rented the *Liebre* and *Paz* to FitzRoy and been acting as pilot on the *Paz*, agreed to accompany Darwin as far as the Rio Colorado, eighty miles to the north, where Rosas was encamped. On the eve of their departure, five gauchos, also on their way to Rosas's camp, joined them. As Darwin wrote, "every body seemed glad of companions in this desolate passage."

CHAPTER NINE

El Naturalista Don Carlos

Traveling light with just "hammer, pistol and shirt," Darwin galloped with his companions over flat plains of withered brown grass and low spiky bushes so dry they were grateful for puddles deposited in wagon ruts by heavy rain the previous day for their horses to drink. Except for the occasional rabbit, agouti, rhea, or guanaco, wildlife was scarce. As the track began to rise to higher ground, Darwin saw on the horizon a lone tree that the gauchos told him the "Indians" revered and to which they made offerings. Drawing nearer he saw cigars, pieces of bread, meat, and cloth, even threads that poor people with nothing else to give had pulled from their ponchos, dangling from the leafless branches. The surrounding ground was piled with the bleached bones of horses sacrificed by "Indians" who believed this would make their remaining horses tireless and themselves prosperous.

That evening the "lynx-eyed" gauchos spotted, hunted down, lassoed, and killed a cow. The party halted for the night at a spot offering "the four necessaries for life 'en el campo,'—pasture for the horses,—water (only a muddy puddle)—meat and firewood." To Darwin, this ability "at any moment to pull up your horse, and say here we will pass the night" seemed admirably free-spirited and independent. "The death-like stillness of the plain, the dogs keeping watch, the gipsy-group of Gauchos making their beds around the fire," using their *recados*—saddles—to sleep on were unforgettable images.

On the third day they reached the reed- and willow-fringed Rio Colorado at a point where it was some sixty yards wide or, Darwin estimated, using his favorite reference point for rivers, "once and half as wide as the Severn at Shrewsbury." They held back while a group of Rosas's soldiers swam "immense troops of mares" across the river. They were following a division of soldiers into the interior since, Darwin learned, "mare's flesh is the only food of the soldiers" when on the move. "A more ludicrous spectacle I never beheld, than the hundreds of heads, all directed one way, with pointed ears and distended nostrils, appearing just above the water like a great shoal of some amphibious animals."

After crossing the Colorado himself, Darwin entered Rosas's nearby camp, which was laid out in a square bounded by wagons, artillery, and straw huts. Rosas's men were such a "villainous Banditti-like army [as] never before collected together," Darwin decided. In his war on the Patagonian peoples, Rosas was relying on a personal army of mercenaries answerable only to him, not the Buenos Aires government. Needing approval for his onward journey, Darwin sought out Rosas's secretary. Clearly suspicious, the man rigorously cross-examined him until Darwin remembered he was still carrying a letter of recommendation from the authorities in Buenos Aires—provided in recognition of his status as "Naturalist of the *Beagle*"—to the governor of Patagones. The secretary bore it off to Rosas, quickly returning "all smiles and graciousness." Darwin was given accommodation in the rancho—a thatched hut he described as "a hovel"—belonging to "a curious old Spaniard" who had fought in Napoleon's army during his disastrous Russian campaign.

Rosas had some six hundred "Indian allies," and Darwin watched some of them buying supplies from his host. They were an "exceedingly fine race" but "it is easy to see the same countenance, rendered hideous by the cold, want of food and less civilization, in the Fuegian savage." With their "high colour," sparkling eyes, graceful limbs and plaits of black hair hanging to their waists, some of the women "deserved to be called even beautiful" and were excellent riders. However, their lives seemed hard. They, not the men, had to load and unload the horses and set up tents for the night—"They are in short, like the wives of all Savages useful slaves."

On his second day in the camp, Darwin was surprised to learn that Rosas himself wished to see him. By now he was well briefed about the forty-year-old general from a wealthy Buenos Aires family—a man of "extraordinary character . . . He is said to be owner of 74 square leagues of country and has about three hundred thousand cattle . . . He

General Juan Manuel de Rosas

first gained his celebrity by his laws for his own Estancia and by disciplining several hundred workmen or Peons, so as to resist all the attacks of the Indians.—He is moreover a perfect Gaucho:—his feats of horsemanship are very notorious; he will fall from a doorway upon an unbroken colt as it rushes out of the corral, and will defy the worst efforts of the animal. He wears the Gaucho dress and . . . has gained an unbounded popularity . . . and in consequence despotic powers—A man a short time since murdered another; being arrested [and] questioned he answered, 'the man spoke disrespectfully of General Rosas and I killed him'; in one week's time the murderer was at liberty." Darwin had also been told that Rosas kept a pair of "mad buffoons" as his jesters, was at his most dangerous when he laughed, and inflicted an agonizing punishment on those who displeased him: "four posts are driven into the ground, and the man is extended by his arms and legs horizontally, and there left to stretch for several hours.—the idea is evidently taken from the usual method of drying hides."

The interview "passed away without a smile," although "in the most obliging and ready manner," Rosas gave Darwin a passport and papers allowing him to use the chain of small military posts—*postas*—defended by soldiers and well supplied with horses that Rosas had set up at intervals across the pampas to protect supply and communication lines for his campaign against the "Indians." Though Rosas's desire to see him had meant postponing his departure by a day, Darwin was "altogether pleased with my interview with the terrible General. He is worth seeing, as being decidedly the most prominent character in S. America."

Early the next morning, August 16, Darwin set out for the two-day ride of some ninety miles to Baia Blanca, this time without Harris, who was unwell, but with gaucho companions. Following the course of the Rio

Colorado they reached the first of Rosas's *postas*. Darwin thought the surrounding land looked fertile enough to grow corn, while the many willows on the riverbanks could provide timber for the *estancias* Rosas planned to build. Darwin was conflicted about Rosas and his activities, writing in his diary, "This war of extermination, although carried on with the most shocking barbarity, will certainly produce great benefits; it will throw open four or 500 miles in length of fine country for the produce of cattle."

Between the second and third *postas*, Darwin noted how the landscape was changing with what he understood to be the start of a "grand geological formation" extending six hundred miles north beyond Buenos Aires. He concluded a belt of red sand dunes was "invaluable . . . for resting on the clay the[y] cause small lakes in the hollows and thus supply that most rare article, fresh water." Beyond the dunes was the fourth *posta*, this time "commanded by a negro Lieutenant born in Africa," with a small room for travelers, a corral of sticks and reeds for their horses, all protected by a defensive ditch, which Darwin decided would be little obstacle to attacking "Indians." Darwin thought the lieutenant was very obliging and polite which, he wrote in his diary, made it "the more painful . . . that he would not sit down and eat with us." He did not explain the reason for the man's reluctance, but the implication is that race was a factor.

Early the next morning Darwin and his companions were off again, enjoying an "exhilarating gallop" until they reached swamps and saltpeter marshes that extended all the way to Baia Blanca. Darwin's horse fell, leaving him "well souzed in black mire, a very disagreeable accident, when one does not possess a change of clothes." Not long afterward, they encountered a man who told them he had heard the firing of a great gun—a warning that "Indians" were near. They immediately left the road to hug the edge of the marshes, so that, if attacked, they could try to escape into them. However, it proved a false alarm. Riding into Baia Blanca, Darwin learned that though "Indians" had been spotted, they were "friendly ones, who wished to join General Rosas."

Darwin was disappointed to receive no news of the *Beagle*: "I had nothing to do, no clean clothes, no books, nobody to talk with.—I envied the very kittens playing on the floor." In frustration, he set out with a

soldier as his guide to ride some twenty-five miles around the bay to a
point close to where he expected the *Beagle* to anchor. Reaching their
destination safely, Darwin saw no sign of the ship. As their horses were
too exhausted for them to return that night, they camped out, sleeping fit-
fully on ground "thickly encrusted with salt petre and of course no water."
After breakfasting on an armadillo they killed and roasted in its shell,
they headed back to Baia Blanca. However, their thirsty horses were still
weak, particularly that of Darwin's guide—"As a Gaucho cannot walk,
I gave up my horse and took to my feet." At last Darwin himself "could
walk no more" and retrieved his horse, "which was dreadful inhumanity
as his back was quite raw." They eventually regained Baia Blanca, after
twenty hours without water—"I do not know whether the poor horse
or myself were most glad."

The *Beagle* had still not arrived by August 22, two days later. "So
tired of doing nothing," Darwin set out with the same soldier guide for
nearby Punta Alta, that "perfect catacomb for monsters of extinct races"
where he had dug up his trove of fossils a year earlier and with good views
of the entire bay in case the *Beagle* appeared. This time he was careful
to bring "bread and meat and horns with water" to avoid the problems
of his journey around the bay. But it seemed their luck at evading raid-
ing parties had run out. Just as they were approaching Punta Alta, they
saw riders in the distance. The guide immediately dismounted, watched
them intently, then said "they don't ride like Christians." When one
rider disappeared out of sight over a hill, he told Darwin, who had also
dismounted, "'We must now get on our horses, load your pistol' and he
looked to his sword.—I asked are they Indians.—'Quien sabe? (who
knows?), if they are no more than three it does not signify'"—in other
words, they were too few to pose a threat.

Darwin suggested "that the one man had gone over the hill to fetch
the rest of his tribe . . . but all the answer I could extort was, Quien
sabe?—His head and eye never for a minute ceased scanning slowly the
whole horizon." Disconcerted by such unusual coolness, Darwin asked
whether they should turn back, which they did. However, when they
reached the foot of a hill, the guide again dismounted, handed Darwin the

reins of his horse, then crawled on hands and knees up the hill to recon-
noiter. After some time, he suddenly burst out laughing and exclaimed
"'*Mugeres* (women).'" He identified the riders as the wife and sister-in-law
of an officer from the settlement probably out searching for rhea eggs.

Reaching Punta Alta unmolested, Darwin found places where fossilized bones protruded and left markers so he could retrieve the remains at
a future date. Then, as a brilliant sunset streaked the sky and "everything
became quiet and still," they settled down for the night. Returning to
Baia Blanca the next day in heavy rain, Darwin was intrigued by a "very
curious" creature, the zorilla—or skunk, unknown in Britain—noting
how the noxious fluid it could eject made every animal avoid it. Back in
the settlement, Darwin found that Harris, now recovered, had arrived
from Rosas's camp with grisly news that just a few days before, "Indians"
had murdered "every soul" in one of the *postas*. Rosas suspected members
of a tribe supposedly allied to him and had threatened their chief that if
he "failed to bring the heads of the murderers, it should be his bitterest
day, for not one of his tribe should be left in the Pampas."

Next day, Darwin finally saw the *Beagle* in the distance, beyond some
wide mud banks, "its figure curiously altered by the refraction" of the
light on the water. However, the winds were too strong for FitzRoy to
send a boat to the settlement. While Darwin waited, one of Rosas's commanders arrived with three hundred soldiers and orders to track down
those who had attacked the *posta*, establish whether they were from
the suspected tribe, and, if so, to massacre them all. Darwin observed
the new arrivals with fascinated horror. Many were themselves "Indians," and "nothing could be more wild or savage than the scene of their
bivouacking.—Some of them drank the warm, steaming blood of the
beasts which were slaughtered for supper."

Finally, Edward Chaffers arrived in one of the *Beagle*'s boats, bringing
a letter from FitzRoy: "My dear Philos, Trusting that you are not entirely
expended—though half starved,—occasionally frozen, and at times half-
drowned . . . I do assure you that whenever the ship pitches . . . I am
extremely vexed to think how much *sea practice* you are losing—and how
unhappy you must feel upon the firm ground." FitzRoy told Darwin to

send word "when *you* want a boat . . . Take your own time—there is abundant occupation here for *all* the *Sounders* [surveyors]—so we shall not growl at you when you return." In a postscript, he added, "I do not rejoice at your extraordinary and outrageous peregrinations because I am envious—jealous,—and extremely full of uncharitableness. What will they think at home of 'Master Charles' 'I do think he be gone mad'— Prithee be *careful* . . ."

Darwin returned with Chaffers to the *Beagle,* where his "travellers tales" were in high demand. Three days later with Sulivan, Covington, and others in the yawl, he went back to Punta Alta to fetch the bones he had previously marked. They moored overnight among the mud flats, well provided with tea, food, and a quarter of a pint of rum per man. The next morning, Sulivan watched Darwin find "the teeth of animals six times as large as those of any animal now known in this country, also the head of one about the size of a horse, with the teeth quite perfect and totally different from any now known, and just at low-water mark he found the remains of another about six feet long, nearly perfect, all embedded in solid rock." Delighted with his discoveries, Darwin left Covington and another man to continue "digging out the bones" while he returned with Sulivan to the *Beagle* to arrange to continue his overland expedition northward.

Intent on finding yet more fossils, Darwin now planned not only to ride the four hundred miles to Buenos Aires, but, if possible, a further two hundred beyond to the Carcarana River, where, he had read, the banks were "so thickly strewed with great bones, that they build part of the Corral with them." His information came from a book in the ship's library by Thomas Falkner, an English Jesuit missionary, who in 1760 had unearthed the fossilized skeleton of a huge armadillo-like animal from cliffs along the Carcarana. Conscious of the cost to his family, Darwin wrote to one of his sisters that he hoped his only real expense on the journey would be paying a trusty guide, since "in that depends your safety, for a more throat-cutting gentry do not exist than these Gauchos."

Meanwhile, Rosas's "bloody war of extermination" seemed to be suc- ceeding. Darwin noted that Baia Blanca was in "great excitement" as reports of victory after victory against the "Indians" flooded in. A witness of a recent battle told him, "The Indians are now so terrified that they offer no resistance in body; but each escapes as well as he can, neglecting even his wife and children,—The soldiers pursue and sabre every man.—Like wild animals however they fight to the last instant.—One Indian nearly cut off with his teeth the thumb of a soldier, allowing his own eye to be nearly pushed out of the socket.—Another who was wounded, pretended death with a knife under his cloak, ready to strike the first who approached."

Darwin thought this "a dark picture; but how much more shocking is the unquestionable fact, that all the women who appear above twenty years old are massacred in cold blood.—I ventured to hint that this appeared rather inhuman. He [the witness] answered me, 'Why what can be done, they breed so'—Every one here is fully convinced that this is the justest war, because it is against Barbarians. Who would believe in this age in a Christian civilized country that such atrocities were committed? . . . in another half century, I think there will not be a wild Indian in the Pampas North of the Rio Negro." Instead "the country will be in the hands of white Gaucho savages instead of copper-coloured Indians. The former being a little superior in civilization, as they are inferior in every moral virtue." As in the case of slavery in Brazil, Darwin seems to have felt unable to protest further to local people, perhaps feeling it futile or, in this case, dangerous to do so.

Despite Rosas's success against the "Indians," Darwin struggled to find a suitable guide in Baia Blanca: "The father of one man was afraid to let him go, and another, who seemed willing, was described to me as so fearful, that I was afraid to take him, for I was told that even if he saw an ostrich at a distance, he would mistake it for an Indian, and would fly like the wind away." At last Darwin identified a reliable man and his assistant to accompany him, and on September 8, 1833, set out.

Darwin and his companions camped the first night beneath a saddle linking two of the four peaks of the Sierra de la Ventana, a range that Darwin had been able to see from Baia Blanca. The air was so cold that they woke to find the coarse saddle blankets they slept beneath filmed with ice and the water in their kettle frozen solid. Darwin spent the morning scrambling up a rock-strewn ridge, hoping to reach the summit of one peak only to find a precipitous valley barring his way. Suddenly spotting two horses grazing below, he flung himself to the ground, grabbed his telescope, and looked for any sign of "Indians," but there were none. Later that day, he struggled up the steep, rugged slopes of another peak to the summit, but the effort brought on such agonizing cramps in his thighs that he abandoned further exploration. In any case, he was "much disappointed in this mountain; we had heard of caves, of forests, of beds of coal, of silver and gold etc etc, instead of all this, we have a desert mountain of pure quartz rock . . . The scene however was novel, and a little danger, like salt to meat, gave it a relish."

Two days later, real danger seemed imminent as a vast, swirling cloud of dust approached. It could mean only one thing—horsemen. Though the riders were some distance away, Darwin's guide could tell "by the streaming hair, that they were Indians." To Darwin's relief they were not on the attack but heading for a *salina* to collect salt, of which, he noted— with his characteristic desire to compare—they ate large amounts, "the children sucking it like sugar," whereas gauchos, despite "living the same life, eat scarcely any." The group "gave us good-humoured nods as they passed at full gallop, driving before them a troop of horses, and followed by a train of lanky dogs."

That night, Darwin's party halted at a *posta* where, Rosas had told him in a letter that reached him in Baia Blanca, he should await the arrival of a small military escort to accompany him to Buenos Aires. The lieutenant commanding the *posta* was hospitable, but Darwin thought his four soldiers ill-assorted. The first was "a fine young negro" while the second was "half-Indian and Negro." The remaining pair were "quite non descripts, one an old Chilean miner of the color of mahogany, and the other partly a mulatto; but two such mongrels, with such detestable expressions I never saw before." Around the men squatting by the fire playing cards lay "dogs, arms, remnants of Deer and Ostriches, and their long spears were struck in the ground." Their tethered horses were nearby in the darkness, "ready for any sudden danger.—If the stillness of the desolate plain was broken by one of the dogs barking, a soldier leaving the fire, would place his head close to the ground and thus slowly scan the horizon."

Just as he had pitied the solitary existence of Dickson, the keeper of the British flag in the Falkland Islands, Darwin was sorry for these men, leading "a life of misery" in discomfort and constant fear of attack. The "little hovel" of thistle stalks where they slept kept out neither wind nor rain. Their only food was what they could catch—principally rheas, deer, and armadillos—while their sole fuel was a small aloe-like plant. Their only pleasures were smoking "little paper cigars and sucking mattee [herbal tea]."

After three nights with no sign of the promised military escort, on September 14, Darwin decided to join two soldiers traveling the same route. With his own two companions, this made a party of five, which seemed safe enough as they were all armed. They set off at a gallop across the pampas until, reaching low swampy country that reminded Darwin of the Cambridgeshire fens, they camped. The next day, they passed the deserted *posta* where "Indians" had recently killed all the soldiers—stabbing the lieutenant eighteen times with their *chusas*—to arrive at a much larger *posta*, "the most exposed on the whole line," and protected by twenty-one soldiers. It was beside a shallow lake teeming with wild fowl and black-necked swans. That night, "brilliant conflagrations" lit the horizon. The purpose of the fires, Darwin learned, was less to deter "any stray Indians" than to improve the pasture: "In grassy plains unoccupied by the larger ruminating quadrupeds, it seems necessary to remove by fire the superfluous vegetation, so as to render serviceable the new year's growth."

Darwin was still enjoying the life: "I am become quite a Gaucho, drink my Mattee, and smoke my cigar, and then lie down and sleep as comfortably with the Heavens for a canopy as in a feather bed.—It is such a fine healthy life, on horseback all day, eating nothing but meat, and sleeping in a bracing

Argentinian gaucho, 1868

air . . ." He decided that the gauchos' diet explained why "like other carnivorous animals [they] can go a long time without food and can withstand much exposure."

Approaching Buenos Aires they passed through small settlements of Rosas's native allies. Darwin admired both the young women, riding two or three to a horse, and the embroidered horse blankets they were taking to market, which were of such high quality they could have been made in England—or so an English merchant in Buenos Aires thought when he later inspected one of Darwin's purchases. Soon they passed the first cattle ranch and Darwin saw his first "white woman" since his journey began. By September 19, he was in the little town of Guardia del Monte, a pretty place with quince and apple orchards and fennel growing in the hedgerow. Nearby, he unearthed "a perfect piece of the case [carapace] of the Megatherium," which he thought connected "the Geology of the different parts of the pampas."

While he waited for a change of horses, townspeople quizzed him about Rosas's progress. "I never saw anything like the enthusiasm for Rosas and for the success of this 'most just of all wars, because against Barbarians.'" Their attitude seemed natural enough given that "even here neither man, woman, horse or cow was safe from the attacks of the Indians." Journeying on, Darwin arrived after dark and in torrential rain at the eighteenth *posta* since Baia Blanca. The commandant at first thought he was a robber, but on checking his papers stating he was "El Naturalista Don Carlos," his "respect and civility were as strong as his suspicions had been before.—What a Naturalista is, neither he or his countrymen had any idea; but I am not sure that my title loses any of its value from this cause."

The next day, September 20, twelve days after leaving Baia Blanca, Darwin arrived in Buenos Aires, where olives, peaches, and willow trees were coming into leaf with the arrival of the Southern Hemisphere spring. He stayed with a hospitable merchant, Edward Lumb, and was soon enjoying "all the comforts of an English house," including watching "a lady making tea." A few days later Syms Covington arrived from Montevideo, where the *Beagle* had just anchored. Darwin dispatched him to an *estancia* on the banks of the River Plate to shoot and skin birds for his collection while he himself obtained the usual permits and letters of introduction for his onward journey to the Carcarana and wrote to FitzRoy asking where and when to rendezvous with the *Beagle*.

FitzRoy replied, in the letter that would not catch up with Darwin for some time, that he should be in Montevideo no later than early November, adding teasingly, "my good Philos why have you told me nothing of your hairbreadth scapes and moving accidents? How many times did you flee from the Indians? How many precipices did you fall over? How many bogs did you fall into?——How often were you carried away by the floods? And how many times were you kilt [*sic*]?" He also praised Conrad Martens, Augustus Earle's successor as the *Beagle*'s artist, whom Darwin had not yet met: "By my faith in Bumpology, I am sure you like him . . . His landscapes are *really* good . . . though perhaps in *figures* he cannot equal Earle. He is very industrious, and gentlemanlike in his habits——(not a *small* recommendation)."

A week after reaching Buenos Aires, Darwin set off again with his guide, galloping northwest over "sea-like pampas" covered with clumps of bright green thistles, which, he learned, in summer grew so thick and high they formed "an excellent . . . home for numerous robbers, where they can live, rob and cut throats with perfect impunity." That evening he was ferried across the Rio Arrecife on a raft of empty barrels lashed together and slept at the *posta* on the other side where he hired fresh horses. In his diary he complained the distances between *postas* were universally exaggerated——"a man who pays for 50 leagues by the post by no means rides 150 English miles"——the equivalent distance.

Following the course of the Parana River, at 3,030 miles long, the second longest in South America after the Amazon, Darwin reached the province of Santa Fe, where he had been warned "all the good people . . . are most dexterous thieves." Soon afterward, he discovered his pistol had been stolen. In Rozario, a large clifftop town overlooking the Parana, "a most hospitable Spaniard," to whom Darwin had a letter of introduction, lent him "this most indispensable article"——a replacement pistol. Starting out from the town by moonlight, by dawn Darwin had reached the Carcarana, the reputed treasure house of fossils, which flowed into the Parana north of Rozario. He soon found "a curious and large cutting tooth." Later, taking a canoe down the Parana itself, he saw "two immense skeletons"—which he thought must be those of mastodons—projecting from the perpendicular cliffs but "so completely decayed and soft" he could extract and preserve only "small fragments of one of the great molar teeth." During his voyage on the *Beagle*, Darwin would designate nearly all the fossilized remains of large mammals he found as belonging to either the Megatherium or mastodon class—the only large classes then known for South America—though some would prove to be entirely new creatures when examined by experts.

The next day, feeling weak and feverish, which he attributed to over-exertion in the heat searching for fossils, Darwin continued north toward the settlement of Sante Fe itself. Pleasure at the beauty of birds, flowers, some of which he had never seen before, and new types of cacti—the result, he decided, of the warmer temperatures north of Buenos Aires—was tempered by the knowledge that he was in a sparsely populated area where "the Indians sometimes come down and kill." His guide showed him the skeleton of one "Indian" dangling from a tree, "the dried skin hanging to the bones."

Still feeling ill, after crossing an arm of the Parana to Santa Fe, built on a large island in the river, Darwin rented a room and took to his bed, resisting "a good-natured old woman" who urged him to try several odd remedies to cure his headache. He later wrote, "Many of the remedies used by the people of the country are ludicrously strange, but too disgust-ing to be mentioned. One of the least nasty is to kill and cut open two puppies and bind them on each side of a broken limb. Little hairless dogs are in great request to sleep at the feet of invalids." During his recovery, Darwin heard stories about the local governor—a tyrant whose favorite pastime was hunting and killing "Indians" and selling off their children.

Darwin had originally intended to journey eastward to the Uruguay River and sail downstream toward Buenos Aires. However, by October 5 he still felt so unwell he decided instead to return at once down the Parana. Unable to hire a private boat, he booked a berth in a *balandra*, a single-masted, one-hundred-ton communal riverboat. To his annoyance, bad weather and "the indolence of the master" delayed departure for five days, which Darwin used to examine the geology along the Parana. Probing beds of sand, clay, and limestone at the base of the cliffs, he found fossil shells, shark's teeth, and also the "teeth of the Toxodon and Mastodon."

To his surprise, nearby he also found a stained, decayed, fossilized tooth, which "I at once perceived . . . had belonged to a horse . . . I carefully examined its geological position, and was compelled to come to the conclusion, that a horse . . . lived as a contemporary with the various great monsters that formerly inhabited South America." A few weeks earlier he had picked up another seemingly fossilized horse's tooth close to some ancient cream-colored flint arrowheads. He had not then thought it sufficient proof that horses had ever existed in the pampas before the conquistadores brought them. However, when sent to Britain, these teeth would provide conclusive evidence that a horse—*Equus curvidens*—indeed existed in the Americas in pre-Columbian times but had become extinct before the conquistadores arrived.

The *balandra* finally departed on October 12, but almost immediately, strong winds and heavy showers decided its cautious captain to moor it to a tree on one of the Parana's many islands. Darwin's stroll ashore ended abruptly when he came across what he called the "most indubitable and recent sign" of a jaguar, "a much more dangerous animal than is generally supposed." Instead, he spent his time trying to catch fish or lying in bed as the low ceiling of his tiny cabin made it impossible to sit up. By October 15 the weather had improved and the *balandra* sailed rapidly on with the current, but not for long. Toward sunset, "from a silly fear of bad weather," the captain halted again, this time tying up in a creek that Darwin explored in one of the *balandra*'s boats. In the fading light he saw "a very extraordinary bird, the scissor-beak," similar in size to a tern, with a beak "flat and elastic as an ivory paper-cutter" and with a lower mandible an inch and a half longer than the upper. It ploughed through the water, lower mandible submerged as it seized small fish. As darkness fell, a mass of bright fireflies appeared, but also "very troublesome" mosquitoes. Exposing his hand for just five minutes, Darwin observed with scientific detachment some fifty batten on to it, "all busy with sucking."

As the *balandra* continued downstream, Darwin pondered the few other vessels on the Parana, deciding the region's commercial potential was being wasted: "One of the best gifts of nature seems here wilfully thrown away, in so grand a channel of communication being left unoccupied. A river in which ships might navigate from a temperate country, as surprisingly abundant in certain productions as destitute of others, to another possessing a tropical climate, and a soil . . . unequalled in fertility in any part of the world. How different would have been the aspect of this river, if English colonists had by good fortune first sailed up the Plata! What noble towns would now have occupied its shores!"

Frustrated by the slow progress, when the *balandra* neared the mouth of the Parana, Darwin disembarked with his guide and, after some miles on horseback, eventually hired a canoe in which they were paddled down a narrow channel to the northern outskirts of Buenos Aires. Stepping from the canoe, to his astonishment he found himself "a sort of prisoner" in the hands of "a furious cut-throat set" of rebel soldiers. They informed him a violent revolution had broken out against the city's governor and all the ports were closed, so that "I could not return to my vessel, and as for going by land to the city it was out of the question." Finally, the officer in charge gave grudging permission for him to visit the nearby camp of General Rolor, one of the rebel leaders.

Reaching the encampment to find it milling with "great villains," Darwin sought out the general. He could only persuade him to provide papers allowing him to travel to the camp of the rebel general in chief at Quilmes on the south side of the city. Procuring horses with difficulty, Darwin set off with his guide, making a great sweep around Buenos Aires to Quilmes. At the camp, the rebels again told him that on no account could he enter the city. However, by a stroke of good fortune, Darwin learned that Rosas's brother was among the rebels and made great play of General Rosas's civility to him when they had met on the Rio Colorado. "Magic could not have altered circumstances quicker than this." The rebels told Darwin he could, after all, enter Buenos Aires, but that he must travel on foot, leaving behind his guide, horses, and baggage. He duly set off down a deserted road and "was exceedingly glad when I found myself safe on the stones of B. Ayres."

In the city he learned more about the uprising, including that its leaders were Rosas's supporters, though Rosas himself did not yet know about it. Darwin thought this "quite consonant with his schemes . . . It is clear to me that Rosas ultimately must be absolute Dictator . . . of this country." Indeed, in 1835 Rosas would became dictator of Argentina and remain in power until ousted by his own generals in 1852.[1]

The rising left Darwin "in a pretty pickle." He had lost his baggage, including his fossil finds, and had no way of contacting the *Beagle*. Furthermore, Syms Covington was still outside the city at the *estancia*, where Darwin had sent him to collect bird specimens. Darwin felt "obliged to bribe a man to smuggle him in through the belligerents." Even so, Covington wrote in his journal, to get past the sentries into Buenos Aires was not easy. Darwin soon learned Covington had "nearly lost his life in a quicksand and my gun completely," while the rebels had sacked the house where he was staying and stolen his spare clothes.

While the two men lay low—Covington, as a seaman, feared being press-ganged—the mood in Buenos Aires remained tense, with all the shops shut and constant fear of the town being ransacked. To Darwin the greatest danger "lay with the lawless soldiery." Knowing the *Beagle* was probably at Montevideo, on November 2, Darwin managed to find spaces for himself and Covington on a packet boat "crowded with men, women and children, glad to escape . . ." bound there. He went aboard in baleful mood, hoping "the confounded revolution gentlemen" would "like Kilkenny Cats, fight till nothing but the tails are left."

Two days later, Darwin arrived with Covington in Montevideo, where he indeed found the *Beagle* and after an absence of nearly three months' "shore-roving," as FitzRoy called it, was able to tell his story.

---

1. Rosas spent the remaining twenty-five years of his life in modestly comfortable exile on a four-hundred-acre farm in Swaythling in Hampshire in the south of England, where he met Darwin again.

CHAPTER TEN

"Great Monsters"

It was a blow to Darwin to discover the *Beagle* would not sail until early December. The reason was to allow charts incorporating the survey information collected by the *Paz* and *Liebre* to be completed and dispatched to London. With the already cluttered poop cabin filled by workers at the chart table, Darwin found lodgings ashore and began planning a trip to the Uruguay River. He took until mid-November——he blamed "true Spanish delay"——to obtain the necessary official papers. While waiting he wrote to Edward Lumb, his host in Buenos Aires, asking his help in retrieving the possessions and specimens he had been forced to abandon with the rebels there. Perhaps surprisingly, Lumb managed to have them located and forwarded to Montevideo.

To Henslow, Darwin described his adventures as "un grande galopeador," traversing wild terrain with bands of gauchos, and detailed the further specimens he was sending him——plants, seeds, fossil shells, birds, and animal skins, including "a fine collection of the mice of S. America." He added somewhat diffidently that he was dispatching separately to a contact in Plymouth "an immense box of Bones and Geological specimens . . . as they do not want any care it does not much signify where kept.——another reason is not feeling quite sure of the value of such bones as I before sent you." Darwin had not heard from Henslow about the specimens he had already sent, and, as he wrote to his cousin Fox, was finding his silence "a mortification . . . It is disheartening work to labour with zeal and not even know whether I am going the right road."

In fact, Henslow had written to him twice, in January 1833 and again in August 1833. The first letter would not catch up with Darwin until nearly eighteen months later in July 1834, while the second would reach him before then in March 1834 in the Falkland Islands. In the latter, Henslow told him his Megatherium bones had "turned out to be extremely interesting as serving to illustrate . . . parts of the animal which the specimens formerly received in this country and in France had failed to do" and had already been exhibited, adding, "Send home every scrap of Megatherium skull you can set your eyes upon.—and *all* fossils."

In mid-November 1833, Darwin finally set out northwest with a *vaqueano* (cowhand) as his guide. His route lay across several flooded rivers. He admired one gaucho's skill at getting himself and his horse across a fast-flowing river: "He stripped off his clothes, and jumping on its back rode into the water till it was out of its depth; then slipping off . . . he caught hold of the tail . . . As soon as the horse touched the bottom on the other side, the man pulled himself on, and was firmly seated" before the horse reached the bank. "A naked man on a naked horse" made "a very fine spectacle," like a figure from "the Elgin marbles."

Three days after setting out, Darwin reached Colonia del Sacramiento, an old town set among peach and orange groves on the northern bank of the Plate, where a wealthy local man invited him to stay. Darwin learned that Colonia had been a military headquarters during Uruguay's recent conflict with Brazil. Its church had served as a powder magazine until a lightning strike detonated it. Darwin thought the ruins "a shattered and curious monument of the united powers of lightning and gunpowder." Reflecting on the war's political consequences for Uruguay, he decided it had created "a multitude of generals, and all other grades of officers" who "have learned to like power and do not object to a little skirmishing. Hence arises a constant temptation to fresh revolutions." He found some hope for stability in people's evident interest in forthcoming elections but noted they did not require their representatives to be highly educated: "I heard some men discussing the merits of [candidates] for Colonia, 'that although they were not men of business, they could all sign their names.' With this every reasonable man was satisfied."

Darwin's host invited him to his *estancia* on the Plate, a prosperous place with plenty of cattle and sheep but also eight hundred mares valued only for breeding and for their skins. Darwin watched gauchos compete to kill and flay the greatest number in a day: "One individual will stand 12 yards from the gate of the Corral, and will bet that he will catch every horse by the legs as it rushes by him.——Another will enter on foot a Corral, catch a mare, fasten its front legs, drive it out, throw it down, kill, skin and stake the hide (a tedious job) and this whole operation he will perform on 22 mares in one day."

On November 20, Darwin reached Punta Gorda, where the Uruguay and the Parana Rivers meet before draining into the Plate. The Uruguay was "a noble body of water," clearer and swifter flowing than the Parana—"when the sun shines, the two colors of the water may be seen quite distinct." Along its thickly wooded shores, he found fresh jaguar tracks and tree bark grooved by yard-long claw marks but failed to see one. Riding on through acacia forests he reached Mercedes on a tributary of the Uruguay called—like the river in Patagonia—Rio Negro and "nearly as large as its namesake to the South." He begged accommodation at a large *estancia*, where his hosts expressed "unbounded astonishment" that the world was round, not flat, and that English farmers twirled neither lassos nor bolas to control their horses and cattle. However, his views on the women of Buenos Aires interested them far more. Had he ever seen such beauty or such large combs in ladies' hair? His polite assurances that he had not elicited delight. Darwin recorded their reaction: "Look there, a man, who has seen half the world, says it is the case; we always thought so, but now we know it."

Continuing across a forbidding landscape of densely growing thistles, Darwin reached the *estancia* of an Englishman to whom he had a letter of introduction and who showed him a point where the Rio Negro snaked beneath precipitous cliffs. Knowing that giant bones had been found nearby, Darwin poked about and "with much trouble extracted a few broken fragments" of "a Megatherium." He also watched a gaucho *domidor*, or horse-breaker, at work. After a troop of young colts had been driven into a corral, the horse-breaker picked out a target, caught its front legs in his lasso, and brought it crashing down. He then quickly roped one of its hind legs tightly to the front legs and, sitting on the horse's neck, fixed a bit-less bridle to the lower jaw. Loosening his lasso, he allowed the horse to stagger up and, bridle tightly gripped, flung a cloth and saddle on its back and jumped on. "During these operations the horse throws himself down so repeatedly and is so beaten, that when his legs are loosed and the man mounts him, he is so terrified as hardly to be able to breathe, and is trickling . . . with sweat.—Generally however a horse fights for a few minutes desperately, then starts away at a gallop . . . till the animal is quite exhausted.—This is a very severe but short way of breaking in a colt."

Like the mass killing and flaying of mares, Darwin did not find it comfortable viewing: "Animals are so abundant in these countries, that humanity and self-interest are not closely united; therefore . . . the former is here scarcely known." A gaucho once urged him to spur a horse he had lent him, shouting, "Why not? . . . . spur him—it is *my* horse." Darwin had "some difficulty making him comprehend that it was for the horse's sake . . . that I did not choose to use my spurs. He exclaimed, with a look of great surprise, 'Ah, Don Carlos, que cosa!' It was clear that such an idea had never before entered his head."

Setting off back to Montevideo, Darwin called at an *estancia* that, he had heard, possessed "some giant's bones." For the equivalent of eighteen pence, he purchased "a part, very perfect, of the head of a Megatherium," minus only the lower jaw and a few teeth knocked out by gauchos shying stones. The remarkable number of giant fossils seemingly littering the landscape convinced him that "the whole area of the Pampas is one wide sepulchre of these extinct gigantic quadrupeds," and all South America "must have swarmed with great monsters."

Galloping on with his prize across hilly, rock-strewn terrain, Darwin paused for the night at a post house kept by "a man, apparently of pure Indian blood," clearly drunk and who—Darwin's guide later told him—had in Darwin's very presence called him "a Gallego"—"an expression synonymous with saying he is worth murdering." Darwin, who had already noticed the man's companions laughing oddly, decided he must have been trying to suborn his guide, who "luckily for me was a trust worthy man.—Your entire safety in this country depends upon your companion." Not for the first time Darwin had had a lucky escape.

Safely back in Montevideo, Darwin reflected on what the past, sometimes dangerous, six months had taught him about those he had met. Gauchos were invariably "obliging, polite and hospitable," spirited and bold. Their swift galloping and diet of beef were "exhilarating to the highest pitch," but he condemned the habitual bloodshed for which he blamed their carrying of knives: "It is lamentable to hear how many lives are lost in trifling quarrels. In fighting, each party tries to mark the face of his adversary by slashing his nose or eye" causing "deep and horrid-looking scars." Robbery was equally commonplace—"a natural consequence of universal gambling, much drinking and extreme indolence."

The police and justice authorities were at best inefficient, at worst corrupt. If a man charged with murder was "rich and has friends he may rely on it nothing will happen." Equally, attitudes toward authority were ambivalent: "the most respectable people . . . will invariably assist a murderer to escape.—They seem to think that the individual sins against the government and not against the people." The only certain protection for travelers like himself were their firearms.

Darwin's verdict on "the higher and more educated classes" in the towns was crushing. The only positive aspect was the remarkable social equality—the humble and ill-educated mingled on equal terms with the wealthy and privileged. Otherwise, "stained by many other crimes" and with few of the virtues of the gaucho, the town dwellers were profligate sensualists who scoffed at religion, thought it weakness not to cheat their own friends, and never told the truth if lying was more expedient: "The term honor is not understood; neither it, nor any generous feeling, the remains of chivalry, have survived the long passage of the Atlantic." With so much overt corruption, he predicted people would soon "be trembling under the iron hand of some Dictator," which—supporter of the British Reform Act though he was—he thought might be better for them: "If I had read these opinions a year ago, I should have accused myself of much illiberality; now I do not."

As the *Beagle*'s departure from Montevideo finally approached, Darwin dashed off a letter to his sister Susan, asking her to tell their father he had drawn a further £17, making a total of £217 over the past seven months of the voyage: "I can offer no excuse. But the Captain believes that instead of passing the summer in Tierra del Fuego we shall also winter there." He added, "The Captain has been exerting himself to a degree which I thought no human being was capable of . . ." FitzRoy had indeed been pushing himself and others. By December 6, as the newly repainted *Beagle* and *Adventure* left the muddy River Plate for the last time, he had deposited the finished survey charts of the region with the British consul to be sent to Beaufort on the first available vessel.

FitzRoy planned to survey further stretches of the Patagonian coastline and Tierra del Fuego before returning to Woollya Cove in the Beagle Channel. As well as checking how Jemmy Button, York Minster, and Fuegia Basket were faring, he was contemplating "a second attempt to place Matthews among the natives of Tierra del Fuego." FitzRoy ordered Wickham, commanding the *Adventure*, to lead because he was worried how well she would sail after her refit and wanted to observe her. His concern was justified. The two ships took seventeen days to reach Port Desire, a safe natural harbor on a river estuary nearly nine hundred

Port Desire, by C. Martens

nautical miles south on the Patagonian coast, because in the blustery conditions the *Adventure* was "found not to sail well on a wind."

Both ships anchored a few miles within the estuary, opposite the ruins of a long-abandoned Spanish fort. Darwin, as usual, was quickly ashore. On an arid, gravelly plain 250 feet above sea level, he found "shells of the same sort which now exist.—and the mussels even with their usual blue colour.—It is therefore certain, that within no great number of centuries all this country has been beneath the sea."[1] His later exploration of nearby cliffs also suggested "the usual geological story" of "being upheaved in modern days." Such findings helped him conclude that "everything in this southern continent has been effected on a grand scale: the land from the Rio Plata to Tierra del Fuego, a distance of 1200 miles, has been raised (and in Patagonia to a height of between 300 and 400 feet), within the period of the now existing sea-shells."

On Christmas Eve, while seamen worked on rerigging the *Adventure* to improve her performance, Darwin shot a guanaco. Even when gutted it weighed 170 pounds, "so that we shall have fresh meat for all hands on Christmas Day." Next day, as well as feasting, most of the crew competed ashore in the *Beagle*'s own Olympic games—wrestling, racing, jumping in sacks, and a naval game called "slinging the monkey," which involved suspending a crewman by his heels, with FitzRoy awarding the prizes. Darwin described "with what schoolboy eagerness the seamen enjoyed them; old men with long beards and young men without any playing like so many children . . . a much better way of passing Christmas than the usual one, of every seaman getting as drunk as he possibly can." Martens painted a lively watercolor of the revels, with the *Adventure* and *Beagle* in the background. Appraising it, FitzRoy told him the *Beagle*'s main mast was too far forward, while the angle of the mizzen mast was wrong.

Whenever he could, Darwin walked into the bleak, treeless interior, finding a "high pleasure . . . which I can neither explain or comprehend" in the "stillness and desolation." On New Year's Day, behind two enormous stone slabs propped against each other on a hilltop, he found

---

1. More recent thinking is that the elevation from the ocean happened 120,000 years ago.

an "Indian" grave consisting of "a heap of large stones placed with some care." The remains of several small fires and piles of horse bones nearby suggested the site was still frequented. A party of officers returned with him next day "to ransack the . . . grave in hopes of finding some antiquarian remains." Beneath successive layers of rocks and stones they found a foot-deep bed of earth but no artefacts or bones. Darwin was surprised since he had read that although "Indians" buried their dead immediately, they later dug up "the less perishable part"—the bones—and carried them to ancient burial grounds since, like many societies, they preferred to lie among their ancestors. He decided "water and changes in climate had utterly decomposed every fragment" of any bodies.

Darwin meanwhile was collecting specimens of "new birds and animals." Always welcoming specimens brought by others, Darwin was given by Martens a small rhea he had shot. In northern Patagonia, gauchos had told Darwin about a rare rhea they called an Avestruz Petise—*petiso* meaning "little." However, "forgetting . . . in the most unaccountable manner . . . the whole subject of the Petises," Darwin dismissed Martens's kill as merely "a not full-grown bird of the common sort." The rhea had been skinned and cooked before he suddenly remembered the gauchos' comments and rushed to retrieve what was left. Only the head, neck, wings, larger feathers, and a big piece of skin remained, but enough to enable a later reconstruction in London that established it as a species separate from the rhea of northern Patagonia— the larger Rhea americana—a distinction that would later influence Darwin's evolutionary thinking by causing him to ponder whether and why two distinct species, found in such close proximity, might share a single common ancestor.

With the *Adventure* still needing adjustments to her masts and rigging, including a new square topsail, FitzRoy

A Rhea Petise (*Rhea darwinii*)

decided to leave her at Port Desire and take the *Beagle* a further one hundred nautical miles south to survey Port St. Julian. On January 4, 1834, as a strong ebb tide bore the *Beagle* out of the estuary, she struck a rock so heavily "as to shake her fore and aft," FitzRoy wrote. He was certain it was the same rock the *Beagle* had hit in 1829, soon after he had assumed command of her. This time the surging tide carried the *Beagle* clear. Sulivan "dived down under the keel, and, having ascertained things were not so bad, came up the other side, bleeding from several scratches received from the jagged copper" of the hull's sheathing to report the hull itself was intact. "Wishing to make doubly sure," FitzRoy also dived down to take a look before, finally satisfied, he ordered the *Beagle* to sail on.

Five days later the ship anchored in Port St. Julian—the spot where, in 1520, two of Ferdinand Magellan's captains, Luis de Mendoza and Gaspar de Quesada, had led a failed mutiny, resulting in Mendoza's death in the fighting, Quesada's subsequent execution, and the impaling of both their dismembered bodies on stakes on the shore. Before leaving, Magellan raised a large cross on a hill four miles northwest of where the *Beagle* anchored. In 1578 a further execution had occurred there. The aristocrat Thomas Doughty, one of Francis Drake's captains, had attempted to usurp Drake's position. After trying him for mutiny, Drake had Doughty beheaded—very unfairly, FitzRoy thought—and his name inscribed in Latin on a rock to be seen "by all that should come after us." FitzRoy found it remarkable two such tragedies had happened in the same place.

FitzRoy and a small group including Darwin set out by boat along the shore to search for a place marked *pozos de agua dulce*—"sweet water springs"—on an old Spanish map. With the wind blowing hard, after a while FitzRoy beached the boat and continued some distance inland on foot with Darwin and four of the boat's crew, all carrying instruments, ammunition, and weapons—in FitzRoy's case "a heavy double-barreled gun." They failed to find fresh water, and as their own ran out, they became exhausted. "Too tired and thirsty to move," FitzRoy stretched out on a hilltop from which Darwin saw the shimmer of what he thought might be a lake about a mile or two away. "More accustomed than the men, or myself, to long excursions on shore," FitzRoy watched Darwin set off only to see him "stoop down at the lake, but immediately leave it." The "lake" proved to be "a snow-white expanse of salt, crystallized in great cubes!"—a glistening *salina*.

The thirsty men trudged slowly back toward their boat. By dusk, after eleven hours without water, FitzRoy could not continue. One man, "the most tired next to me," stayed with him, while in gathering darkness Darwin and the others eventually managed to reach the boat. Fresh men immediately set out with water for FitzRoy and his companion while Darwin and the others lit a fire on the shore to signal to the *Beagle* that they needed help. By mid-morning all were safely back on board. "No one suffered afterwards from the over-fatigue, except Mr Darwin, who had had no rest during the whole of that thirsty day—now a matter of amusement, but at the time a very serious affair," wrote a somewhat sheepish FitzRoy.

Indeed feeling a little feverish, Darwin spent two days in his hammock but then resumed his wanderings ashore. Coming across "a large Spanish oven built of bricks" and a wooden cross on a hilltop, he wondered whether they were relics of navigators of old. Just conceivably, the cross could have been Magellan's, though it's doubtful it could have survived three centuries of Patagonian weather. Beneath a layer of red mud on a gravel plain, he found "some very perfect bones of some large animal."

By January 20, 1834, the *Beagle* was back in Port Desire. Satisfied the *Adventure* was ready to sail, FitzRoy ordered Wickham to take her to complete earlier survey work in West Falkland, where they would rendezvous on the *Beagle*'s return from the far south. On January 22, both ships stood out to sea, the *Adventure* for the Falklands and the *Beagle* for the Strait of Magellan.

On January 26, the *Beagle* entered the strait and, after three days beating against strong westerly winds, reached St. Gregory Bay on the strait's north shore where, watched by Patagonian "Indians," whose *toldos*, or tents, lined the shore, FitzRoy anchored. Accustomed to frequent visits from sealing vessels, the "Indians" spoke Spanish and a little English and gave FitzRoy and his men "a very kind reception" as they came ashore. The men were tall—the personification of Magellan's "Patagones"—with

A Patagonian, by Philip
Parker King

an average height, Darwin reckoned, of over six feet. With their guanaco cloaks, long hair streaming about their faces, and "rather wild" appearance they reminded him of the "Indians" at Rosas's camp though more painted—"many with their whole faces red, and brought to a point on the chin, others black," while one was "ringed and dotted with white like a Fuegian."

As FitzRoy and his party climbed back into their boat, many "Indians" jostled to join them. FitzRoy insisted he could take only three back to the *Beagle* but found it "a very tedious and difficult operation" to reduce the number. Eventually three were chosen who, once aboard the *Beagle*, "behaved quite like gentlemen," using knives and forks and helping themselves with spoons. Darwin noticed they particularly relished sugar. Next day, some of the *Beagle* men bartered ashore for items such as guanaco skins and rhea feathers for which the "Indians" wanted firearms in exchange. When these were refused, their next request was tobacco, which they valued more highly than knives or axes. "An amusing scene," Darwin wrote, "and . . . impossible not to like these . . . giants, they were so thoroughly good-humoured . . ."

Continuing along the strait, on February 2 the *Beagle* arrived in ill-omened Port Famine where, in 1826, the *Beagle*'s previous captain Pringle Stokes had committed suicide and that FitzRoy needed to survey. The Elizabethan sailor Thomas Cavendish had named the place in 1587 after discovering famished survivors of a failed Spanish settlement clinging to life there. For the first few days, torrential rain restricted Darwin to brief walks along the beach. When the weather improved a little, though still "blowing a gale" he set off to climb the highest point in the immediate vicinity—the 2,600-foot-high Mount Tarn. Struggling up and along deep, dark ravines blocked "by great mouldering trunks" that collapsed beneath his weight as he tried to climb over them, he finally emerged in piercing cold onto a bare ridge leading to the summit and "a true Tierra del Fuego view; irregular chains of hills, mottled with patches of snow; deep yellowish-green valleys; and arms of the sea running in all directions . . ."

By February 10, surveying complete, FitzRoy headed northeast back out of the strait to begin charting the northeastern shores of Tierra del Fuego. En route, from the crew of a passing sealer, he and Darwin heard shocking news about the Falkland Islands: "The Gauchos had risen and murdered poor Brisbane and Dixon [sic] . . . and it is feared several others," and some English sailors had managed to escape and were hiding out on West Falkland. The British warship HMS *Challenger* had subsequently called at the islands and left a handful of marines to restore order—"six (!)" Darwin wrote, clearly thinking the number inadequate, adding "in my opinion the Falkland islands are ruined . . ."

Despite the many dangerous sand and mud banks, FitzRoy completed his survey of eastern Tierra del Fuego within a week. Darwin was only able to land once but found other distractions such as an acrobatic display by a pod of sperm whales "jumping straight up out of the water; every part of the body . . . visible excepting the fin of the tail. As they fell sideways into the water, the noise was as loud as a distant great gun." FitzRoy now set a southwesterly course for the Beagle Channel and Woollya Cove and on February 22 brought the *Beagle* into Le Maire Strait between Tierra del Fuego's eastern tip and rugged, castellated Staten Island, which Darwin vividly recalled from the previous year as "one of the most desolate places . . . the mere backbone of a mountain forming a ridge in the ocean."

Here the *Beagle* ran into serious difficulties. Covington wrote "the ship plunged and would not answer her helm for a short time . . . looked . . . as if going down." Darwin blamed "a very great and dangerous tide rip" for the heavy pitching that "in a weak vessel . . . would almost have been sufficient to have jerked out her Masts." Eventually FitzRoy brought the *Beagle* safely out of the strait only for the wind to drop so completely that for a while the ship was at the mercy of the high swell and risked being pushed dangerously close to the western shore of Staten Island. "What a great useless animal a ship is, without wind," Darwin wrote.

On February 24, the *Beagle* reached Woollaston Island, surveyed by FitzRoy the previous year. In contrast to the engaging "giant" Patagonian "Indians" of St. Gregory Bay, Darwin thought the Fuegian inhabitants "poor savages . . . I never saw more miserable creatures; stunted in their growth, their hideous faces bedaubed with white paint and quite naked.— One full aged woman absolutely so, the rain and spray . . . dripping from her body; their red skins filthy and greasy, their hair entangled, their voices discordant, their gesticulation violent and without any dignity . . . one can hardly make oneself believe that they are fellow creatures placed in the same world . . . What a scale of improvement is comprehended between the faculties of a Fuegian savage and Sir Isaac Newton . . ."

He wondered, "What could have tempted, or what change compelled a tribe of men, to leave the fine regions of the north, to travel down the Cordilleras [parallel mountain ranges] or backbone of America, to invent and build canoes . . . and then to enter on one of the most inhospitable countries within the limits of the globe?" Anthropologists today confirm that thousands of years before, the Fuegians' ancestors crossed the Bering Strait from Asia into Alaska and headed south toward Tierra del Fuego, which rising temperatures at the end of the Ice Age had made more hospitable. A land bridge created by falling sea levels had enabled them to cross to Tierra del Fuego before, about eight thousand years ago, further climate change caused the Strait of Magellan to flood, cutting the Fuegians off from the north.

The reason for their primitive state, Darwin decided, summing up his two visits to Tierra del Fuego, was the conditions in which the Fuegians lived. The cold climate, poor diet, and other environmental limitations had stifled their development, although it had hardened their bodies to enable them, though scantily clothed, to survive the cold, and it had enhanced their hearing and vision: "How little can the higher powers of the mind be brought into play: what is there for imagination to picture, for reason to compare, for judgement to decide upon? To knock a limpet from the rock does not require even cunning, that lowest power of mind. Their skill in some respects may be compared to the instinct of animals; for it is not improved by experience . . . Nature by making habit omnipotent . . . has fitted the Fuegian to the climate and the productions of his miserable country."

Beneath a leaden sky, the wind began blowing hard from the west and white foam flecked the churning sea: "Dear Tierra del has recollected her old winning ways.—The ship is now starting and surging with her gentle breath.—Oh the charming country." As the weather continued to worsen, FitzRoy moved the *Beagle* to a more sheltered mooring in a small cove inside the eastern entrance to the Beagle Channel. While the crew went ashore to search for fresh water and wood, three canoes filled with Fuegians came out to the ship to barter—"very quiet and civil and more amusing than any Monkeys," with "a fair idea of barter and honesty," Darwin wrote. They begged for every item they saw, sometimes pointing to young women or children "as much as to say, 'If you will not give it me, surely you will to them.'"

Still battling strong westerlies, FitzRoy headed along the Beagle Channel. As the ship entered Ponsonby Sound, more and more canoes appeared, filled with Fuegians eager to barter fish and crabs for pieces of cloth. Darwin found it "very amusing to see with what unfeigned satisfaction one young and handsome woman with her face painted black, tied . . . several bits of gay rags round her head.—Her husband, who enjoyed the very unusual privilege in this country of possessing two wives, . . . became jealous of all the attention paid to his young wife, and after a consultation with his two naked beauties was paddled away by them." As he himself admitted, Darwin was more comfortable interacting with the Fuegians from the security of the *Beagle* than while sitting in a small boat: "Last year I got to detest the very sound of their voices; so much trouble did it generally bring to us . . . But now we are the stronger party, the more Fuegians the merrier . . ."

Pursued by Fuegian canoes that "vainly strove to follow us in our zigzag course," FitzRoy tacked the *Beagle* into the northern reaches of Ponsonby Sound. The magnificent views all around included the mountain FitzRoy had just named for Darwin in honor of his twenty-fifth birthday a few days before. "There is such splendour in one of these snow-clad mountains, when illuminated by the rosy light of the sun," and at seven thousand feet Mount Darwin was the highest in Tierra del Fuego, Darwin exulted.

The next day, March 5, the *Beagle* reached an apparently deserted Woollya Cove on Navarin Island. The wigwams built for Jemmy, York, and Fuegia still stood, and FitzRoy found turnips and potatoes growing in the neglected vegetable garden, but of Jemmy, York, and Fuegia there was no sign until finally a canoe "with a flag hanging up" approached. Sitting in it was Jemmy Button, though until the canoe came close, neither FitzRoy nor Darwin recognized him. Before leaving Montevideo, Darwin had written to Susan Darwin that he expected to find Jemmy and the others "naked and half starved—if indeed they have not been devoured," and at least some of his predictions had come true. When last seen, Jemmy was "very fat, and so particular about his clothes, that he was always afraid of even dirtying his shoes, scarcely ever without gloves and his hair neatly cut." Now it was "quite painful" to see him so "thin, pale, and without a remnant of clothes, excepting a bit of blanket round his waist: his hair, hanging over his shoulders . . ." Jemmy himself seemed embarrassed, turning his face away as his canoe drew near.

However, according to FitzRoy, once he had taken Jemmy belowdecks and clothed him, Jemmy assured everyone he had "plenty fruits," "plenty birdies," and "too much fish" to eat, was not cold, had built his own canoe, and that "very good people" were his friends. He presented fine otter skins to FitzRoy, two spearheads he had fashioned to Darwin, and he had even brought a bow and quiver filled with arrows for his old schoolmaster in Walthamstow. That evening "a young and very nice looking squaw," as Darwin put it, arrived, who Jemmy eventually and bashfully admitted was his wife.

Jemmy surprised FitzRoy and Darwin by insisting he had no desire to return to England. Strangest of all to Darwin was that Jemmy could still talk only a little of "his own language" but appeared "to have taught all his friends some English." Jemmy related how several months earlier, York Minster had persuaded him and his family to accompany him and

Fuegia to his own land. The party had set out in four canoes, but in "an act of consummate villainy," one night York deserted Jemmy and his relations, taking all their possessions, including their clothes and tools. Fuegia, Jemmy claimed, was "quite contented with her lot" and had assisted York in the robbery.

That night Jemmy slept ashore, but the following morning came aboard and "shared my breakfast," FitzRoy wrote, after which "we had a long conversation by ourselves." Everything Jemmy told him, including accounts of attacks on Woollya by "the dreaded Oens-men" from across the Beagle Channel, convinced him a second attempt at establishing a mission would be futile, even in the unlikely event Matthews could be persuaded to stay after his previous experiences. Nevertheless, FitzRoy hoped "some benefit, however slight, may result from the intercourse of . . . Jemmy, York and Fuegia, with other natives of Tierra del Fuego. Perhaps a shipwrecked seaman may hereafter receive help and kind treatment from Jemmy Button's children . . . by an idea, however faint, of their duty to God as well as their neighbour."

Jemmy remained on board until the ship got underway, which so frightened his wife, waiting in their canoe, that she did not stop crying until he was safely off the *Beagle* with all the presents he had been given. Darwin thought that "every soul on board was as sorry to shake hands with poor Jemmy for the last time, as we were glad to have seen him.—I hope and have little doubt he will be as happy as if he had never left his country; which is much more than I formerly thought."

After paddling back to the shore, Jemmy lit a fire, the smoke curling up in "a last and long farewell" to his friends as the *Beagle* stood out of Ponsonby Sound to begin her voyage to the perennially troubled Falkland Islands to complete her survey work there.

CHAPTER ELEVEN

The Furies

As soon as the *Beagle* anchored in Berkeley Sound, East Falkland, Lieutenant Henry Smith, left by the captain of HMS *Challenger* as the official British Resident in the islands, hurried aboard to give FitzRoy a fuller report of the recent troubles than the crew of the sealing vessel had been able to provide. He cataloged "such complicated scenes of cold-blooded murder, robbery, plunder, suffering, such infamous conduct in almost every person who has breathed this atmosphere" that Darwin reckoned it would take two or three pages to do them justice.

The murders of Brisbane, Dickson—the Union Jack keeper—and others had occurred the previous August. The *Challenger*'s commander, Captain Seymour, had sent marines ashore, but despite searching for four days they failed to locate the murderers who had fled inland. Unable to stay longer, Seymour left Smith with six marines and some sailors to continue the search and restore order. Smith arrested all but one of the murderers, whom, with the exception of a man who agreed to testify against his fellows to save himself, he dispatched by cutter to the British naval station in Rio de Janeiro. Shortly before the *Beagle* arrived, Smith finally caught the remaining and "principal murderer," Antonio Rivero—"a desperate character"—and confined him on an islet in the sound. Nevertheless, Smith still feared for his own safety if Rivero should escape and asked FitzRoy to put Rivero in irons on the *Beagle* until the cutter returned. FitzRoy agreed. Also at Smith's request, he took on board the prisoner willing to give evidence against the others, together with a third man now suspected of complicity in the plot.

FitzRoy moved the *Beagle* closer to Port Louis, where on March 13, 1834, Wickham joined him with the *Adventure*. Aboard was William Low, the Scottish captain from whom FitzRoy had purchased the vessel. Low had returned to the Falkland Islands from a sealing expedition shortly after the murder of his friend Brisbane. Realizing his own life was in danger, he had fled west by boat and eventually been picked up by the *Adventure*. By now "in great distress," he had offered Wickham his services as a pilot, which Wickham had provisionally accepted. FitzRoy confirmed the appointment, "trusting that the Admiralty would approve of my so engaging a person who, in pilotage and general information about the Falklands, Tierra del Fuego, Patagonia and the Galapagos Islands, could afford us more information than any other individual . . ."

After the *Adventure* departed to continue her survey of the islands, Fitz-Roy visited Port Louis, which seemed "more melancholy than ever; and at two hundred yards' distance from the house in which he had lived, I found, to my horror, the feet of poor Brisbane protruding above the ground. So shallow was his grave that dogs had disturbed his mortal remains, and had fed upon the corpse. This was the fate of an honest, industrious and most faithful man . . . murdered by villains, because he defended the property of his friend; he was mangled by them to satisfy their hellish spite; dragged by a lasso, at a horse's heels . . . and left to be eaten by dogs."

Darwin wanted to explore inland, but with so many "villains" around worried about his safety. He decided to entrust himself to "the only two Spaniards . . . not directly concerned" with the killings, even though he suspected they had known what was going to happen. In the event, "they had no temptation to murder me and turned out to be most excellent Gauchos." With cold winds rippling the withered grass and bursts of drenching rain, Darwin rode out with them to explore southwest of Berkeley Sound.

Beyond "some little geology," he found nothing interesting until, with some effort, they picked their way through the broken peaks of a two-thousand-foot-high range of "quartz rock hills." One obstacle was the massive, wide "stone-runs" of angular chunks of quartz—some two miles long—extending along and crossing the intersecting, steep-sided valleys. On his first visit to the Falkland Islands eleven months earlier, Darwin had noticed these tumbled "streams of stones," some fragments "like the ruins of some vast and ancient cathedral" and large enough to give shelter from the driving rain. Now examining them more closely, he wondered whether they were once white lava flowing down from the mountains that "when solidified . . . had been rent by some enormous convulsion"—perhaps by earthquakes.

Riding on, Darwin and the gauchos crossed swampy, peaty moorland strewn with decomposing carcasses of cows killed for food by the murderers of the previous August as they retreated inland. Darwin probed inside several, searching for what had been his early love in the natural world—beetles. Under a "dead old calf" was a *Catops* beetle. A large number of bulls were wandering the wilderness, savage but so magnificent with their powerful heads and shoulders that they reminded him of Greek marble sculptures. To revenge themselves on one old bull for blocking their path, the gauchos violently felled and callously castrated it. In his diary, which he knew his sisters would read, Darwin wrote euphemistically that they rendered the animal "for the future innocuous."

Darwin's companions also pursued and caught a cow with their *bolas*. After dispatching it with a swift knife thrust into the spine, they hacked off enough flesh to last the entire expedition, then constructed a fire using bullock bones for fuel and roasted some. Darwin had developed a taste for "carne con cuero"—meat cooked in its skin: "A large circular piece taken from the back, is roasted on the embers with the hide downwards and in the form of a saucer, so that none of the gravy is lost. If any worthy alderman had supped with us that evening, 'carne con cuero,' without doubt, would soon have been celebrated in London."

After three days, the weather was so bad—almost constant hail and snow—that Darwin decided to return to the *Beagle*. The journey was difficult. "From the great quantity of rain this boggy country was in a very bad state.—I suppose my horse fell at least a dozen times and sometimes [all of us] were floundering in the mud together. All the little streams have their sides soft, so that it is a great exertion for the horses to jump over them and . . . they repeatedly fall." Forced to ford an inlet where "waves from the violent winds broke over us," he was sodden and exhausted when he finally reached the *Beagle*.

With FitzRoy still awaiting the return of the cutter to take Rivero and the others off his hands, Darwin spent his time geologizing, making notes on barnacles and examining kelp. "The immense quantity and number of kinds of organic beings which are intimately connected with the Kelp"—small fish, crabs, sea-eggs, star fish, flat worms, and snails that in turn attracted "all the fishing quadrupeds and birds (and man) [to] haunt the beds"— amazed him. He decided "the zoology of the sea is . . . generally the same here as in Tierra del Fuego" and speculated about the interdependence of living organisms for food, shelter, and hence life itself—anticipating the science of "ecology"—"One single plant form is an immense and most inter- esting menagerie," he wrote. "If this Fucus [the kelp] was to cease living, with it would go many: the seals, the cormorants and certainly the small fish and then sooner or later the Fuegian man must follow.—the greater number of the invertebrates would likewise perish . . ."

The jackass penguin, so named for "throwing its head backwards, and making a loud strange noise, very like the braying of [an ass]," both amused and intrigued him. He noticed how "in diving, its little plumeless wings are used as fins; but on the land as front legs." It was also brave. When he blocked the path of one trying to reach the sea, "it regularly fought and drove me backwards . . . every inch gained he firmly kept, standing close before me, erect and determined."

Darwin pondered the islands' quadrupeds and how introduced spe- cies like cattle, horses, and rabbits had fared. Earlier visitors, including de Bougainville, whose book was in the *Beagle* library, had described the islands' only native quadruped, the "wolf-like" Falkland fox (also known as the warrah), which lived on both East and West Falkland. Of the four speci- mens brought aboard, Darwin noted, "the three larger ones are darker and come from the East; there is a smaller and rusty coloured one which comes from the West Island; Low states that all from this island are smaller and of this shade of colour." Darwin also believed the Falkland fox was "confined to this archipelago" and the product "of what

A Falkland fox

appears to me to be a centre of creation." This was because "many sealers, Gauchos, and Indians, who have visited these islands, all maintain that no such animal is found in any part of South America . . . As far as I am aware, there is no other instance in any part of the world, of so small a mass of broken land, distant from a continent, possessing so large a quadruped peculiar to itself." The differences and similarities between the foxes of East and West Falklands would—like those he observed between the two types of South American rheas—prompt him to consider the possibility they shared a common ancestor.

Because of their tameness and curiosity, fox numbers were already declining. In the notes he made on the island, Darwin presciently predicted that "within a very few years after these islands shall have become regularly settled, in all probability this fox will be classed with the dodo, as an animal which has perished from the face of the earth." By 1880, the Falkland fox would indeed be extinct throughout the islands. Darwin attempted to bring one home on the *Beagle*, but after several months it died, and only its skin reached Britain.[1]

During his final days in the Falkland Islands, Darwin also caught up with his letter writing. On arrival he had found his first letter from Henslow, whose enthusiastic response to the specimens he had sent home, especially the giant fossil bones, was an enormous relief. "Nothing for a long time has given me so much pleasure," Darwin wrote back, ". . . your account of the safe arrival of my second cargo and that some of the specimens were interesting has been . . . most highly satisfactory to me." He told Henslow, "I am quite charmed with geology . . . When puzzling about stratification etc, I feel inclined to cry a fig for your big oysters and your bigger Megatheriums.—But then when digging out some fine bones, I wonder how any man can tire his arms with hammering granite." He described the new type of "ostrich" (rhea)—the *Avestruz petise*—he had

---

1. The Natural History Museum in London possesses two skins of the Falkland fox presented by FitzRoy. Darwin preserved another to allow a check to be made of whether it was indeed a fox or a wolf.

identified and enthused about the future. "The prospect before me is full of sunshine . . . glorious scenery, the geology of the Andes; plains abounding with organic remains . . . and lastly an ocean [the Pacific] and its shores abounding with life." He assured Henslow—who had urged him, even if "sick of the expedition," to think carefully before abandoning it— "if nothing unforeseen happens I will stick to the voyage; although . . . this may last till we return a fine set of white-headed old gentlemen."

Also awaiting Darwin were letters from his sisters Catherine and Caroline relaying six-month-old gossip about his brother Erasmus's ill-concealed devotion to the wife of his cousin Hensleigh Wedgwood and "how prettily and coquettishly" Fanny Biddulph had inquired, "Has Charles quite forgotten me?" He wrote back describing parting from Jemmy Button and his pleasure in his work. There was "nothing like geology"—not even the first day of partridge shooting or fox hunting could compare with the joy of finding fossil bones, "which tell their story of former times with almost a living tongue."

His pen portrait of "this little miserable seat of discord" and "scene of iniquity"—the Falkland Islands—was damning—"the whole country is . . . an elastic peat bog." He castigated British government policy: "Here we, dog-in-the-manger fashion seize an island and leave to protect it a Union jack; the possessor has been of course murdered; we now send a lieutenant, with four sailors [sic], without authority or instructions." Yet, if administered properly, the Falkland Islands could become "a very important halting place in the most turbulent sea in the world." He asked Catherine to let him know—"in more than one letter" in case correspondence got lost or delayed—whether the latest installment of his diary, dispatched the previous July, had reached Shrewsbury safely, requested more gossip, sent thanks to Fanny Biddulph for remembering him, and sent love to his childhood nurse, Nancy, warning that if she saw him with his "great beard," she would not recognize him. He concluded he would have written more, but "I have a host of animals, at this minute, surrounding me, which all require embalming and numbering."

The cutter had still not returned from Rio, but FitzRoy found a way of disposing of Rivero and the two others in his custody, presumably in return for a fee, dispatching them to Rio on a French whaling vessel with a mutinous crew that put into the Falkland Islands for repairs. Darwin wrote, "Having thus removed . . . the worst prisoners there are little fears for [Lieutenant] Smith's safety."[2]

FitzRoy's final task in the islands was to bury a lieutenant from HMS *Challenger*, drowned three months earlier when his boat overturned. Martens had found his body lying at the high-water mark in an unfrequented part of Berkeley Sound while shooting geese for the *Beagle*'s pantry. FitzRoy interred the lieutenant "not far from the tomb of our regretted shipmate Hellyer." Later that day, April 6, 1834, they sailed from the Falkland Islands, "depressed more than ever by the numerous sad associations connected with their name."

Battling stiff westerlies—Darwin had "never seen His Majesty's vessel under a greater press of sail or much closer to a lee-shore"—six days later they reached the mouth of the Santa Cruz River just north of the Strait of Magellan. FitzRoy was still worried about possible damage to the *Beagle*'s keel after hitting the rock at Port Desire and wanted to confirm the copper sheathing was intact since, in the warmer Pacific waters, parasites like Teredo worms would burrow into exposed planking. He therefore decided to beach the *Beagle* at high tide to inspect the hull below the waterline—a "rather ticklish operation," Darwin thought. Martens sketched the ship tilted over on the wet sand, supported by beams, with figures looking up at the curved hull. In fact, damage was slight—a piece of the false keel had been knocked off, and a few sheets of copper sheathing were damaged. Hard work by carpenter Jonathan May and his team meant that "one tide was sufficient to repair her and after noon she floated off and was again moored in safety."

FitzRoy next turned to the exploration of the Santa Cruz River, intending if possible to reach its unknown source, believed to be in the Andes. On the *Beagle*'s previous voyage, the late Captain Pringle Stokes

---

2. Rivero became something of a national hero in Argentina, symbolizing resistance to the British in the Falkland Islands, admired by President Perón, among others, in more recent times.

The *Beagle* laid ashore, River Santa Cruz, by C. Martens

had traveled thirty miles upriver before his provisions ran out and he turned back. Darwin thought FitzRoy's plan "a glorious scheme . . . I cannot imagine anything more interesting."

FitzRoy assembled a party of twenty-five, including Darwin, John Lort Stokes, Bynoe, Martens, and Covington, along with a number of seamen and marines, all well armed and capable of defying "a host of Indians." On April 18, taking three weeks' supply of "beef, pork, preserved meats, rice . . . and biscuit" and a large collection of instruments, including three chronometers, a sextant, a theodolite, and Darwin's two aneroid barometers for measuring the height above sea level, they set out in three whaleboats. For FitzRoy, usually preoccupied with coastal surveying, this was the closest he had yet come to the adventurings of his shipboard "philosopher." He had written a little enviously to his sister Fanny, "While I am pottering about in the water, measuring depths and fixing positions, he wanders over the land, and frequently makes long excursions where I cannot go, because my duty is Hydro-not Geo-graphy."

The wide, milky-blue Santa Cruz at first flowed seaward at a brisk four to six knots through seemingly boundless flood plains strewn with smooth pebbles and shells. Rowing upstream against such a powerful current became impossible. Fitzroy left two men in each boat, ordered the rest ashore, and had the boats made "fast to one another . . . a few yards apart, in . . . line a-head." Then he divided those ashore into teams to haul the boats, rather like horses towing canal barges. After about an hour and a half, one team replaced another. Everyone, FitzRoy and Darwin included, took their turn in the hard work. "Many were the thorny bushes . . . If the leading man could pass, all the rest were bound to follow. Many were the duckings, and serious the wear-and-tear of clothes, shoes and skin," FitzRoy wrote.

Darwin noticed how the officers lived with their teams, eating the same food, and sleeping in the same tent. Every night, as soon as the sun began to sink, FitzRoy chose a place to camp, and they took turns to cook. While the appointed cook lit a fire, two men pitched the tent while others fetched items from the boat and collected firewood. Throughout the night two men and an officer took hourly shifts to keep up the fire and watch for "Indians." Next morning, half an hour after sunrise, they were off again. Martens thought "it would surprise the most experienced manager of picnics in England to see with what dispatch our tents were taken down and stowed and our breakfasts prepared and eaten." One day, the party came across a dead guanaco. Most readily ate the meat, though a few balked at eating carrion.

Progress was slow, especially where they had to negotiate narrow channels between close-packed islands. Darwin calculated they were managing no more than ten miles a day if measured in a straight line, "perhaps 15 or 20 as we were obliged to go." Distant spirals of smoke and the occasional skeleton of a horse told them native people were about. On the third night, after finding an old boat hook lost on Pringle Stokes's expedition, they reached the point where he had turned back.

Though it was sunny by day, the nighttime temperature dropped to 25°F so that no one undressed to sleep. Like polar explorers, by morning they found "the inside of the tents completely covered with a heavy frost arising from the breath of those within." Darwin's wet fishing net froze solid during the night.

One morning, they woke to find fresh horse tracks and the marks of long spears—the *chusa* "which trails on the ground"—near their camp, suggesting "Indians [had] reconnoitred us during the night" without alerting the sentries. In case they were indeed being watched, FitzRoy ordered "the stragglers . . . to keep close," and "we now formed a small compact body with the exception of two . . . sent forward by way of advanced piquet or scouts." Pushing on, they found fresh footprints of people, dogs, and horses and the remains of a fire, where a large party had clearly achieved the "difficult and dangerous" feat of crossing the Santa Cruz. A Spaniard who lived with the indigenous people had told Darwin how they used a *pilota*—"a sort of canoe" made by tying the corners of a hide together, "which generally is pulled over by catching hold of the horse's tail." However, from then on, apart from occasional large, distant columns of smoke, this was the last sign of "Indians." Martens wondered whether they were "retreating into these barren and desolate regions" to escape Rosas's "war of extermination."

The characteristic Patagonian dry shingle plains dotted with stunted bushes through which they were passing disappointed Martens, who had not yet seen anything worth sketching. Observing the few fish in the river or waterfowl on its banks, Darwin grumbled, "The curse of sterility is on the land." The only animals that appeared to thrive were great herds of llama-like guanacos—food for the pumas whose paw prints were almost everywhere on the riverbanks—and the condors, which pounced on the pumas' leavings. Darwin also noticed "several species of mice" with large, thin ears, luxuriant fur, and cannibal tendencies—any mouse caught in one of his traps was immediately eaten by others.

As the days passed and the river wound higher above sea level, expectation grew. By the seventh day, April 24, Darwin thought they were "like the old navigators approaching an unknown land," everyone watching "for the most trivial signs of a change"—"the drifted trunk of a tree, a boulder of primitive rock," or anything suggesting they might be nearing "the stony ridges of the Andes." They were disappointed when what at first looked like mountains turned out to be only clouds piling the horizon. However, they were in fact "a true harbinger"—clouds of vapor condensing around the Andes's icy summits.

From this point, Darwin found the geology some of the most interesting in his time in South America. He began noting the direction and composition of the ever more striking lava flows and making comparisons with what he had seen in Port Desire to Port St. Julian to the east coast of Tierra del Fuego. The pebbles of basalt—a hard, igneous rock formed by the cooling of magma brought to the Earth's surface by volcanoes—in the gravel of the riverbed that he had noticed from the start "suddenly became abundant" and much larger. Then ahead the party saw "the angular edge of a great basaltic platform"—dark cliffs capped with a layer of basalt 120 feet thick, with tumbled blocks of basalt at their base.

The river now became "encumbered with these basalt masses," and, as Martens noticed, "the whole country was gradually becoming more elevated." Darwin decided, correctly, that the basalt was lava, "which at some remote period when these plains formed the bottom of an ocean, was poured forth from the Andes" into the water in eruptions "on the grandest scale." It surprised him since, as he noted in his diary, the most southerly volcanic Andean rocks "hitherto known" were hundreds of miles to the north, near the Chilean island of Chiloe.

Halting at noon on April 26, they explored what FitzRoy described as "a wild looking ravine, bounded by black lava cliffs . . . a kind of jungle at the bottom . . . Lions or rather pumas shelter in it, as the recently torn remains of guanacos showed . . . Condors inhabit the basaltic cliffs . . . imperfect columns of basalt give to a remarkable rocky height, the semblance of an old castle . . . a scene of wild loneliness." Enthused by the "fine bold crags," Martens at last got out his sketchpad.

As they continued to manhaul the boats, the river narrowed and the current flowed ever faster over "great blocks of lava," which knocked

Basalt glen, Santa Cruz River, Patagonia, by C. Martens

two holes in the side of one boat, which they only just prevented from sinking. While it was being repaired, Darwin shot a condor measuring eight and a half feet from wingtip to wingtip and four feet from beak to tail. He thought the condors magnificent, especially "when seated on a pinnacle over some steep precipice, sultan-like they view the plains beneath them" or when they rose heavily from their nests to "wheel away in majestic circles."

On this occasion, the "steep precipice" from which the condors were launching themselves were black lava cliffs that rose ever higher above the Santa Cruz River. On April 29, Darwin and Stokes scaled the cliffs and finally glimpsed the snowy peaks of the Andes "peeping through their dusky envelope of clouds" to the west. The next day, the views were clearer. Yet their progress was slowing as the tortuous Santa Cruz became ever more choked with great blocks of stone——one of which was "five yards square and five feet deep"——which Darwin thought "in former periods of commotion have come from the Andes." By May 3, the snow-covered peaks of the Andes stretching away on the western horizon seemed to

FitzRoy as far distant "as on the day we first saw them." The next day he decided reluctantly to turn back.

Not only was everyone exhausted, but supplies were running low. In recent days FitzRoy had halved the biscuit ration, which Darwin found "very unpleasant after our hard work . . . It was very ridiculous how invariably the conversation in the evening turned upon all sorts, qualities and kinds of food." Martens noted, "he was a rich man who had still some tobacco or a few cigars yet left," and everyone's shoes "were in the last stage of consumption"—guanaco skins would have made fine moccasins, he thought, if they only had more time.[3]

Despite the hardships, Darwin was frustrated to turn back with the mountains only about thirty miles away: "We were obliged to imagine their nature and grandeur, instead of standing . . . on one of their pinnacles and looking down . . ." He joined those who had the energy on a westward walk led by FitzRoy for "a farewell look at the Cordilleras which probably in this part had never been viewed by other European eyes." They halted at a point they calculated was "about 140 miles from the Atlantic and 60 from the nearest inlet of the Pacific." FitzRoy had suspected the source of the Santa Cruz to be a lake and, though he could not know it, he was correct. It flowed eastward out of the glacial Lake Argentino in the Andean foothills, from which they were only about twelve miles away.

Before sunrise the following day, the party began the descent, shooting downstream in the boats, Darwin wrote, "with great rapidity." Their speed of around ten knots meant that in a single day they traveled as far as five and a half days' worth of laborious hauling. Three weeks after setting out, they were back aboard the *Beagle*—"fresh-painted and as gay as a frigate." Though the expedition had been hard work and they had failed to reach the Andes or discover the river's source, there were compensations. Soon after his return to England, FitzRoy would describe the expedition in a paper to an impressed Royal Geographical Society. Darwin found the

---

3. The large basalt blocks Darwin noted are today interpreted as indications of glacial expansion in Patagonia.

journey "most satisfactory, from affording so excellent a section of the great modern formation of Patagonia" and his best opportunity so far to confirm his view that the Patagonian plains had once been seabed—the smoothness of the pebbles the result of the movement of the ocean. In coming weeks in his cramped cabin, he would write an essay, "Elevation of Patagonia," suggesting how the plains had risen.

This unpublished essay formed the basis for Darwin's later published work on the subject, including in his *Journal of Researches*. In the essay, he built on the detailed and precise measurements of the height of the land made by FitzRoy and the *Beagle*'s officers as part of their survey work and his own observations to analyze how and over what area the land might have moved. His conclusion was that successive distinct elevations propelled by geological forces acting over a wide area had raised the Patagonian plains and that the Santa Cruz River valley had originally been a channel like the Strait of Magellan beneath the sea. On the basis of his observations he believed that the elevation of the vast area from Cape Horn to Baia Blanca, and quite possibly beyond, had been caused by movement from deep within the Earth. This view, as Darwin noted, differed from that of Lyell, who proposed that such elevations were caused by movements much closer to the Earth's surface and occurred more gradually. Darwin's willingness to deviate in his conclusion from the work of Lyell, whom he admired, is perhaps indicative of his growing confidence in his own analytical abilities. On his return from the *Beagle* voyage he would draw the first geological map of Patagonia.[4]

At this stage FitzRoy seems to have shared some of Darwin's thinking regarding the likelihood that the geological changes they had observed could not have been caused by a biblical flood. In a chapter titled "A Very Few Remarks with Reference to the Deluge" in his narrative of the *Beagle* voyage recanting this view, he would recall that while crossing vast plains composed of "sea-worn" stones "bedded in diluvial detritus some hundred feet in depth," he had remarked to "a friend"—who could only

---

4. In his published work, Darwin revised his opinion somewhat, suggesting that the Patagonian elevation had been somewhat more continuous than he had previously thought.

have been Darwin—that "this could never have been effected by a forty days' flood."

While FitzRoy prepared for the next leg of the journey, back south to Port Famine, then westward into the Pacific, Darwin took long walks "collecting for the last time on the sterile plains of this Eastern Side of S. America." On May 12, the *Beagle* sailed and almost immediately ran into cold, buffeting weather that rendered Darwin "sick and miserable." Ten days later, off the entrance to the Strait of Magellan, the *Adventure* hove in view from the Falkland Islands. Wickham reported that all now seemed quiet. Of even greater interest were the letters and parcels the *Adventure* was bringing from the Falkland Islands. Darwin's sisters had sent him a new pair of walking boots and a supply of books, including *Poor Laws and Paupers Illustrated* by social reformer Harriet Martineau—"Erasmus knows her and is a very great admirer and everybody reads her little books and if you have a dull hour you can, and then throw them overboard, that they may not take up your precious room." A letter from Fanny Biddulph—now with a six-month-old daughter—teased Darwin about settling with a "*little wife*" in his "*little* parsonage" on his return and flirtingly reminded him of "the good old times of the *housemaid* and *postillion*" they had once shared.

While the *Adventure* bore away to survey a final stretch of the Strait of Magellan, FitzRoy headed for Port Famine on the strait's northern shore, where the *Beagle* anchored on June 1. Darwin had never seen "a more cheer-less prospect" with "the dusky woods, pie-bald with snow" barely visible through fog and rain. FitzRoy sent parties to collect firewood and water—the latter not a moment too soon, as the water they had recently been drinking "contained so much salt that brackish is almost too mild a term to call it," Darwin thought, and did nothing to satisfy thirst. However, Fuegians harassed the shore parties so much that FitzRoy ordered the firing of one of the *Beagle*'s cannon. When that failed to frighten them away, he ordered his men to fire musket balls wide of them. The Fuegian response, Darwin noticed, was to hide behind trees and every time a musket was discharged to fire an arrow. Seeing FitzRoy's men laughing as their arrows fell short, the Fuegians "frantic with rage . . . shook their very mantles with passion" before running off.

Darwin meanwhile examined small plants and animals he had fished from the strait. Ever since studying flustrae in Edinburgh, he had been interested in the tiny plantlike creatures he called "corallines" or "zoophytes"—soon to be known as *bryozoans*—and that for a long time people had classified as plants, not animals. They were, Darwin realized, "an enormous branch of the organized world; very little known . . . and abounding with most curious, yet simple, forms of structures." Placing a reddish-colored "elegant little coralline" under his microscope, he saw its body was covered with long bristles. Uncertain about their purpose, he tested the creature's response to different stimuli and saw how the bristles moved—sometimes independently, sometimes in concert—suggesting that, primitive though it was, it could interact to its benefit with the surrounding environment. He called it a "perfect transmission of will."

Early on June 8, rejoined by the *Adventure*, the *Beagle* set course for the Magdalena Channel—a route out of the Strait of Magellan recently discovered by William Low, now aboard the *Adventure*, which took them past 7,370-foot-high Mount Sarmiento. Through ragged clouds, Darwin glimpsed "jagged points, cones of snow, blue glaciers." A solitary deserted wigwam in a little cove was a reminder "that man sometimes wandered amongst these desolate regions," though it was difficult to imagine "a scene where he seemed to have less claims or less authority . . ." The true rulers of these lands were to Darwin "rock, ice, snow, wind and water." When the clouds lifted next morning, Darwin had a clear view of Sarmiento's glaciers, "cataracts of blue ice" descending like "great frozen 'Niagaras.'"

By the following evening, the two ships were in the westerly reaches of the Magdalena Channel, but finding safe anchorage amid the East and West Furies—islets that were only the summits of steep, partially submerged mountains—proved difficult. FitzRoy sent the *Adventure* to safety in a tiny cove with only room for one vessel. For "a long, pitch-dark night of 14 hours," the *Beagle* beat back and forth within an area of only four square miles and hazardously close to rocks. The next morning, the two ships continued their passage, running a gauntlet of islands and innumerable rocks on which, Darwin wrote, "the long swell of the open Pacific incessantly rages.—We passed out between the 'East and

Fuegian wigwams, Magdalena Channel, by Philip Parker King

West Furies' [into the Pacific]; a little further to the North, the captain from the number of breakers called the sea the 'Milky way.'—The sight of such a coast is enough to make a landsman dream for a week about death, peril and shipwreck."

The crew congratulated themselves on entering the Pacific, but relief proved premature as northerly gales pummeled the ships. "Never has the *Beagle* had such ill luck; night after night, furious gales . . . when the wind ceased, the great sea prevented us from making any way," Darwin lamented. FitzRoy, who had sailed along the Chilean coast as a young midshipman, steered for San Carlos [Ancud] on the island of Chiloe, only some seven hundred miles north of the Strait of Magellan. Shortly before reaching it, the thirty-eight-year-old purser George Rowlett, who had been "gradually sinking under a complication of diseases," died. Forty miles off Chiloe next morning, FitzRoy read the funeral service on the quarterdeck and Rowlett's body was consigned to the sea—"an aweful and solemn sound, that splash of the waters over the body of an old ship-mate," thought Darwin.

CHAPTER TWELVE

"The Very Highest Pleasures"

Darwin spent three nights in San Carlos, "a small straggling dirty village" with houses made of cedar planking, while exploring the surrounding forests. After barren Patagonia, "the teeming luxuriance" of Chiloe's evergreen trees, elegant ferns, and groves of bamboo—the result, Darwin thought, of a wet, temperate climate and volcanic soil—delighted him. Early one morning he saw standing out in stark relief the conical 8,700-foot-high Osorno volcano across the strait separating Chiloe from the mainland. What he had seen of Chiloe's geology had already convinced him, correctly, that, like the Patagonian plains, it was "only an appendage to the Andes . . . formed of the debris of its rocks and of streams of lava" as it had risen from the ocean bed.

With dense forest covering most of the 120-mile-long, 40-mile-wide island, most Chilotans—as the islanders, then as now, called themselves—lived around the coast's creeks and bays. Though they had "an abundance of plain food, coarse clothes, and fire-wood," they lacked money for even "the smallest luxuries." One cause of their poverty was that during the struggle for independence from Spain, Chiloe had been loyal to the Spanish throne. Though Chile claimed its independence in 1818, the government still penalized the islanders. Darwin pitied their subsistence living but deplored their wet climate, which "nothing but an amphibious animal could tolerate." Constant rain rendered Chiloe such "a miserable hole" that even the inhabitants complained, "No es muy mala?" "Is it not a miserable place?" He had, in fact, arrived in one of the wettest months of the year in one of the wettest places in Chiloe—the average rainfall in San Carlos is around ninety inches a year.

By July 13, FitzRoy was ready to take his ships northward, well reprovisioned with potatoes, apples, chickens, pork, and beef, which Martens found

San Carlos (Ancud), Chiloe, by
C. Martens

most welcome since supplies had been "reduced to a very disagreeable state of simplicity." A week later they arrived in what seemed another world, the wide harbor of Valparaiso, Chile's principal port. After so long in remote places, Darwin enjoyed seeing other ships "like great animals of the sea, come up and reconnoitre each other." With its jumble of low whitewashed, red-tiled houses beneath 1,600-foot-high hills, Valparaiso itself looked beautiful, while the climate was "quite delicious; the sky so clear and blue, the air so dry and the sun so bright." To the northeast the rugged outlines of the Andes, including snowcapped Aconcagua, at over 22,800 feet the highest mountain in the Americas, were clearly visible.

Waiting for Darwin were further pairs of new walking shoes, little gifts—a chain, pencil case, and purse—and letters from his "sisterhood." Having received nothing since his letters from Buenos Aires, his sisters were eager to know how he had escaped the revolution there and—as practical, thrifty women—what had become of his baggage. They told him they had all read his diary. Susan thought it would make "a nice amusing book of travels" but listed "several little errors in orthography." "Cannabal," for example, should be spelled "cannibal." She also recommended *Peter Simple*—"the best novel that has come out a long time," which she planned to send him. Since the author, Captain Frederick Maryatt, was a naval officer, "the sea terms which puzzled us you will understand and relish."

Turning to politics, Susan wrote somewhat sneeringly that Daniel O'Connell, an Irish nationalist politician and prime mover behind the recently passed Catholic Emancipation Act enabling Catholics to sit in the House of Commons, was "boring the House with Ireland's wrongs till one is quite sick of him and his country." The other great topic of the day was the campaign to repeal the Corn Laws—statutes that protected British farmers against cheaper foreign imports but artificially inflated the price of grain to the detriment of the poor. They would be repealed a decade after the *Beagle* arrived home.

Adding to the "long straggling" letter he had begun on the voyage to Valparaiso, Darwin told his family he was well and was glad they liked his diary while acknowledging it was a little slapdash: "My geological notes and descriptions of animals I treat with far more attention: from knowing so little of Natural History, when I left England, I am constantly in doubt whether these will have any value . . . Of one thing, I am sure; that such pursuits, are sources of the very highest pleasures I am capable of enjoying." Since no one was officially allowed ashore alone "except in civilized ports," being able to employ Covington had already given him much needed independence. He thought his servant "an odd sort of person; I do not very much like him; but he is, perhaps from his very oddity, very well adapted to my purposes."

To an old school and Cambridge friend, Charles Whitley, Darwin revealed some anxiety about the future. Suggesting that Whitley might now be married, perhaps even "nursing, as Miss Austen says, little olive branches, little pledges of mutual affection," he confessed such thoughts conjured a vision of "retirement, green cottages and white petticoats.— What will become of me hereafter, I know not; I feel, like a ruined man who does not see or care how to extricate himself.—That this voyage must come to a conclusion, my reason tells me, but otherwise I see no end to it . . ." He longed for when "ideas gained during the voyage can be compared to fresh ones." Darwin wrote to Henslow in similar weary vein that it was a comfort to know a ship made of wood and iron "cannot last for ever and therefore this voyage must have an end." He excused his egotism in talking so much about himself by saying Henslow was his "father in Natural History, and a son may talk about himself to his father."

As usual, Darwin was quickly ashore. Valparaiso, he discovered, had only one main street, parallel with the coast and intersected by steep ravines. Nevertheless, compared to recent ports of call it was "a sort of London or Paris." The only drawback was that for the first time in months

Valparaiso harbor, watercolor, by C. Martens

he felt obliged to shave and dress respectably. Casting an artist's eye over Valparaiso, Martens also found it picturesque, especially the "innumerable cottages hanging as it were one over the other to a considerable height . . . accessible only by narrow winding and zigzag paths."[1]

Valparaiso had a large community of foreign merchants, many British. Darwin was "much struck by the great superiority in the English residents over other towns in S. America. Already I have met with several people who have read works on geology . . . and actually take interest in subjects in no way connected with bales of goods and pounds shillings and pence.—It was as surprising as pleasant to be asked, what I thought of Lyell's geology." Among the books Darwin had found waiting in Valparaiso was the third and final volume of Lyell's *Principles of Geology*.

Richard Corfield, who had overlapped with Darwin at Shrewsbury School, was among the British merchants and invited him to stay in his airy house in the elegant Alemendral suburb east of the town. Darwin found it "most pleasant to meet with such a straightforward, thorough Englishman, as Corfield is, in these vile countries." Exploring Valparaiso's

---

1. In 1883, the first of thirty *ascensores*—funicular elevators—would be built to help citizens reach the upper areas of their vertiginous city, some of which are still in use.

hilly hinterland from his new quarters, Darwin was surprised to see few birds and even fewer animals. Finding beds of seashells, colors still bright, at an altitude of 1,300 feet, he speculated whether the scarcity of animals might "be owing to none having been created since this country was raised from the sea," a comment that rejected the concept of the single creation of an unchanged world.

While Darwin rambled and hypothesized, FitzRoy had to complete the survey charts of Patagonia, Tierra del Fuego, and the Falkland Islands for dispatch to England. Realizing the amount of work involved, rather than go himself, he dispatched the Spanish-speaking Wickham to the Chilean capital, Santiago, to present the *Beagle*'s credentials to the authorities. Meanwhile, taking Stokes, King, and Usborne with him, he moved into spacious lodgings ashore to finalize the charts.

With time on his hands, in mid-August Darwin set out with a *huaso*—the Chilean equivalent of a gaucho—toward the Andes. His first stop was the Hacienda de Quintero, an estate once owned by Admiral Thomas Cochrane, the British naval officer who commanded the newly formed Chilean navy during the war of independence with Spain, where Darwin had heard great quantities of seashells were being dug up from the earth to be ground into lime. From here he headed for the fertile Quillota Valley—a pleasant patchwork of gardens planted with peach, orange, and olive trees and occasional date palms—between the tall bare mountains.

Arriving at the base of the 6,400-foot Campana (Bell) mountain, so named for the shape of its summit, Darwin hired a second *huaso* as his mountain guide, who led the way up through the forested lower slopes toward the summit, past several places where, Darwin noted, people had attempted to mine for gold. As night fell, they camped by the Agua del Guanaco, a spring high on the mountainside in air so clear Darwin made out the masts of vessels anchored in Valparaiso bay, twenty-six miles away. The sunset was "glorious . . . the valleys being black whilst the snowy peaks of the Andes yet retained a ruby tint." The *huasos* lit a fire beneath "a little arbor of bamboos" over which they fried *charqui*—dried strips of beef—which they washed down with *mate*. Again, as on the pampas,

Darwin was seduced by "the inexpressible charm in . . . living in the open air.—The evening . . . so calm and still . . ."

The next morning, leaving their horses, Darwin and his two companions clambered up "the rough mass of greenstone" capping the summit. Examining the enormous broken rock fragments, some covered with lichen, Darwin was certain the region's "constant earthquakes" were the cause of the fracturing of the rocks and spent the entire day geologizing. Looking down from the summit of the Campana he saw "Chili [*sic*] and its boundaries the Andes and the Pacific . . . as in a Map." The Andes themselves—"more like a wall" than a range of separate mountains— differed from his expectations and were an awesome sight: "Who can avoid admiring the wonderful force which has upheaved these mountains, and even more so the countless ages which it must have required to have broken through, removed and levelled whole masses of them?" The idea that not only the Andes but the surrounding plains had been gradually uplifted from the sea was already in his mind.[2]

Continuing northward, Darwin reflected on what he had learned so far of Chilean society. Travelers could not expect the unbounded hospitality offered with no thought of payment they received in Argentina. While almost every Chilean house owner would offer a bed for the night, payment was expected next morning—"even a rich man will accept of two or three shillings." As for the *huaso*, he could not compare with the gaucho who, though "a cut-throat, is a gentleman . . . seems part of his horse and scorns to exert himself excepting when on its back." Though an accomplished horseman too, the *huaso* was "a vulgar, ordinary fellow" prepared to work as a hired hand in the fields. Also, with his "black and green worsted leggings" and "absurdly large" clanking spurs, the *huaso* was far less picturesque than the gaucho with his white boots, wide trousers, and scarlet *chiripa*—a woolen garment tucked in at the belt and wound round the hips and thighs.

———

2. A plaque, erected on the summit of the Campana in 1935 by "The Scientific Society of Valparaiso, the British community and admirers" commemorates Darwin's visit and quotes his description of the dramatic view.

Arriving at the Jajuel copper mines high in an Andean ravine—"the rage for mining has left scarcely a spot in Chili [sic] unexamined, even to the regions of eternal snow"—Darwin met the manager, "a shrewd but ignorant Cornish miner," who asked him, "Now that George Rex was dead [George IV], how many of the family of Rex's were yet alive." Amused that the manager was unaware "Rex" was Latin for "king" and not a surname, Darwin suggested in his diary that the miner probably also thought "this Rex certainly is a relation of Finis who wrote all the books"—in other words that he was similarly ignorant that "Finis" at the conclusion of a book was Latin for "end" and not the author's name.

A child of the British industrial revolution, Darwin was surprised to see no "smoke or furnaces or great steam-engines" in this country of mines. The reason was that the raw copper ore was shipped to Swansea in Britain for profitable smelting. Clearly shocked by the conditions of the workers, Darwin recorded that those who labored from first light till dark carrying the ore from the mines on their backs received only their food—figs, bread, boiled beans, and roasted wheat grain—and twelve pounds a year to support themselves and their families. Those who dug in the mine itself were paid only a little better, though allowed some *charqui* in their rations. However, the conditions were perhaps closer to those in British mines than Darwin realized. In Britain, though wages were higher, men, women, and children commonly worked twelve hours a day, six days a week in hazardous conditions, and deaths or serious injuries were common. The family of his uncle Jos's wife Bessy—the Allens of Cresselly in Pembrokeshire—were mine owners.

Darwin spent five days in the Jajuel area, "scrambling in all parts of these huge mountains" amid huge, branching cacti, some as high as fifteen feet. The chaos of "shattered and baked rocks traversed by dykes of formerly melted greenstone showed what commotion has taken place during the formation of these mountains." However, a heavy snowstorm prevented him exploring further, and by late August he was on the road to Santiago at the base of the Andes. Comfortably settled in an English-run hotel, he spent an enjoyable week, writing to FitzRoy that, though neither so large nor so grand, Santiago reminded him of Buenos Aires. His

friend Corfield joined him "to admire the beauties of nature, in the form
of signoritas." Darwin himself dined with various British merchants, galloped over the surrounding plain, and several times climbed a "little pap
of rock"—Cerro Santa Lucia, a rocky hillock in the middle of the city.[3]

In early September, he began his return to Valparaiso by a different route that took him over one of the "famous suspension bridges of
hide"—bundles of twigs and branches bound together by leather straps
and suspended from ropes—across the Maipo River. He did not enjoy
the experience since "the bridge oscillates rather fearfully with the weight
of a man leading a horse." One night at a *hacienda* he shocked some "very
pretty signoritas" by confessing he had visited a church to look around it
rather than to pray. "They asked me, why I did not become a Christian . . .
I assured them I was a sort of Christian."

Continuing south, Darwin roused further interest, indeed suspicion,
while staying with an American called Nixon, owner of the large Yaquil
gold mines. An elderly Spanish lawyer who came to call seemed unconvinced by Darwin's explanation of the purpose of his travels in a foreign
land, commenting, "It is not well . . . I do not like it; if one of us was to
go and do such things in England, the King would very soon send us out
of the country." Nixon offered Darwin "some Chichi, a very weak, sour
new made wine" that "half-poisoned" him. For several days he was too
sick to leave. When he finally set out, riding across a great treeless plain,
he again began to feel ill, though he still summoned enough energy to
collect some fossilized shells. However, growing ever weaker, he had to
halt frequently to rest. By September 24—five days after leaving Nixon's
house—he was so exhausted all he could think of was sleep. He found
lodgings where he had "the uncommon luck of obtaining some clean straw
for my bed. I was amused afterwards by reflecting how truly comparative
all comfort is. If I had been in England and very unwell, clean straw and
stinking horse cloths would have been thought a very miserable bed."

He managed to ride on a little farther but, as soon as he was close
enough to Valparaiso, dispatched a message to Richard Corfield, who sent

---

3. As on the Campana mountain, a plaque commemorates his visit to Cerro Santa Lucia.

Covington in a carriage to fetch him. Tottering "very miserable" into Corfield's house, Darwin went straight to bed, where he remained for nearly five weeks until the end of October. Bynoe took charge of him, prescribing calomel—a purgative popular in the nineteenth century but now banned because of its high mercury content—and complete rest. Darwin fretted at time lost when he could have been adding to his specimen haul, though he did experience "one little earthquake . . . on a sudden I heard such a hubbub in the dining room . . . at the same moment I felt my bed *slightly* vibrate . . ."

A fortnight passed before Darwin felt strong enough even to sit up and write home. Some suggest this was the start of the ill health that would dog the rest of his life, and also that, rather than being poisoned by tainted wine as he thought, he had been bitten by the bloodsucking *Triatoma infestans* bug, vector of Chagas disease that causes serious heart and digestive problems. Hiding by night in cracks in walls and roofs to emerge at night, the bugs look for exposed skin to bite and suck blood from and often defecate or urinate close to the wound. The parasite enters the body when the victim instinctively rubs the bite, smearing the feces or urine into it. Others, however, suggest Darwin caught typhoid.

In letters home, the invalid Darwin confessed that this latest excursion, as well as making him ill, had been his most expensive so far and—beyond the chance for "some more hammering at the Andes" and finding further fossil shells—probably not worth it. He also related more dramatic news that reached him while still on his sick bed in Corfield's house. FitzRoy had suffered a nervous breakdown. Already under stress from his heavy workload of surveying and taking longitudinal measurements, FitzRoy had learned that "the Lords Commissioners of the Admiralty did not think it proper to give me any assistance" in purchasing, refitting, and operating the *Adventure* over the past two years. They also criticized the length of time he had taken to survey the east coast of South America. As a result, FitzRoy had become "very thin and unwell. This was accompanied by a morbid depression of spirits, and a loss of all decision and resolution. The Captain was afraid that his mind was becoming deranged (being aware of his hereditary predisposition [the suicide of his uncle

Lord Castlereagh]." Bynoe tried to convince FitzRoy his problems were physical not mental—he was simply exhausted—but he would not listen and was insisting on relinquishing his command to Wickham.

Darwin interpreted the Admiralty's coldness toward FitzRoy as political spite—the government was Whig, while FitzRoy was a former Tory parliamentary candidate from a well-known and largely Tory family. He also realized the crisis had implications for his own plans. If FitzRoy resigned his command there would be no crossing of the Pacific since the Admiralty's orders were explicit. If anything happened to FitzRoy, the new commanding officer was to complete the section of the survey on which the *Beagle* was currently engaged but then to return straight home via the Atlantic. Darwin lamented this prospect: "We shall see nothing of any country, excepting S. America." He contemplated leaving the *Beagle*, crossing the Andes, and returning to England from Buenos Aires.

FitzRoy himself recalled, "At this time I was made to feel and endure a bitter disappointment; the mortification it caused preyed deeply . . ." He knew he had no option but to dispose of the *Adventure*—he had written to his banker that "if her expenses exceed what I can pay I must sell her, and enjoy the satisfaction of having done my little all for the good of my country." Yet without her, he was convinced "all my cherished hopes of examining many groups of islands in the Pacific, besides making a complete survey of the Chilian [*sic*] and Peruvian shores, must utterly fail."

As soon as he could find a buyer, FitzRoy sold the schooner for nearly fourteen hundred pounds—more than he paid but not nearly enough to cover all he had spent on refitting her and paying and provisioning her crew. Even for a man whose attitude toward money was far more cavalier than Darwin's, it was a heavy blow. Just a few weeks before FitzRoy had written to his sister Fanny, "I have ideas of money different from those of many persons—and I cannot see the wisdom of hoarding money with a view to the latter part of a life which is as precarious as the wind," but he had badly depleted his capital. To his friend Beaufort he wrote, "I am in the dumps. It is heavy work—all work and no play . . . Troubles and difficulties harass and oppress me so much that I find it impossible either to say or do what I wish . . . Continual hard work—and heavy expense—These and many other things have made me ill and very unhappy."

Darwin saw FitzRoy's overwrought state for himself when he was well enough to visit the *Beagle*. There they quarreled—an incident still so vivid he recorded it for his family many years later: "[FitzRoy] complained bitterly to me that he must give a great party to all the inhabitants. I . . . said that I could see no such necessity on his part . . . He then burst out into a fury, declaring that I was the sort of man who would receive any favours and make no return." Darwin left without a word and returned to Corfield's house. A few days later, he came back to the *Beagle*, where FitzRoy greeted him "as cordially as ever, for the storm had by that time quite blown over."

FitzRoy, however, had taken time to calm down. Wickham complained to Darwin, "Confound you, philosopher, I wish you would not quarrel with the skipper; the day you left the ship I was dead-tired . . . and he kept me walking the deck till midnight abusing you." Darwin decided "the difficulty of living on good terms with a captain of a Man-of-war, is much increased by its being almost mutinous to answer him as one would answer anyone else; and by the awe in which he is held . . . by all on board."

In the end, the sensible, good-natured Wickham rescued the *Beagle*'s circumnavigation. According to Darwin, "Very disinterestedly giving up his own promotion," he pointed out to FitzRoy that the Admiralty's instructions obliged him only "to do as much of the west coast as *he [had] time* for and then proceed across the Pacific . . . and then asked the Captain, what would be gained by his resignation. Why not do the more useful part and return, as commanded by the Pacific. The Captain, at last, to every one's joy consented and the resignation was withdrawn." Although Darwin thought FitzRoy quickly regained "his cool inflexible manner, which he had quite lost," FitzRoy remained inwardly troubled, confessing to his sister shortly before leaving Valparaiso, "I am so surrounded with troubles and difficulties . . . My brains are more confused even than they used to be in London."

One unavoidable consequence of the *Adventure*'s sale—as Darwin perceived—was that the *Beagle* again became very crowded. FitzRoy paid off the *Adventure*'s crew, but Wickham and the ship's other officers had to be accommodated. Though FitzRoy had dispatched two casks, a box, and a large jar filled with specimens on a ship bound for Portsmouth on his behalf, Darwin fretted about storage for his collections. For Martens, who had filled four sketchbooks during his year aboard and completed several watercolors, matters were more serious. For lack of space, FitzRoy now dismissed "our little painter," as Darwin labeled him, "to wander about the world."

FitzRoy had delayed sailing until Darwin was fully recovered but on November 10 set course back to the island of Chiloe to survey it and the little-known Chonos archipelago to its south. Eleven days later, the *Beagle* again anchored off San Carlos. As darkness fell, "torrents of rain and a gale of wind" reminded Darwin that even though it was now summer he was indeed back in an island fit for amphibians. Three days later, when FitzRoy dispatched Sulivan with the yawl and a whaleboat to check the accuracy of existing charts of Chiloe's eastern coastline, Darwin went too. However, rather than starting out with the boats, he decided to ride to Chacao on Chiloe's far northeastern tip and rendezvous there with Sulivan on the first evening.

Following a coastal trail through forests so dense the sun never penetrated the interlaced foliage and over ground so waterlogged the Chilotans had built a boardwalk from squared tree trunks so riders could pass, Darwin reached Chacao toward nightfall and joined Sulivan as planned. Soon after, the local governor's son arrived barefoot to ask who the new arrivals were. The appearance of what were obviously naval boats had astonished the inhabitants, who at first hoped they were "the forerunners of a Spanish fleet coming to recover the island from the patriot government of Chili [*sic*]."

From Chacao, the survey team headed down the east coast past numerous islets covered with "impervious blackish-green forest." Despite frequent rain storms, one morning dawned so clear that Darwin again saw Osorno, this time "spouting out volumes of smokes [*sic*]; [a] most beautiful mountain, formed like a perfect cone and white with snow." A second volcano with a saddle-shaped summit was also emitting puffs of steam, and Darwin made out a third—"the lofty peaked Corcobado, well deserving the name of 'el famoso . . .' Thus we saw at one point of view three great active volcanoes, each . . . of about seven thousand feet high."

Ashore, Darwin encountered "a family of pure Indian extraction." The father's features reminded him strongly of York Minster, while the ruddy-cheeked younger boys "might be mistaken for Pampas Indians."

Osorno Volcano

Though they spoke a different language, their appearance convinced him they were closely related to the tribes on the mainland. Charles Douglas, a longtime British resident on the islands whom Sulivan had hired as his pilot, told Darwin Chiloe's population numbered around forty-two thousand, mostly "of mixed blood," of whom eleven thousand still retained their Indian surnames. Though all were Christian, some still held "superstitious communication in caves with the devil," and in previous centuries people had been arrested on this account and sent for trial by the Inquisition in Lima.

Douglas also confirmed what Darwin had noticed on his first visit to Chiloe—that though food was plentiful, most Chilotans were very poor. Almost no one had a regular income, and if anyone did accumulate a substantial sum "it would be stowed away in some secret place, for each family generally possesses a hidden jar or chest buried in the ground." A major obstacle to prosperity was the difficulty in purchasing cultivatable land. Not only was buying land complicated and heavily taxed but the climate was too dank to clear the huge forests by burning. Instead, trees—many of them alerce trees, members of the cypress family—had to be felled, a difficult, time-consuming, and expensive process.

Five days after leaving the *Beagle*, Sulivan's survey party sailed up a wide inlet to Castro, founded in 1567 and once Chiloe's capital but now virtually deserted, where a few fishermen lived in *pilotas*, stilt houses built out over the water. What had once been the central plaza and the surrounding streets were overgrown with green turf grazed by sheep. However, on the plaza was a "picturesque and venerable" church built entirely of wood, without a single iron nail, by the Jesuits. Darwin found it "wonderful that wood should last for half a century in so wet a climate."[4]

The arrival of strange boats was so rare that nearly all Castro's inhabitants congregated on the beach to watch the *Beagle* men land and pitch their tents. They were, Darwin thought, civil, and one man even gave the new arrivals a cask of cider. Castro's governor was "a quiet old man, who in his appearance and manner of life was scarcely superior to an English cottager." As darkness fell, despite heavy rain, the large circle of onlookers remained by the tents, still watching.

The next day, the survey party sailed south to Lemuy Island where Darwin had heard there was a coal mine. Again they found themselves surrounded, this time by "a large group of nearly pure Indian inhabitants" who seemed astonished by their arrival, exclaiming, "This is the reason we have seen so many parrots lately; the Cheucau . . . has not cried 'beware' for nothing." Darwin realized that the islanders had a superstitious fear of the bird but thought they had "chosen a most comical little creature for their prophet." He had already observed the cheucau—"an odd red-breasted little bird, which inhabits the thick forest and utters very peculiar noises." Sometimes, "although its cry may be heard close at hand, let a person watch ever so attentively, he will not see the cheucau; at other times, let him stand motionless, and [it] will approach within a few feet, in the most familiar manner. It then busily hops about the entangled mass of rotting canes and branches, with its little tail cocked upwards."

As usual the people wanted to barter; "their esteem and anxiety for tobacco was something quite extraordinary: after tobacco, indigo came next in value, then capsicum, old clothes and gunpowder; the latter

---

4. Sixteen of Chiloe's remarkable wooden churches still survive and form a UNESCO World Heritage site, while some of the old *pilotas* have also survived and, restored, serve as boutique hotels.

article . . . required for a very innocent purpose; each parish has a public musket, and the gunpowder was wanted to make a noise on their Saint or Feast days." For tobacco worth "three half-pennies," Sulivan's men acquired a chicken and a duck, while three shillings' worth of cotton handkerchiefs bought three sheep and a large bunch of onions. Meanwhile, Darwin's rumored coal mine turned out to be lignite—brown coal—and of little interest to him, but examining some yellow sandstone, he discovered a fossilized tree with "a great trunk (structure beautifully clear), throwing off branches: main stem much thicker than my body . . ."

Though Darwin loathed Chiloe's wet climate, on the rare occasions the weather was good, the island's beauty moved him. As they continued south, he wrote, "I cannot imagine a more beautiful scene, than the snowy cones of the Cordilleras seen over an inland sea [the channel between Chiloe's east coast and the Chilean mainland] of glass, only here and there rippled by a porpoise or logger-headed duck. And I admired this view from a cliff adorned with sweet-smelling evergreens, where the bright colored, smooth trunks, the parasitical plants, the ferns, the arborescent grasses, all reminded me of the Tropics . . ."

In squally weather on the evening of December 6, the party reached the island of San Pedro, at Chiloe's southeastern tip, where FitzRoy had arranged to rendezvous with Sulivan after he himself had surveyed the exposed, wilder western coast. They found the *Beagle* already at anchor because bad weather had frustrated the survey work. Accompanying two of Sulivan's men ashore again to take measurements, Darwin spotted a small, shaggy fox—"a rare animal" in Chiloe—sitting and watching them so intently that he was able to sneak up behind it "and actually kill him with my geological hammer."[5] Still commonly known as Darwin's fox, today it is even rarer and listed as an endangered species.

FitzRoy, Darwin, and others attempted to climb to San Pedro's highest point. As they struggled to find a way up over, through, or beneath

---

5. Darwin's foxes are dark gray with paler gray underbellies. They are about twenty-one inches long, their tails are approximately nine inches long, and they weigh between 4.4 and 6.6 pounds.

the "confused mass of dead and dying trunks," the steep-sided hills and slate rock reminded Darwin of Tierra del Fuego. "Oftentimes for quarter of an hour our feet never touched the ground, being generally from 10 to 20 feet above it; at other times, like foxes, one after the other we crept on our hands and knees under the rotten trunks." Finally, feeling "like fish struggling in a net," they gave up the attempt.

While Sulivan and his team departed in their boats to continue surveying the coast of Chiloe, Darwin rejoined the *Beagle* as FitzRoy set course south for the Chonos Archipelago, a string of more than 150 low, densely forested islands separated by a deep channel from the Chilean mainland. Three days sailing brought them to the archipelago's southern reaches, where FitzRoy found safe anchorage just as "a real storm of T. del Fuego" was about to break—"With white massive clouds . . . piled up against a dark blue sky and across them black ragged sheets of vapor were rapidly driven . . . successive ranges of mountains appeared like dim shadows . . . a most ominous, sublime scene."

Despite the bad weather, in subsequent days Darwin tried to explore ashore, scrambling across the rocky beaches but failing to penetrate the dark woods and badly scratching and scraping his face, hands, and shins in the attempt. As the *Beagle* sailed on, heading for the peninsula of Cabo Tres Montes at the archipelago's extreme southern point, before turning north again, Darwin landed when he could, always looking for the highest point to climb—"in these wild countries it gives much delight to reach the summit of any high hill; there is an indefinite expectation of meeting something very strange." There was also "a little vanity of distinction, that you perhaps are the first man who ever stood on this pinnacle, or admired this view." However, beyond some volcanic rock and signs that someone—perhaps a shipwrecked sailor—had lit a fire and made a grass bed beneath a stone ledge, destroying Darwin's illusions of being the first man there, little caught his eye.

Sometimes gales held the *Beagle* back and time hung heavy on everyone's hands. Darwin thought Christmas Day 1834, spent surveying between thirty and forty miles of coast, "not such a merry one" as the previous year ashore at Port Desire when FitzRoy had awarded prizes to winners of the

Beagle Olympics. Fitzroy blamed their subdued Christmas celebrations on the setting. The archipelago was just "another Tierra del Fuego, a place swampy with rain, tormented by storms, without the interest even of population." Darwin too was disappointed and surprised not to have seen any "Indians." He wondered whether their absence on these islands where seal meat—"the Indians' highest luxury"—as well as oysters, mussels, and wild potatoes were so plentiful signaled yet another "step to the final extermination of the Indian race in S. America," evidence of which he had witnessed in Patagonia during his encounters with Rosas's men.

They did, however, spot a figure on the shore frantically waving a shirt, and FitzRoy sent a boat to find out what he wanted. The man turned out to be one of six American whalers who had deserted their ship fifteen months earlier. Since then five of "the poor wretches," as Darwin called them, had survived on seal meat and shellfish, but one had tumbled to his death from a cliff. Darwin realized the camp he had found a few days before was theirs. The party came gratefully aboard the *Beagle*.

On an island off Chiloe, Sulivan's party spent a more enjoyable Christmas, sheltering from a heavy storm in a priest's house and dining handsomely on "one side of a sheep roasted, another side boiled, [and] twelve pounds of English fresh roast beef heated," followed by "two immense plum puddings" for which they had flour and raisins brought with them from the *Beagle* and foraged for the eggs. The bad weather had given them a holiday since, Sulivan wrote, "we could only afford to knock off work when it rained too hard . . . which happened on Christmas Day . . ."

FitzRoy headed next for Yuche Island in the Chonos Islands, where Darwin found large herds of wild goats, descendants he assumed of goats left by Spanish missionary expeditions. The island's 2,400-foot peak was composed of granite looking as if it had been "coeval with the very beginning of the world . . . Granite to the Geologist is a classic ground . . . the fundamental rock, and however formed, we know it to be the deepest layer in the crust of this globe to which man is able to penetrate." Granite is igneous rock—rock formed by the slow cooling beneath the Earth's surface of hot, molten magma. The granite Darwin examined during the *Beagle* voyage had, as he knew, been gradually uplifted from where it was formed. His observations would provide one of the first detailed descriptions of the nature, composition, and diversity of igneous rocks. His use of the word *geologist* indicated that this is how he then saw himself.

New Year's Day 1835 brought boisterous gales. Darwin thanked his stars that by the end of that year they should be "where a blue sky does tell one there is a heaven, a something beyond the Clouds . . ." After finding secure anchorage for the *Beagle*, despite wind and rain, FitzRoy set out with Darwin by boat to explore a deep creek. They found every foot of beach and every rock packed with seals—probably South American fur seals—snuggling together like pigs. Their stench was overpowering, but Darwin enjoyed watching "the heap of seals, old and young," impetuously flinging themselves into the water as their boat passed by to follow them "with outstretched necks, expressing great wonder and curiosity."

Turkey buzzards patiently watched and waited for a chance to feed on a dead seal. Darwin thought the buzzard a "disgusting bird, with its bald scarlet head formed to wallow in putridity." More appealing were the terns, gulls, cormorants, "beautiful black-necked swans," and small sea otters.

On a further reconnoitering trip with FitzRoy, their boat passed the floating carcass of a newly dead whale, its flesh and blubber pink and crawling with parasitical crabs. The huge carcass reminded Darwin of "the great fossil animals; he appears altogether too big for the present pigmy race of inhabitants. He ought to have coexisted with his equals, the great reptiles . . ." A few days later, near the archipelago's northern tip, Darwin made several new finds, including a sea slug, later called *Thecacera darwinii*, and a tiny parasitical barnacle that bored into larger mollusks and lived inside them.

In mid-January, finished with the Chonos, FitzRoy set course back to Chiloe to anchor again near San Carlos. The night after their arrival, the Osorno volcano on the Chilean mainland put on a display. At midnight, "the sentry observed something like a large star, which . . . gradually increased in size . . ." wrote Darwin. Those on deck trained their telescopes on "a very magnificent sight . . . in the midst of the great red glare . . . dark objects in a constant succession might be seen to be thrown up and fall down.—The light was sufficient to cast on the water a long bright shadow . . ."

While FitzRoy took the *Beagle* on a second attempt to survey Chiloe's western coast, Darwin and King set out on horseback to explore across the island. They made first for Castro, riding over log roads through dank forest where flowering trees perfumed the air, but dead trunks loomed "like great white skeletons." Before long a woman and two boys joined them on the journey. Darwin thought the woman rather good-looking. However, brought up with sisters and female cousins who rode decorously sidesaddle, he was shocked that though from "one of the most respectable families in Castro" she rode astride like a man and wore neither shoes nor stockings.

From Castro, Darwin and King followed the coast south, Darwin again admiring the "large barn-like chapel built of wood" that every hamlet seemed to possess. From the settlement of Chonchi, they struck inland through magnificent forests and undulating countryside like "the wilder parts of England." Reaching a lake, they continued by *periagua*, a large canoe rowed by "Indians" who stopped to pick up a cow. Darwin watched in amazement as, within barely a minute, they "brought the cow alongside . . . and heeling the gunwhale towards her, placed two oars under her belly and resting on the gunwhale; with these levers they fairly tumbled the poor animal heels over head into the bottom of the boat."

Arrived in Cucao, the only settlement on the west coast, Darwin and King spent the night in "an uninhabited hovel," where they lit a fire and cooked their supper. The local population of some thirty or forty lived by making oil from seal blubber. Darwin thought them both "discontented, yet humble to a degree which it was quite painful to behold" and for which he blamed the "harsh and authoritative" behavior of the Spanish. On his travels through Chiloe he had noticed that though officials spoke courteously to him, they addressed "the poor Indians as if they were slaves rather than free men," commandeering their food and horses "without ever condescending to say how much, or indeed if the owners should . . . be paid." He and King gave the people cigars, *mate*, and sugar.

After exploring a little of the rugged western shore where the breakers crashed with such force Darwin was sure the roar could be heard in Castro, he and King returned to the *Beagle*. Darwin spent his final days on Chiloe making short excursions to see such curiosities as "a bed of oyster [shells] out of which large forest trees were growing at an elevation of 350 feet" above sea level. By February 4, the *Beagle* was ready to leave. Darwin wrote, "I believe every one is glad to say farewell . . . Yet if we forget the gloom and ceaseless rain . . . Chiloe might pass for a charming island."

These past weeks had yielded a fine haul of specimens. As well as the tiny barnacle Darwin had discovered in the Chonos, he had been intrigued by some giant barnacles relished by the Chilotans for the cooking pot. A detailed study of barnacles, prompted by the specimens he brought back from Chile, would in later years be a critical step in the development of his thinking about the relationship between organisms and their environment and convince him that all animals, including humans, had hermaphrodite ancestors.

He had also gathered insects. As well as the usual beetles and butterflies, they included fleas and lice picked from the Chilotans themselves. "These disgusting vermin are very abundant in Chiloe: several people have assured me that they are quite different from the lice in England. They are said to be much larger and softer (hence will not crack under the nail) [and] they infest the body even more than the head . . ."

Darwin had, in addition, discovered a tiny, fingernail-size frog— subsequently named *Rhinoderma darwinii* after him—and, of course, bagged the Chiloe fox. However, he was leaving Chiloe unaware of the diminutive, nocturnal *monito del monte*, "little bush monkey" (*Dromociops gliroides*). This mouse-sized marsupial is a descendant of a time when Australia, South America, and Antarctica formed the Gondwana supercontinent. An encounter with it might later have prompted Darwin, after the *Beagle*'s visit to Australia, where the majority of the world's marsupials live, to consider the reason both why marsupials with their distinct characteristics developed and why their locations had become separated.

Meanwhile, FitzRoy's survey work had achieved what Darwin called a "most singular result" by proving that earlier Spanish maps had seriously overestimated Chiloe's length—"hence it will be necessary to shorten the island 1/4 of its received size." As FitzRoy wrote to his sister Fanny, the discovery nearly provoked a diplomatic incident. When a member of the *Beagle* crew told a Chilotan their surveys had "cut off twenty-five miles" of their island, he interpreted this as meaning Britain had occupied a part of Chiloe as a prelude to an attack on the Chilean mainland and local men rushed to arm themselves to support what they thought would be an imminent British attack.

CHAPTER THIRTEEN

"Skating on Very Thin Ice"

On February 8, 1835, the *Beagle* entered Corral Bay, 150 miles north of Chiloe, to anchor off one of several forts built to protect Valdivia, ten miles away up the Valdivia River. Darwin joined others curious to see the town in one of the ship's boats. The riverbanks were "one unbroken forest" and only the occasional "Indian" family paddled quietly by. Valdivia itself was so "buried" in apple trees that its streets were mere "paths in an orchard." Darwin could scarcely believe this quiet little hamlet was one of the oldest settlements on the west coast of America. Founded by conquistadores in 1552, for a while it had been Chile's second most important city. When destroyed during bloody and protracted wars with the local Mapuche and Huilliche people, the Spanish rebuilt and refortified it. By the eighteenth century, Valdivia was a base for the colonization of southern Chile. However, in the independence struggle it had sided with Spain and been overwhelmed by an assault led by Admiral Cochrane with three hundred men of the Chilean navy. By the time the *Beagle* arrived, Valdivia had declined into a quiet backwater.

Valdivia's small British community included seven transported convicts who had escaped from the British penal colony on Tasmania. Reaching Valdivia, they had settled down and married. Just as in Buenos Aires, where he and FitzRoy had met another escaped convict, Donna Clara, Darwin was indignant and pontificated that "in all these Spanish colonies . . . the commital of enormous crimes lessens but very little the public estimation of any individual." In the case of the convicts, "their being such notorious rogues appears to have weighed nothing in the Governor's opinion, in comparison with the advantages of having some good workmen."

Darwin rode with a guide into the forests beyond Valdivia, finding them as impenetrable as Chiloe's but, with fewer evergreens, "a brighter

Spanish fort at Valdivia, March 2020

and more lively green." The bamboos from which the "Indians" cut their *chusas*, or spears, swayed in great clumps twenty feet high. The only available lodging house was so filthy Darwin slept outside. Even so, he woke to find "not the space of a shilling on my legs which had not its little red mark where the flea had feasted." Journeying on, he felt hemmed in by "the wilderness of trees" and longed for "the free, unbounded plains of Patagonia."

He spent the next night in greater comfort as the guest of a hospitable Catholic priest from Santiago who answered his questions about the local "Indians" who, with their long, dark hair, dark complexions, and grave expressions, reminded Darwin of "old portraits of Charles the First," the English king. The priest told him the tribes around Valdivia were Christian, despite not "much like coming to mass" or observing "the ceremonies of marriage." Those farther north were "yet very wild and not converted," taking as many wives as they could support. The "besetting sin" of all was drunkenness from the sour cider Darwin had tasted on Chiloe, which, the priest observed, made them "very dangerous and fierce." The priest complained of his own isolation. Darwin reflected that "with no particular zeal for religion, no business or pursuit, how completely must this man's life be wasted."

Returning to the *Beagle*, Darwin found unusual gaiety aboard as local bigwigs visited, on one occasion bringing a boatload of ladies whom "bad weather compelled to stay the night, a sore plague both to us and them." Valdivia's signoritas were certainly "very charming; and what is still more surprising, they have not forgotten how to blush, an art . . . quite unknown in Chiloe." With Wickham, Darwin toured the Niebla fortress in Corral Bay, finding its defenses in ruins, which he thought emblematic of Spain's "fallen greatness." Using his knowledge of Spanish, Wickham remarked tactlessly to the officer commanding the fort that the wooden gun carriages were so rotten that "with one discharge they would all fall. The poor man trying to put a good face on it, gravely replied, 'No I am sure Sir they would stand two!'"

February 20 brought a dramatic event and, for Darwin, one of the most significant of the voyage because it enabled him to see with his own eyes the effects on land and sea levels of violent geological phenomena of which so far he had only read. While he and Covington were resting from specimen gathering in the woods, an earthquake "came on suddenly and lasted two minutes (but appeared much longer). The rocking was most sensible; the undulation appeared both to me and my servant to travel from due East. There was no difficulty in standing upright; but the motion made me giddy.—I can compare it to skating on very thin ice or to the motion of a ship in a little cross ripple . . . a breeze moved the trees, I felt the earth tremble . . ." Covington thought the sensation "something like [a] ship in a gentle seaway" and noticed the trees "wave a lot to and fro."[1]

Aboard the *Beagle* "the motion was very perceptible; some below cried out that the ship . . . was touching the bottom." In Valdivia, where Fitz-Roy was, the wooden houses were "shaken violently and creaked much, the nails being partially drawn," though none collapsed and few people were injured. What struck Darwin most when he arrived in the town was the "horror pictured in the faces of all the inhabitants." He reflected that "an earthquake like this at once destroys the oldest associations; the

---

1. The earthquake of February 20, 1835, has since been estimated at its epicenter as magnitude 8.5 on the Richter Scale.

world, the very emblem of all that is solid, moves beneath our feet like a crust over a fluid; one second of time conveys to the mind a strange idea of insecurity."

Darwin noted that the earthquake, which had occurred at low tide, had a "very curious" effect on the tides. An old woman who had been on the beach told him how "the water flowed quickly but not in big waves to the high water mark, and as quickly returned to its proper level . . . like an ordinary tide, only a good deal quicker." In the evening, a series of aftershocks produced "the most complicated currents, and some of great strength in the bay." Darwin wondered whether the Villa Rica volcano, visible on the horizon, might be connected to the quake even though it "appeared quite tranquil." Though damage in Valdivia was slight, Darwin predicted correctly that "we shall hear of damage done at Concepcion," the *Beagle*'s next port of call.

Two days later, the *Beagle* departed to continue surveying northward. Anchoring off the "dangerous coast," the swell was so heavy the anchor cable snapped and the anchor itself was lost—"the sixth anchor since leaving England!" Darwin recorded. Six days later, another "snapped right in two," leaving the *Beagle* with only one bower anchor—the main anchor located at the bow. Knowing it was not safe to continue surveying without first obtaining a replacement in case of further losses, FitzRoy decided that after Concepcion, he would have to sail on to Valparaiso. Darwin grumbled—"Nobody but those on board . . . can know how vexatious these petty misfortunes are." The loss was more than a minor irritant to FitzRoy, who had taken the precaution of sailing from England with more than the usual number of spare anchors.[2]

On March 3, as the *Beagle* neared Concepcion, 179 nautical miles due north of Valdivia, Darwin felt beneath the ship "a very smart shock of an earthquake: some compared the motion to that of a cable running out, and others to the ship touching on a mud bank." The next day, he

2. The *Beagle* carried three types of anchors—large bower anchors, medium-sized stream anchors to stabilize the ship in fast-flowing tidal waters and river channels, and lighter kedge anchors to help turn the ship while in motion.

Remains of the cathedral at Concepcion, ruined by the great
earthquake of 1835, by J. C. Wickham

discovered what had happened on February 20, the day the quake had
shaken Valdivia: "not a house in Concepcion or Talcahuano (the port) was
standing . . . seventy villages were destroyed, and . . . a great wave had
almost washed away the ruins of Talcahuano."

Touring the damage, Darwin found it "the most awful yet interesting spectacle." In Talcahuano half an hour after the quake, according to
FitzRoy, "the sea having retired so much, that all the vessels at anchor,
even those which had been lying in seven fathoms water, were aground,
and every rock and shoal in the bay was visible,—an enormous wave,"
spotted while still some five or six miles away, and followed by two
others, had crashed into the town with great force. Receding again,
the waves of the tsunami had left the shore strewn "with timber and
furniture as if a thousand great ships had been wrecked," Darwin wrote.
Chairs, tables, bookshelves, even entire roofs, together with sodden
sacks of cotton, yerba leaves used for mate, and other merchandise
from the now shattered warehouses lay everywhere. A schooner that
had been swept two hundred yards inland remained stranded, while
in the fort the water had carried a four-ton gun and its carriage fifteen
feet higher. People told FitzRoy how "during the remainder of the day,

and the following night, the earth was not quiet many minutes . . . Frequent, almost incessant tremors, occasional shocks more or less severe, and distant subterranean noises kept everyone in anxious suspense." Darwin was quickly convinced "the permanent level of the land and water is . . . altered."

To Darwin, Concepcion itself resembled the classical ruins of Ephesus or Palmyra rather than a modern town inhabited until just a few days ago. Though it had escaped the tsunami, nearly every house had collapsed or partially collapsed, leaving only a few half-shattered and tottering walls standing. Though not as completely devastated as Talcahuano, Darwin found it "the more terrible, and if I may so call it, picturesque sight." Only "the *constant* habit of these people of running out of their houses *instantly* on perceiving the *first* trembling" had saved them. "The inhabitants scarcely passed their thresholds before the houses fell in."

The quake had happened without warning. Henry Rouse, British consul in Concepcion who had been breakfasting, had rushed into his courtyard just before "one side of his house came thundering down." He had scrambled onto the rubble just as "the other side of the house fell . . ." Dust darkened the sky as "shock succeeded shock . . . no one dared approach the shattered ruins; no one knew whether his dearest friends or relatives were perishing . . . The thatched roofs fell over the fires, and flames burst forth in all parts . . ."

Many survivors had lost their livelihoods and could barely find food. Rouse himself was living beneath a makeshift shelter in his apple orchard, struggling to protect what little he had saved from looters. A shocked Darwin reflected that an earthquake was enough "to destroy the prosperity of a country." Even in wealthy, powerful Britain, if "a volcanic focus should reassume its power; how completely the whole country would be altered. What would become of the lofty houses, thickly packed cities, the great manufactories . . . If such a volcanic focus should announce its presence by a great earthquake, what a horrible destruction there would be of human life.—England would become bankrupt; all papers, accounts, records . . . would be lost: and Government could not collect the taxes . . ."

About one hundred people were known to have died, though Darwin
suspected many more lay buried in the ruins. On the shore of a small
island in the harbor, numerous rocks once beneath the sea had been
hurled high on the beach. Great fissures in the land convinced him the
quake had done more "in degrading or lessening the size of the island,
than 100 years of ordinary wear and tear."

Both Darwin and FitzRoy gathered information about the earthquake. Most towns between Concepcion and Santiago to the north had
been destroyed, and the Antuco volcano, east of Concepcion, was active,
though not actually erupting. Some claimed that an old witch with a
grudge against Concepcion had caused the quake by sealing up Antuco's
vents. While this was a "silly belief," Darwin realized experience had
taught local people to associate "the suppressed activity of volcanoes"
with "tremblings of the ground."

Drawing on Lyell's theories and on the physical evidence he himself
had seen that seemed to support them, Darwin suggested in his diary
that vibrations from the rumbling, spitting Antuco might have caused
the quake: "Many geological reasons have been advanced for supposing
that the earth is a mere crust over a fluid melted mass of rock and that
volcanoes are merely apertures through this crust. When a volcano has
been closed for some time, the increased force . . . which bursts open
the orifice might well cause an undulation in the fluid mass beneath the
earth; at each successive ejection of lava a similar vibration would be
felt over the surrounding country . . . till at last the expansive force is
counterbalanced by the pressure in the funnel of the volcano." When
earthquakes occurred without any volcanic activity, "we may either imagine that melted rock is injected in the inferior strata, or that an abortive
attempt at an eruption has taken place beneath the volcano."

During two sharp aftershocks in Concepcion, Darwin again felt solid
ground transform beneath his feet into "a partially elastic body over a fluid
in motion." Despite all the earthquake's horrors, he thought that "since
leaving England we have scarcely beheld any one other sight so deeply
interesting. The earthquake and volcano are parts of one of the greatest phenomena to which this world is subject." To his sister Caroline he

wrote, "The three most interesting spectacles I have beheld since leaving England [were] a Fuegian savage.—Tropical vegetation—and the ruins of Concepcion."

Leaving Stokes and Usborne behind in tents ashore to work on survey charts until his return, on March 7, FitzRoy set course back to Valparaiso—238 nautical miles north of Concepcion and unaffected by the recent earthquakes—from where he promised to bring the citizens of Concepcion a supply of cash, of which there was "a great dearth." When the *Beagle* again anchored in Valparaiso, letters from his sisters awaited Darwin. Caroline was grateful he had "always been prudent and fortunate enough to get into no difficulties in your several adventures" but cautioned, "Do not let the having escaped so long make you careless and daring for the time to come." She also reported a "blundering" family friend had told them the Admiralty had ordered the *Beagle* home, but they assumed this was a false report.

Replying, Darwin confessed that he was still not fully recovered from the stomach illness that had confined him to bed in Valparaiso but that "some good rides will make another man of me." What he actually had in mind was a major expedition. With FitzRoy occupied over coming weeks with obtaining replacement anchors, further surveying and securing cash for the stricken people of Concepcion, Darwin planned to travel to Santiago and on through the Andes across the Portillo Pass to Mendoza in Argentina.

Traveling the seventy miles to Santiago by covered gig, Darwin stayed with Alexander Caldcleugh, the British owner of a copper mine and a naturalist whose book about his travels through South America was in the *Beagle* library. By March 18, armed with "a strong passport from the President of Chile"—just as Darwin had found in Argentina, the authorities were ready to assist the *Beagle*'s naturalist—and plenty of "horse cloths, stirrups, pistols and spurs," Darwin left Santiago. With him were Mariano Gonzales, his guide on his previous expedition to the Andean foothills, a good supply of food "in case of being snowed up, as the season was rather late," and an *arriero*, or muleteer, in charge of ten mules and their *madrina*—an "old steady mare" with a small bell round her neck. Darwin thought the mare "a sort of step-mother to the whole troop.—It is quite curious to see how steadily the mules follow the sound of the bell . . . The affection of the mules for the Madrina saves an infinity of trouble; if one is detained for several hours and then let loose, she will like a dog track out . . . the Madrina." Of the mules, six were for riding and four for baggage. Darwin was astonished that on a level road, a mule could carry more than four hundred pounds—"that a hybrid should possess more . . . muscular endurance, than either of its parents, seems to indicate that art has here out-mastered nature."

His route first lay southward across a "great burnt up plain" to the fertile valley of the Maipo—the 160-mile-long river Darwin had nervously crossed on a swaying suspension bridge six months before. In the valley's orchards were trees "bending and breaking" beneath ripe apples, nectarines, and peaches—the product, Darwin thought, of minerals and nutrients carried downriver from the foothills of the Maipo volcano, high in the Andes. Approaching the border with Argentina, Darwin decided Chile was "better guarded by the Cordilleras than by so much sea; the mountains on each side of the few narrow valleys where there are Custom-houses, are far too steep and high for any beast of burden to pass over." The border officials, no doubt impressed by Darwin's presidential passport, were extremely polite. In a rare criticism of Britain, he decided "the contrast is strong with the same class of officers in England," while Chileans in general seemed more courteous than Europeans. When he and his guides met "a very little, fat, poor Negress, with so enormous a goitre, that ones eyes almost involuntarily were fixed with surprise," he noticed his companions "after looking for a short time took off their hats as an apology."

As the party wound up the Maipo valley to the tinkling of the *madrina*'s bell, Darwin again reveled in the freedom and independence of his overland expeditions. Each evening the party found simple lodgings, hired pasture for the mules, purchased firewood, and, after cooking their supper in an iron pot, ate beneath "the cloudless sky [knowing] no troubles." The valley grew ever narrower until they emerged on to a shingle plain, a hundred feet above the Maipo, by now "rather a great mountain torrent than a river," roaring with such force over great chunks of stone that Darwin heard them rattle.

The shingle itself consisted of "rudely-stratified" terraces. "No one fact in the geology of South America, interested me more," Darwin wrote on his return from the voyage.

> I am convinced that the shingle terraces were accumulated, during the gradual elevation of the Cordillera, by the torrents delivering, at successive levels, their detritus on the beach-heads of long narrow arms of the sea, first high up the valleys, then lower and lower down as the land gradually rose. If this be so . . . the grand and broken chain of the Cordillera, instead of having been suddenly thrown up, as was till lately the universal, and still is the common opinion of geologists, has been slowly upheaved in mass, in the same gradual manner as the coasts of the Atlantic and Pacific . . .

Though lines of mules still zigzagged up steep tracks to silver mines high above, herds of cattle being driven down from the higher valleys——a reminder that winter was coming——made Darwin hurry on "more than was convenient for geology" through a landscape where the bright reds and purples of "utterly bare and steep hills" amazed him and the dramatic stratification created "the wildest and most picturesque groups of peaks."

On March 21, Darwin reached the base of a steep ridge, and the climb began in earnest. Breathing became harder, and Darwin experienced "a slight tightness over the head and chest; a feeling which may be known by leaving a warm room and running violently on a frosty day." Even the hardy, acclimatized mules paused every fifty yards or so. His companions explained the Chileans called this shortness of breath in thin air *puna* and showed him graves of several *punado*——people who, they claimed, had died from it. Unaware of the effects of altitude sickness, a skeptical Darwin decided they must have had heart or chest problems and wondered whether *puna* was more in the mind than the body since "upon finding fossil shells . . . in my delight I entirely forgot the Puna." His finds included oyster shells "and a piece of an ammonite as thick as my arm."

Halfway up, they passed a large group with seventy heavily laden mules. It was "a pretty sight to see the long string descending, and hear the wild cries of the muleteers; they looked so diminutive; no bushes, nothing but the bleak mountains with which to compare them." Reaching the snow-covered summit of the ridge in a freezing, biting wind, Darwin noticed that where the mules' hooves had crushed it, the snow was stained red—"the famous Red Snow of the Arctic Countries." Wondering whether this might be red porphyry dust blown from surrounding rocks, he placed a little snow to dry between the pages of his note book. Examining the residue later beneath his microscope he saw "groups of minute red balls" resembling "eggs of small molluscous animals."[3]

Looking back from the ridge, Darwin thought the view "glorious . . . The atmosphere so resplendently clear, the sky an intense blue, the profound valleys, the wild broken forms, the heaps of ruins piled up during the lapse of ages, the bright coloured rocks . . . produced a scene I never could have imagined . . ." With only a few condors circling the higher pinnacles to distract attention "from the inanimate mass," it was as magnificent as "watching a thunderstorm, or hearing in the full orchestra a chorus of the Messiah." All he was seeing further convinced him "that nothing, not even the wind that blows, is so unstable as the level of the crust of this earth."

Descending on the other side, the party camped on the bare slopes. Darwin had such a bad headache—probably the result of being at ten thousand feet—that he retired early to the folding bed he had brought with him. Waking during the night, he noticed the starry sky suddenly cloud over. Aware how quickly blizzards could strike at this time of year, he roused the muleteer, who assured him that "without thunder and lightning" there was no risk. In the morning they found the potatoes they had left to cook overnight over the fire were still rock hard. Darwin's companions declared "the cursed pot . . . did not choose to boil potatoes," but he realized the cause—at the low atmospheric pressure of altitude, water boils at a lower temperature, so food takes much longer to cook.

---

3. Red snow is in fact red-pigmented algae.

After a potato-less breakfast, they continued toward the fourteen-thousand-foot Portillo Pass through the Andes, crossing plains grazed in summer by cattle but from which even most guanaco had now gone. As they climbed toward the narrow cleft of the pass, Darwin noted "bold conical hills of red granite" and the distant snowy mass of the Tupungato mountains, in the middle of which was a "blue patch" he assumed was a glacier. Picking their way over a terrain of icy pinnacles formed by the thawing and refreezing of "perpetual snow," they came across "a frozen horse . . . sticking to one of these points as to a pedestal, with its hind legs straight up in the air." Darwin concluded it must have tumbled headfirst into a hole.

As they neared the summit, a cloud of what Darwin called "minute frozen spiculae"—needles of ice—enveloped them, so dense that as they began their descent, to his disappointment it obscured the view of the plains below. Reaching the tree line, they camped beneath the shelter of some large rocks, and as darkness fell, the sky finally cleared, revealing a sight "quite magical, the great mountains, bright with the full-moon, seemed impending over us from all sides . . . The increased brilliancy of the moon and stars at this elevation is very striking . . . owing to the great transparency of the air." He attributed the latter to the atmosphere's "extreme dryness." Another curious effect was "the facility with which electricity is excited." When he rubbed his flannel waistcoat, it looked "as if washed with phosphorous" while "every hair on a dog's back crackled."

Continuing their descent, Darwin noticed the gradient was much steeper than on the climb up—evidence that the mountains rose more abruptly on the Argentinian side than the Chilean. He also observed a "general difference" in the vegetation and the wildlife "in the valleys on this side and those of the other." Since soil, climate, and latitude were so similar, this surprised him and stimulated ideas about how geological change might influence the geographical distribution of species, an idea originally proposed by Lyell. After his return to Britain, he would write that the Andes "have existed as a great barrier, since a period so remote that whole races of animals must subsequently have perished from the face of the earth. Therefore, unless we suppose the same species to have been created in two different countries, we ought not to expect any closer similarity between the organic beings on the opposite sides of the Andes, than on shores separated by a broad strait of the sea."

After passing through "a brilliantly white sea of clouds," they reached a spot with pasture for the mules and brushwood for a fire and halted for the night. Early the next morning, Darwin had his much-anticipated first view of the extensive pampas below, like "a distant view of the ocean." Most striking were the rivers, which in the rising sun "glittered like silver threads till lost in the immense distance." Reaching a customs post, Darwin handed over his papers for inspection. One of the soldiers was "a thorough-bred Pampas Indian," who Darwin learned, was "kept much for the same purpose as a blood hound, to track . . . any person who might pass by secretly" and apparently had a notable success record.

A week after leaving Santiago, Darwin and his companions rode in baking heat across a flat, dry, salty plain "devoid of all interest" northward toward Mendoza. The appearance the next day of distant rows of poplars and willow trees in the village of Luxan was welcome, but not the "large ragged cloud of a dark reddish brown color" that Darwin spotted moving up behind them. At first he mistook it for "heavy smoke," but it was locusts. Their approach sounded like "a strong breeze passing through the rigging of a ship" and they overtook Darwin at a speed, he calculated, of ten to fifteen miles an hour, flying at a height of between twenty and three thousand feet above ground, he thought. Villagers brandished sticks and burning brands in vain attempts to divert the hungry swarm from their crops.

Halting that night in Luxan, Darwin himself "experienced an attack, and it deserves no less a name, of the benchuca, the great black bug of the pampas . . . before sucking they are quite thin, but afterwards round and bloated with blood, and in this state they are easily squashed." He found it "horribly disgusting, to feel numerous creatures nearly an inch long and black crawling soft in all parts of your person—gorged with your blood." Nevertheless, struggling to be a dispassionate observer, he thought it "good to experience everything once." The benchuca was the insect some suggest might have bitten Darwin some months earlier in Chile and caused his weeks of illness in Valparaiso. He would later catch one in Peru and record how a single feed of blood—"for which the benchuca was indebted to one of the officers [of the *Beagle*]"—kept it fat for four months, though within two weeks, "it was quite ready to have another suck." After keeping it alive for a while, he killed it to preserve as a specimen.

Approaching Mendoza the next day, Darwin admired the well-irrigated plantations of figs, peaches, and vines laden with "very fine grapes."[4] He and his companions bought delicious watermelons "nearly twice as large as a man's head" and "half a wheel-barrow" of ripe peaches. However, Mendoza itself, built around the usual plaza, had "a forlorn and stupid air," while its population were to Darwin "sad drunken raggamuffins" with "the reckless lounging manners of the pampas." He agreed with a British traveler a decade earlier that "the happy doom of the Mendozinos is to eat, sleep and be idle."

With little to detain him, the next day, Darwin began his return to Santiago, this time following the route across the Uspallata range taken by General Jose de San Martin in 1817 on his epic journey to liberate Chile from Spain. The early stages lay across a dry, dusty terrain with a few gold mine workings and dotted with dwarf cacti "armed with formidable spines" called "little lions." However, as Darwin wound into the mountains, a remarkable landscape opened up—"white, red, purple and green sedimentary rocks and black lavas," the strata "broken up by hills of porphyry of every shade of brown and bright lilacs. All together they were the first mountains which I had seen which literally resembled a coloured geological section." Some of the formations reminded him of those he had noted along the Chilean coast, and he began searching for further similarities, in particular pieces of petrified—"silicified"—wood such as he had seen on Chiloe.

"I succeeded," Darwin wrote jubilantly of his discovery on a seven-thousand-foot-high green sandstone escarpment of a whole grove of petrified trees, "snow white" as "Lot's wife," with the patterning of their bark still visible and with trunks still up to seven feet tall. "It required little geological practice to interpret the marvellous story, which this scene at once unfolded; though I confess I was at first so much astonished that

---

4. Mendoza's vineyards today are the basis of its flourishing wine production.

Fossilized trees near the Uspallata Pass, March 2020

I could scarcely believe the plainest evidence . . . I saw the spot where a cluster of fine trees once waved their branches on the shores of the Atlantic, when that ocean . . . approached the base of the Andes. I saw that they had sprung from a volcanic soil which had been raised above the level of the sea, and that this dry land, with its upright trees, had subsequently been let down to the depths of the ocean. There it was covered by sedimentary matter, and this again by enormous streams of submarine lava . . . The ocean which received such masses, must have been deep; but again the subterranean forces exerted their power, and I now beheld the bed of that sea forming a chain of mountains more than seven thousand feet in altitude . . . and the trees now changed into silex were exposed projecting from the volcanic soil now changed into rock, whence formerly in a green and budding state they had raised their lofty heads."

Darwin's "subterranean forces" were a process that would not be identified until the mid-twentieth century—plate tectonics. As far as he was concerned, as in the earthquake in Concepcion, here were Lyell's theories brought to life. These petrified trees—recently dated to around 245 million years ago, the late Triassic period, even older than Darwin thought—showed that natural causes operating over long periods of time could create a chain of mountains, even as mighty as the Andes. Darwin chipped off specimens that now are in Cambridge's Sedgwick Museum.[5]

---

5. Today, sadly, only a few fragments of the petrified forest remain in situ, the rest taken by souvenir hunters or, in one case, dynamited when the road passing by was rebuilt. However, hollows in the sandstone reveal some of their shapes. A plaque securely fastened to a large rock—Creationists removed an earlier memorial—commemorates Darwin's discovery.

Continuing over a series of passes with high winds whipping up dust clouds, Darwin found the dry landscape dotted with "low resinous bushes" monotonous and worried there was nothing for the mules to eat. He reflected on the exaggerations of those who had warned him of the "*awful* dangers" of crossing the Andes and of the need to "carry thick worsted stockings"——"No doubt in very many places if the mule should fall you would be hurled down an enormous precipice; in a like manner if a sailor falls from aloft, it is probable he will break his neck." Despite all the warnings, the only drama of the journey would occur some time later when the party woke to discover that during the night a thief had stolen one of the mules and the *madrina*'s bell.

A week after leaving Mendoza, Darwin was beyond the Uspallata range and descending toward the Puente del Inca——the Bridge of the Incas. Despite its name, the bridge had nothing to do with the Incas but was a natural formation where hummingbirds flitted. Darwin pondered how it

The so-called Inca Bridge, actually a natural formation and now a tourist attraction, March 2020

had been formed, sketching the oblique slab of creamy rock fused with soil and stones and its arch hollowed out by the river flowing beneath. Though dismissing "the bridge" as "miserable . . . not worth seeing"—a view not shared by today's tourists thronging the souvenir stalls—the nearby ruins of Tambillos caught his imagination. Knowing the mighty Inca Empire indeed extended here, he wondered whether the ruins had once been resting houses for Inca kings and their retinues crossing the mountains.

Traveling on, "seated on some little eminence in the wild valleys," Darwin noticed more recent ruins—a chain of *casuchas*, small round towers built by the Spanish as storehouses but long abandoned and "not ill suited to the surrounding desolation." The intense blue of the sky and "the brilliant transparency of the air" again struck him. Nearing Santiago, he crossed a fertile plain with figs and peaches drying on the roofs of the houses and people harvesting grapes in the vineyards. Darwin thought it a pretty scene, though lacking "that pensive stillness which makes the autumn in England indeed the evening of the year."

Before long, however, he began to feel so unwell again—perhaps the result of benchuga bug bites—that he "saw nothing and admired nothing." Reaching Santiago, he again stayed with the hospitable Caldcleugh. Despite his illness, Darwin reflected that he had been "well repaid" for his trouble during his twenty-four days away. The journey had been "such a famous winding up of all my geology in S. America," during which he had scarcely slept from excitement at sights such as "the strata of the highest pinnacles tossed about like the crust of a broken pie." He had "half a mule's load" of specimens and had filled more pages of his notebook than on any other expedition.

Five days later, feeling better, Darwin returned to Valparaiso to stay once more with his friend Richard Corfield until the *Beagle* arrived. While waiting, he wrote a long, buoyant letter to Henslow detailing all that he had seen, from the petrified forest to the complex structure of the mountains—"Some of the facts, of the truth of which I in my own mind feel fully convinced, will appear to you quite absurd and incredible." As with other letters, Henslow would extract passages for publication by the Cambridge Philosophical Society. To his sister Susan, Darwin wrote,

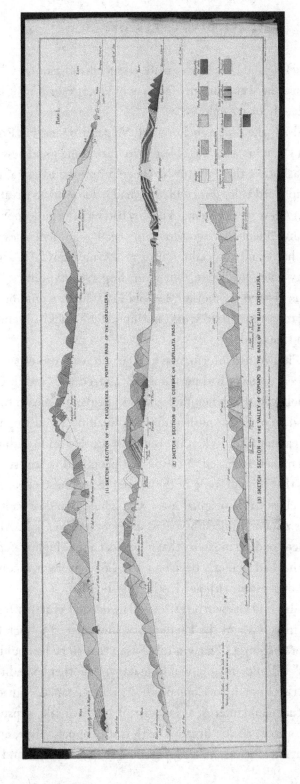

Darwin's geological cross-section of the Portillo Pass, Uspallata Pass, and the Copiapo Valley "since leaving England I have never made so successful a journey." Though very expensive, he was "sure my father would not regret it, if he could know how deeply I have enjoyed it."

On April 23, the *Beagle* anchored off Valparaiso, and FitzRoy sent a boat to fetch Darwin, who was able to give him some good news he had heard while ashore—that FitzRoy had been promoted to captain in rank as well as shipboard title. (Since 1828 he had held only the naval rank of commander.) A few weeks earlier, FitzRoy had written philosophically to his sister Fanny: "Plenty have gone over my head—one feels a *few* wounds but one gets hardened and callous after receiving many." He attributed the Admiralty's delay to "having burn't my fingers with politics." Though pleased by his belated promotion, he told Darwin he would be happier if his efforts to get his "hard working shipmates" Wickham and Stokes promoted had succeeded.

In turn, FitzRoy had information for Darwin, who was eager to know whether the earthquake had permanently altered the land in Concepcion. Lyell believed that when molten rock, heaving beneath the crust, could not find an outlet, it pushed upward, shaking and lifting the land above in an earthquake. When it eventually cooled and solidified, the land was permanently raised. FitzRoy had promised to investigate and told Darwin that though some of the shoreline around Concepcion had settled back close to its original level, the whole island of Santa Maria had been permanently raised by between eight and ten feet. Along the shore were beds of dead mussels "that many feet above high-water mark," while rocks that had formerly been beneath the sea were now completely exposed and encrusted with dead, stinking shellfish.

FitzRoy also told Darwin he estimated he needed a further ten weeks to complete his survey of the Pacific coast and offered to pick Darwin up "at any port I choose." Darwin told Susan that since he could extend his "holidays" till mid-July, he would make yet another expedition. He planned to travel north to Copiapo—"a great distance," but offering "everything that could interest a geologist," from rock salt, gypsum, salt-peter, sulfur mines, rocks threaded with iron, copper, silver, and gold, as well as "curious-formed valleys; petrified shells, volcanos and strange scenery. The country geologically is entirely unknown . . . and I thus shall see the whole of Chile from the Desert of Atacama to the extreme point of Chiloe."

After presenting this dazzling image, he came to the point—the only "black and dismal part of the prospect.—that horrid phantom, money . . . I verily believe I could spend money in the very moon." Though he feared he was taxing his father's patience, he confessed to Susan he had drawn one hundred pounds for the journey since "it is necessary to be prepared for accidents: horses stolen.—I robbed.—Peon sick . . ." but also in case he suddenly heard of "something very wonderful."

CHAPTER FOURTEEN

"Eternal Rambling"

Looking forward to traveling "in the usual independent manner, cooking our own meals and sleeping in the open air," Darwin packed up his "bed and a kettle, and a pot, a plate and basin," spent twenty-five pounds on four horses and two baggage mules, and again engaged Mariano Gonzales as his guide for what would be his last and longest overland journey of the voyage—some five hundred miles. He set out on April 27, 1835, making first for the Campana Mountain to revisit it "for geological purposes." Riding around the bay toward the village of Vina del Mar, named for its vineyards and today a fashionable resort, he looked back one last time at Valparaiso framed against its precipitous backdrop.

Continuing through terrain crisscrossed with streams, Darwin noticed the "numerous scattered hovels" of people panning for gold. "Like all those who gain by chance," they seemed "unthrifty in their habits." From the Campana mountain, he struck north for the coastal town of Coquimbo through an increasingly dry and barren landscape, studded with spiky yucca-like plants. Though sparse, rainfall still produced enough pasture for cattle to graze, as if the grass seeds had "an acquired instinct, what quantity of rain to expect." The distant snow-covered Andes were "a glorious sight."

At the port of Quilimar, where the *Beagle* had recently called as she surveyed north, Darwin found the inhabitants "convinced she was a smuggler" and that FitzRoy had made secret deals with some citizens—"they complained of the entire want of confidence the captain showed in not coming to any terms; each man thought his neighbour was in the secret—I had . . . difficulty in undeceiving them . . . this anecdote . . . shows how little even the upper classes in these countries understand the wide distinctions of manners. A person who could possibly mistake Capt.

FitzRoy for a smuggler, would never perceive any difference between a Lord Chesterfield and his valet."

Darwin next headed inland to Los Hornos, a district noted for its mines, where one large hill was so "drilled with excavations that it was a magnified edition of a large anthill." Darwin thought the miners' habits were like those of sailors on a man-of-war—they lived in close proximity with one another for weeks at a time, and as soon as they had money got drunk and squandered it as fast as they could. He passed one group taking a corpse for burial. "They marched at a very quick trot," four men carrying the body and running as hard as they could for about two hundred yards to be relieved "by four others who had previously dashed on ahead on horseback and so on.——They encouraged each other by wild cries; altogether it formed a most strange funeral."

At Panuncillo, Darwin visited a copper mine acquired by Alexander Caldcleugh, his host in Santiago, at a knockdown price after its previous owners, convinced the type of ore was not worth extracting, had abandoned it. Conditions were, as Darwin had seen before, harsh. Miners carried baskets of ore on their back "up 80 perpendicular yards, by a very steep road" and then "up a zigzag nearly vertical notched pole." They could not pause for breath, even though the mine was more than six hundred feet deep. Darwin tried to pick up a loaded basket but could barely lift it from the ground—it weighed nearly two hundred pounds. The miners, working nearly naked, were expected to haul a dozen such loads a day. They seemed healthy and cheerful enough, but even though "the labour is voluntary, it is yet quite revolting to see the state in which they reach the mouth of the mine.——their bodies bent forward, leaning with their arms on the steps; their legs bowed, the muscles quivering, the perspiration streaming from their faces over their breasts, the nostrils distended, the corners of the mouth forcibly drawn back, and the expulsion of their breath most laborious . . ." Their cries of "ay-ay" sounded "shrill like the note of a fife."

The mine's manager, Don Joaquin Edwards, reinforced what Darwin had already gleaned about people's suspicions of the British. As a schoolboy in Coquimbo, Edwards and his schoolfellows had once been given a holiday to see an English sea captain who came ashore on business. However, nothing had induced any of them, himself included, despite his father being British, to go near the captain "so fully had they been impressed with all the heresy, contamination and evil to be derived from contact with such a person." People still remembered "the atrocious actions of the buccaneers," one of whom had stolen a statue of the Virgin Mary from the church, then returned the following year to help himself to St. Joseph, "saying it was a pity the lady should not have a husband." Caldcleugh had told Darwin of an old lady in Coquimbo who had never forgotten how in her youth the cry of "Los Ingleses" was enough for the entire community to decamp to the hills with their valuables. William Dampier, the later naturalist and hydrographer whose books were in the *Beagle* library, was one of the British buccaneers who terrorized this stretch of coast in the late 1600s.

Darwin arrived in Herradura, just south of Coquimbo, on May 14 to find the *Beagle* anchored in the small bay and all hands living ashore beneath canvas while the ship was refitted for the Pacific voyage. Even the ballast had been temporarily removed. Darwin and FitzRoy rented lodgings together in Coquimbo—a quiet mining town of some eight thousand inhabitants. One evening, as Darwin dined with a young Anglo-Chilean named Jose Edwards, he again felt the "smart shock of an earthquake. I heard the forecoming rumble, but from the screams of the ladies, the running of servants and the rush of several of the gentlemen to the doorway I could not distinguish the motion." A fellow guest who had survived the great earthquake that devastated Valparaiso in 1822 told Darwin the experience had taught him never to sit in a room with the door shut— "the danger in an earthquake is not the time lost in opening a door, but the chance of its being jammed by the movement of the walls."

Darwin understood people's fears but thought their panic excessive and due in part to "a want of habit in governing fear; the usual restraint, shame, being here absent. Indeed the natives do not like to see a person indifferent." He had been told of two Englishmen, who, "sleeping in the open air . . . during a smart shock, knowing there was no danger did not rise," at which the local people "cried out indignantly 'Look at those heretics, they will not even get out of their beds.'"

In a valley near Coquimbo, Darwin examined "step-like plains of shingle" described by a Scottish naval officer and traveler, Basil Hall, some years earlier. Hall's account had caught the attention of Lyell, who thought the shingle formations had "been formed by the sea during the gradual rising of the land." Examining them to find seashells and semi-fossilized shark's teeth some three hundred feet above sea level, Darwin agreed. Further into the hills he visited a silver mine owned by Jose Edwards's father. The economics of mining intrigued Darwin, who had heard it said that in Chile "a person with a copper mine will gain, with silver he may gain, but with gold he is sure to lose." He decided this was untrue—though "a copper mine with care is a sure game" and mining for silver and gold was rather more like "gambling or rather taking a ticket in a lottery," the greatest fortunes were made from "the richer metals." Mine owners lost a great deal through theft, which was hard to prevent since the robbers bribed everyone they thought would betray them. He witnessed this himself a few days later when he encountered "three mules travelling by night loaded with rich ore," whose drivers "very quietly gave a fine specimen" to his guide.

After further geologizing, during which he collected more fossilized shells but found that what he had been told were petrified beans were only small white quartz pebbles, Darwin returned to Coquimbo, where, in his absence, a local man named Don Francisco Vacunan had generously lent FitzRoy his thirty-five-ton schooner, the *Constitucion*, "without any kind of agreement or remuneration." FitzRoy had placed Sulivan in command with orders to survey up the coast of northern Chile and rendezvous with the *Beagle* in Callao, Lima's port. The *Beagle* was to return briefly to Valparaiso to take on nine months of provisions before heading north to Copiapo to collect Darwin before sailing on to Peru.

Before leaving Coquimbo himself, Darwin wrote to his sister Catherine summing up his feelings about his latest travels. In contrast to his previous buoyant anticipations, the journey so far had been "rather tedious," the countryside "very miserable; so burnt up and dry," and sleeping indoors had been impossible because of fleas that left "the skin of my body . . . quite freckled with their bites." His geological work had been a success since on leaving Chile he would have "a very good general idea of its structure," but he was "tired of this eternal rambling, without any rest.—Oh what a delightful reflection it is, that we are now on our road to England."

Setting off again northward, Darwin clearly remained homesick. A rain shower two weeks earlier had brought a little green to the dry land, "just sufficient to remind me of the freshness of the turf and budding flowers" of an English spring. "Travelling in these countries, like to a prisoner shut up in gloomy courts, produces a constant longing for such scenes." The green soon faded, and the rocky desert became so barren there was no grazing for Darwin's horses and mules. An elderly man superintending a copper smelting furnace "as an especial favour . . . allowed me to purchase at a high price an armful of dirty straw, which was all the poor horses had for supper after their long day's work." Darwin was surprised to find the furnace since, as he knew, most Chilean ore was sent to Swansea for smelting—"The poor Chilenos think that England is quite dependent for her copper to Chile; they will scarcely believe that all the quantity . . . imported there must again be exported to other countries."

Traveling on across mountains bare except for a scattering of stunted cacti, Darwin reached the Guasco (Huasco) Valley, where he stayed two nights with the British owner of a copper mine to allow his animals to recuperate. He spent his time exploring the local scenery and calling on the local governor, whose wife, Darwin wrote waspishly, "affected blue-stockingism and superiority over her neighbours. Yet this learned lady never could have seen a map." From Guasco, Darwin headed inland to the town of Ballenar (Vallenar), named after the

Bernardo O'Higgins

birthplace in Ireland "of the family of O'Higgins [who] were presidents and generals in Chile." Bernardo O'Higgins, still then alive, was indeed a hero of the independence fight and Chile's first head of state.

With still nothing for the horses and mules to eat—"I am tired of repeating the epithets barren and sterile"—Darwin was relieved to reach the fertile, clover-scented Copiapo Valley. He found a comfortable *hacienda* with good fodder for his animals since "it is most disagreeable to hear whilst you are eating a good supper, your horse gnawing the post to which he is tied and to know that you cannot relieve his hunger." (He does not say whether he at any time offered the horse some of his own food.) He spent some days exploring the Copiapo Valley, "a mere ribbon of green in a desert," where he found many petrified shells and "great prostrate silicified trunks of trees." Bivouacking one night in a ravine, he again felt the "trifling shock of an earthquake."

On June 22, Darwin reached journey's end, the town of Copiapo—"a miserable looking place" where "every soul appears to be endeavouring to make money and see how soon . . . they can leave it," and mining was all people talked about. Darwin hired a muleteer and a string of eight mules and set off into the southern Atacama desert, examining *salinas*— salt lakes—and ascending high enough into the mountains to the east to experience the discomfort of *puna*, which he no longer doubted was real. When he returned, he learned that the *Beagle* had arrived, and feeling he would be "very glad to be again on board," set off to the coast. After selling his horses and mules for only two pounds less than he had paid and giving his "'adios' with hearty goodwill to my companion, Mariano Gonzales, with whom I had ridden so many leagues in Chile," he embarked with 169 specimens he had collected on a journey, which he complained, except for geology, had been "downright martyrdom."

To his surprise Darwin found FitzRoy absent and Wickham in temporary command. For once, FitzRoy was having a more adventurous time than his "Philos." In mid-June, shortly after the *Beagle*'s return to Valparaiso for supplies, a local British merchant had received a letter reporting the loss near Arauco, south of Concepcion, of HMS *Challenger*—the warship commanded by FitzRoy's friend Captain Michael Seymour that had been sent to restore order in the Falkland Islands the previous year. Recollecting that the master of a recently arrived Swedish ship had reported a large wrecked vessel off that coast, FitzRoy questioned him and inspected his log. The Swede had thought the ship was American, but his description convinced FitzRoy it was the *Challenger*.

Subsequently messages from the *Challenger*'s crew confirmed that on the night of May 19, it had been swept onto the rocks, that all but two men had been saved, and they were now encamped ashore with the stores they had salvaged but in imminent danger of attack by "Indians." FitzRoy tried to rouse the senior British naval officer on the station, the elderly Commodore Francis Mason commanding the flagship HMS *Blonde*, to action. However, Mason feared sailing in winter along "a lee shore"—a shore lying to the leeward of a ship onto which it could be blown in bad weather—and they had a "tremendous quarrel," during which FitzRoy "hinted something about a court-martial to the old commodore for his slowness." Browbeaten by FitzRoy, the commodore finally agreed to take him in the *Blonde* to Talcahuano, Concepcion's recently ruined port, from where he could travel overland to locate the *Challenger*'s stranded men and assess how best to rescue them.

Ordering Wickham to take the *Beagle* north to Copiapo to collect Darwin and then sail via Iquique to Callao where he would rejoin her, FitzRoy took the *Beagle*'s master's assistant, Alexander Usborne; the coxswain James Bennet; and one of the *Beagle*'s whaleboats aboard the *Blonde*. Disembarking at Talcahuano, FitzRoy heard that "a large body of Indians was in motion" toward the *Challenger* crew's camp. Leaving Usborne and Bennet behind, he hired guides and horses and set off at the gallop. On the way, he tried to buy fresh horses but found people reluctant to sell since "every Chilean residing on the frontier endeavours to keep by him a good horse, on which to escape in case of a sudden attack of the Indians." People along the way told him the *Challenger* crew had moved to a more defensible position near the mouth of the Leibu River but that the number of "Indians" converging on them was increasing daily. Riding as "as fast as our tired horses could drag their hooves through deep, loose sand," on the next evening, June 23, FitzRoy approached the bank of the Leibu, where, on the opposite shore, a solitary light pierced the darkness. He shouted, "*Challenger*'s a-hoy." At first no answer came. Then "a faint 'hallo.'"

Seymour sent the *Challenger*'s dinghy—the only ship's boat saved—to fetch FitzRoy. Most of the crew were in good health, albeit hungry and anxious. Henry Rouse, the British Consul in Concepcion to whom Seymour had sent men to seek help, was with them. Seymour told FitzRoy how hard it had been to maintain discipline, with men arguing about what to do "as if they were on equal terms with those whose authority at such a time was more than ever necessary." Some marines had even rifled "chests and boxes belonging to officers." However, he had reimposed order—having one man flogged—and to protect against attack, ordered his men to erect a palisade and dig a ditch around their cluster of tents.

FitzRoy set off back to Talcahuano the next morning to convince Mason that rescuing the *Challenger*'s men, although not difficult, was urgent. Reports that as many as three thousand "Indians" were advancing on Seymour's camp in hopes of plunder, encouraged him to ride "sparing neither whip nor spur" through driving rain. Reaching Talcahuano, Fitz-Roy found that Mason had meanwhile dispatched an American schooner, the *Carmen*, with Usborne, Bennet, and some of the *Blonde*'s crew to search for Seymour.

With FitzRoy aboard, the *Blonde* followed the next day. However, thick fog and heavy gales meant that not until a week later—"the longest . . . I ever passed," FitzRoy recalled—did the *Blonde* finally arrive off the mouth of the Leibu River. Through his telescope, FitzRoy glimpsed smoke rising from the *Challenger*'s camp and realized they were in time. By the following evening, all the ship's men were safely aboard the *Blonde*. Meanwhile, the *Carmen*, having failed to locate the stranded sailors, had gotten into difficulties herself. She was spotted just in time by the *Blonde* and taken in tow. At Seymour's subsequent court-martial—a routine procedure when a ship was lost—he presented a statement in his defense from FitzRoy suggesting changes in the currents resulting from the earthquake had been responsible for the errors in dead reckoning that had caused the *Challenger* to founder. Seymour was acquitted of all blame and assigned a new ship.

At midday, July 6—the day FitzRoy oversaw the embarkation of the *Challenger* men—the *Beagle*, with Darwin aboard, left Copiapo, and after crossing the Tropic of Capricorn, arrived six days later in the small Peruvian port of Iquique (today in Chile, following the war between Chile and a Bolivian-Peruvian alliance in 1879–1884). It lay beneath "a great steep wall of rock about 2000 feet high," and Darwin thought its inhabitants as isolated as those on board a ship. Everything they needed—water, wine, food, firewood—had to be brought from elsewhere. The *Beagle*'s arrival in this remote spot made the inhabitants uneasy since Peru was in a state of anarchy, with Iquique caught between contending factions. To add to the political turmoil, crime was rife. Churches had recently been looted, and enraged inhabitants, convinced "heretics" must be to blame, had "proceeded to torture some Englishmen" and, Darwin recorded, would have shot them had the authorities not intervened.

In the tense situation, Darwin had some difficulty hiring mules and a guide to take him to some local saltpeter works. Ascending a zigzag track of fine, white sand, he reached a desert plateau strewn with the carcasses of pack animals that had collapsed from fatigue. Except for a few vultures gorging on them, he saw "neither bird, quadruped, reptile or insect" and decided this was "the first true desert I have ever seen; the effect . . . was not impressive." Even while traveling from Coquimbo to Copiapo, he had seen some vegetation, however sparse. Here the ground was covered by a thick salt crust and "saliferous" sandstone—"incontestable proof of the dryness of the climate." He slept the night in the house of the owner of one of the saltpeter works, who grumbled about the heavy expense of running it.

Darwin did not enjoy the *Beagle*'s onward voyage to Callao. His consistently negative comments suggest not only homesickness but, as he himself hinted, that any sense of novelty was wearing thin—"custom excludes the feeling of sublimity . . ." Though the steady rolling passage to Callao reminded him of the Atlantic, there was a significant difference. In the Atlantic there was "an ever varying and beautiful sky; the brilliant day is relieved by a cool refreshing evening . . . The ocean teems with life, no one can watch the flying-fish, dolphin and porpoises without pleasure. At night in the clear heavens, the European traveller views the new constellations which foretell the new countries to which the good ship is onward driving.—Here in the Pacific . . . in the winter, a heavy dull bank of clouds intercepts during successive days even a glimpse of the sun . . . in approaching these low latitudes I did not experience that delicious mildness, which is known . . . in the spring of England, or in first entering the tropics in the Atlantic."

The *Beagle* anchored at Callao on July 19, where Darwin soon learned more about the parlous condition of Peruvian politics, convincing him that "no state in S. America, since the declaration of independence, has suffered more from anarchy . . ." Four rivals were currently vying for "supreme government," and if one became too dominant the other three united temporarily against him. The current president had only gained his position by "mutinying against the former president," and at a recent High Mass in church his regiments had unfurled black flags with a death's head, symbolizing their determination to fight to the death for him.

The situation made it unsafe for Darwin to venture far in Callao, which, apart from a handsome fort that had withstood a siege by the ubiquitous Admiral Cochrane, was "a most miserable filthy, ill-built, small sea-port" whose "very atmosphere was loaded with foul smells." Its inhabitants were, Darwin felt, depraved and drunken. Not long before, the British consul general, Colonel Belford Wilson, and two others had been attacked while riding through Callao by soldiers who "plundered them so completely they returned naked, excepting their drawers.—The robbers were actuated by warm patriotism; They waved the Peruvian banner and intermingled cries of 'Viva la Patria'; 'give me your jacket.' 'Libertad Libertad' with 'Off with your trousers.'"

Darwin escaped Callao by buying a seat on one of the stagecoaches that ran twice a day the seven miles to Lima despite the road being "infested with gangs of mounted robbers." There he spent five pleasant days enjoying "much hospitality . . . the conversation of intelligent people in a new and foreign place cannot fail to be interesting. Moreover . . . years in contact with the polite and formal Spaniards certainly improves the manners of the English merchants." He found Wilson, the consul general who had lost his trousers to robbers, particularly good company, having been an aide de camp to the independence leader Simón Bolívar, "El Libertador," and traveled extensively through South America.

Lima itself, founded in 1535 as the Ciudad de los Reyes, the City of Kings, by the conquistador Francisco Pizarro after defeating and murdering the Inca emperor Atahualpa, did not impress Darwin. Cavernous old mansions—now shared between several families—hinted at glory days, as did the many churches, but the city had fallen into "a wretched state of decay," with unpaved streets and "heaps of filth" everywhere. "Better worth looking at than all the churches and buildings," in Darwin's view, were the women in their *tapadas*—a close-fitting "elastic gown" that obliged them "to walk with small steps which they do very elegantly and display very white silk stockings and very pretty feet." They wore a black silk veil, "fixed round the waist behind" which "is brought over the head, and held by the hands before the face, allowing only one eye to remain uncovered.—But then that one eye is so black and brilliant and has such powers of motion and expression, that its effect is very powerful." Darwin felt as if "I had been introduced amongst a number of nice round mermaids, or any other such beautiful animal." Others clearly shared his ogling admiration. Of eight watercolors painted by Syms Covington of scenes in Lima, seven were of women, of which one, titled *Walking Dress of the Females of Lima*, shows a single eye peeping roguishly out just as Darwin described.

After drawing a further fifty pounds—"money for the islands," as he wrote home—Darwin returned to Callao to where Sulivan had now brought the *Constitucion* after completing his survey of northern Chile. Ten days later, on August 9, FitzRoy arrived in excellent spirits aboard HMS *Blonde*. Guessing Darwin would be growing impatient at all the delays, he had written to him, "Growl not at all—Leeway will be made up.—Good has been done unaccompanied by evil—ergo—I am happier than usual."

Impressed with Sulivan's survey work and his praise of the *Constitucion* as "a handy craft and good sea boat," and perhaps reenergized by his recent promotion and role in saving the *Challenger* crew, FitzRoy ignored his previous snubbing by the Admiralty and purchased the schooner from Don Vascunan for four hundred pounds of his own money. His plan was to refit her and, under the command of Alexander Usborne, the master's assistant, assisted by midshipman Charles Forsyth, send her to survey the Peruvian coast. When the survey was complete, Usborne and Forsyth were to sell the ship and return to Britain on a merchant vessel.

Loyal to his friend, Beaufort would later defend FitzRoy's purchase to the authorities in London, arguing that the extra vessel would significantly assist the survey work. Yet again, however, the Lords of the Admiralty would not be moved to refund him: "Inform Capt. FitzRoy that Lords highly disapprove of this proceeding, especially after the orders which he previously received on the subject." By now FitzRoy had spent some seven thousand pounds of his own money, only a few hundred pounds less than the cost to the Admiralty of refitting the *Beagle* for the voyage.[1] As well as the net costs to FitzRoy of the vessels he hired, purchased, and refitted, and, in the case of the *Adventure*, resold, the figure of seven thousand pounds includes the chronometers and other instruments he purchased, his two extra whaleboats for the *Beagle*, the salaries of the artists Earle and Martens, the salary of Stebbings, the instrument maker and repairer, and the two brass cannons he purchased in Rio.

Before setting sail from Callao, somewhat to the annoyance of Darwin—the delay "has been a grievous waste of time for *me*," he somewhat selfishly told his sister Susan—FitzRoy went to Lima for a few days to consult "some old charts and papers, which he thinks of considerable importance." Darwin filled some of his time exploring the ruins of old Callao, destroyed by a great earthquake and tidal wave in 1746, where what he saw convinced him the land had been permanently upheaved in the same way as he had observed at Concepcion and elsewhere.

---

1. Seven thousand pounds is equivalent to more than 915,000 pounds in today's values.

In his geological work in South America, Darwin had collected large numbers of carefully cataloged specimens. In his notebooks he had described in fine and accurate detail and sometimes sketched differing rock strata and formations and their relationships and interactions with one another. He also had the vision, as well as the opportunity, to work on a broad geographical canvas, testing Lyell's theory over much greater areas than Lyell himself had been able to do. He contrasted and compared material and observations from Bahia in Brazil to the vast uniform plains of Patagonia, the boggy Falkland Islands, the broken strata of the high Andes, and the mineral-rich deserts of northern Chile. In his autobiography he later wrote, "The investigation of the geology of all the places visited was far more important [than natural history] as reasoning here comes into play. On first examining a new district nothing can appear more hopeless than the chaos of rocks; but by recording the stratification and nature of the rocks and fossils at many points, always reasoning and predicting what will be found elsewhere, light soon begins to dawn . . . and the structure of the whole becomes more or less intelligible." Analyzing his wide-ranging researches as he prepared to leave South America, he believed Lyell was correct. Geological change was constant. Land gradually rose and fell, producing over large periods of time mountains and ocean trenches.

Finally, on September 7, three and a half years after first arriving in South America, the *Beagle* set course northwest across the Pacific for the Galápagos Islands, 1,060 nautical miles away and the destination most closely associated with both Darwin and the *Beagle* in the public mind.

CHAPTER FIFTEEN

The Enchanted Islands

Spanning the equator some six hundred miles off the coast of Ecuador—known in Darwin's time as the Republic of the Equator—the volcanic Galápagos archipelago consists of thirteen major islands with a total land area of just 3,093 square miles scattered over 23,000 square miles of the Pacific. The then-uninhabited islands were first visited in 1525 by the Spanish Bishop of Panama, Tomás de Berlanga, when driven off-course while sailing to Peru to mediate in a dispute between two leading conquistadores, Francisco Pizarro and Diego Almagar, over control of the former Inca capital, Cuzco. The Bishop named the islands Las Encantadas—the Enchanted Islands. A better English translation might have been the Bewitched Islands, since the bishop was referring to problems in getting between and away from them because of very powerful, fast-moving, and shifting currents and frequent mists. The bishop described the islands as looking "as though God had caused it to rain stones" and marveled at the great numbers of large, saddle-backed tortoises, *galápagos*, living there, hence the name by which Darwin knew them, as we do today.

From then on, Spanish ships began visiting the islands, and also pirates and buccaneers, many of them British, on their quest to intercept Spanish treasure galleons bringing gold from Manila to Panama for transshipment over the Panamanian isthmus and thence to Spain. Both the Spaniards and the buccaneers found the Galápagos tortoises easy to capture and good to eat. A British buccaneer, Ambrose Cowley, produced the first map of the islands in 1684, naming them after various British politicians, royalty, and fellow buccaneers.

William Dampier, who visited the islands on the same buccaneering expedition as Cowley, first described in detail the flora and fauna of the islands in his bestselling account of his voyage, which FitzRoy later referenced in his narrative. To Dampier, the flesh of the tortoise was "so sweet that no pullet eats more pleasantly." The bird life showed no fear and could easily be killed. However, most interesting perhaps was his detailed description of the plentiful and also highly edible green turtles and how they differed from those he had seen in the Caribbean, in the Atlantic around Ascension Island, and elsewhere. Those in the Galápagos were "a sort of bastard green turtle for their shell is thicker than other green turtles in the West or East Indies." In different parts of the world, there were "degrees of them both in respect of their flesh and bigness," but those in the Galápagos were the biggest he had ever seen, with bellies "five feet wide." By contrast other South Sea green turtles were not "so big as the smallest hawkbill turtle."

Dampier's consciousness of geographic variation and of degrees of difference and his use of the word *bastard*—which he used in similar contexts elsewhere—all suggest a willingness to conceive of the characteristics of related creatures being changed by local circumstances and by the selection of those with whom they mated. Dampier also included a description of what he understood to be their migratory pattern.

As the *Beagle* approached the islands, Darwin's greatest enthusiasm about visiting them was not zoology, of which he confessed in a letter to his cousin Fox he had recently been "shamefully negligent," but geology: "I look forward to the Galapagos, with more interest than any other part of the voyage.—They abound with active volcanoes and I should hope contain tertiary strata." On September 15, 1835, battling against the expected strong currents, the *Beagle* neared the easternmost island, Chatham (San Cristóbal). The next day FitzRoy began dispatching small parties in the ship's boats to survey different parts of the archipelago. Then, finding an anchorage protected from the currents and winds, with Darwin he was one of the first ashore on Chatham Island, where, on this first landing, they spent little more than an hour.

Darwin's first impressions were not particularly favorable: "The whole is black lava, *completely* covered by small leafless brushwood and low trees.— The fragments of lava where most porous are reddish and like cinders; the stunted trees show little signs of life.—The black rocks heated by the rays of the vertical sun like a stove, give to the air a close and sultry feeling.

The plants also smell unpleasantly. The country was compared to what we might imagine the cultivated parts of the Infernal regions to be."

FitzRoy's reaction was similar: "We landed upon black dismal-looking heaps of broken lava, forming a shore fit for Pandemonium [the name for the capital of hell coined by Milton in *Paradise Lost*]. Innumerable crabs and hideous iguanas started in every direction as we scrambled from rock to rock. Few animals are uglier than these iguanas, they are lizard-shaped, about three feet in length, of a dirty black colour with a great mouth, and a pouch hanging under it; a kind of horny mane upon the neck and back; and long claws and tail. These reptiles swim with ease and swiftness but use their tails only at that time." He also noticed a wheelbarrow abandoned on the shore by sailors who used such barrows to transport tortoises to their ships for food. A quick search by the *Beagle*'s crew to find tortoises for the same purpose failed.

Over the next days, the *Beagle* began surveying, coasting around Chatham Island from landing place to landing place, sometimes encountering American whaling ships. The Galápagos were surrounded by good whaling grounds full of sperm and humpback whales, and whalers from Nantucket and elsewhere frequently anchored to replenish supplies of tortoise meat.[1]

In St. Stephen's Bay, where they met just such a whaler, Darwin described how "the bay swarmed with animals; fish, shark and turtles were popping their heads up in all parts. Fishing lines were soon put over board and great numbers of fine fish two and even three feet long were caught. This sport makes all hands very merry; loud laughter and the heavy flapping of the fish are heard on every side." Going ashore again on Chatham Island, Darwin took a closer look at marine iguanas: "most disgusting clumsy lizards. They are as black as the porous rocks over which they crawl and seek their prey from the sea.—Somebody calls them 'imps of darkness.' They assuredly well become the land they inhabit." Hunting for botanical specimens, he "obtained ten different flowers; but such

---

1. Among the visitors a few years later was Herman Melville during a whaling voyage on which he gathered material for his novel *Moby-Dick*.

insignificant ugly little flowers, as would better become an arctic than a tropical country." Like Dampier, he found "the birds are strangers to man and think him as innocent as their countrymen the huge tortoises. Little birds within 3 and four feet, quietly hopped about the bushes and were not frightened by stones being thrown at them. Mr. King killed one with his hat and I pushed off a branch with the end of my gun a large hawk."

As he navigated the *Beagle* against the tricky currents and winds, FitzRoy noted how the green turtles were "rather like" those "of the West Indies but not exactly" and that "Honest Dampier" had made similar comparisons during his visit. Landing with Covington in the northern part of Chatham, Darwin studied the geology carefully, climbing one of the island's many currently inactive volcanoes. The surrounding landscape was "studded with small black cones; the ancient chimneys for the subterranean melted fluids." These chimneys reminded him of those of "the iron furnaces near Wolverhampton," a town not far from his Shrewsbury home, which had mushroomed during the industrial revolution in England's West Midlands in what itself became known as "the black country." He counted some sixty of these truncated, chimney-like hillocks rising fifty to a hundred feet above the lava plain: "The craters are all entirely inert; consisting indeed of nothing more than a ring of cinders." The age of the lava streams was "distinctly marked by the presence and absence of vegetation; in the latter and more modern nothing can be imagined more rough and horrid.—Such a surface has been aptly compared to a sea petrified in its most boisterous moments. No sea however presents such irregular undulations,—nor such deep and long chasms."[2]

During this landing, Darwin had his first encounter with the island's tortoises, meeting two, each with a shell of a circumference of "about seven feet. One was eating a cactus and then quietly walked away.—The other gave a deep and loud hiss and then drew back his head.—They were so heavy, I could scarcely lift them off the ground.—Surrounded by the black lava, the leafless shrubs and large cacti, they appeared most

---

2. Interestingly, FitzRoy attributed the original comparison with a petrified sea to Stokes, not Darwin.

Mockingbird, Floreana (Charles Island), Galápagos, January 2022

old-fashioned antediluvian animals; or rather inhabitants of some other planet." He also first observed the island's mockingbirds, or thenca—brownish-gray birds with long tails—which pecked at his shoes. He wrote in his field notebook "the thenca very tame and curious . . . I certainly recognise S. America in ornithology."

On September 24, the *Beagle* crossed to Charles Island (Floreana), where the Ecuadorian government, which had annexed the islands in 1832, had established a penal colony for political prisoners. FitzRoy and Darwin landed at Post Office Bay. Before the penal settlement existed, a box had been placed there in which visiting sailors, often whalers away from port for years, could deposit letters to be picked up by any other vessels heading toward the letters' destination.

The governor of the penal settlement happened to be an Englishman, Nicholas Lawson. He was visiting a whaling ship in the bay when the *Beagle* party landed and was more than happy to conduct them up the four-and-a-half-mile path to the settlement. The rocky trail lay at first through an arid landscape, but as they ascended, the land became more fertile—by a peculiarity of climate, the higher parts of the islands were often shrouded in mist in the rainy season and thus received more precipitation than the lowlands. As a consequence, sweet potatoes, plantains, pumpkins, and melons could be planted in what FitzRoy called "a rich black mould . . . sufficiently clayey for the purpose." Wild goats and pigs were able to find food and in turn formed another source of fresh meat for those in the penal settlement, who Darwin thought "live a sort of Robinson Crusoe life." FitzRoy suggested that rather than, as they did, hunting the pigs and goats with dogs, which would lead to their extinction, the settlers should leave them to multiply for a while, increasing both the supply and sustainability.

Lawson proved a fund of information on the islands' tortoises, which Darwin later used to supplement his own researches. Though Darwin paid little attention at the time, Lawson explained that he could, "on seeing a tortoise[,] pronounce with certainty from which island it has been brought" and that the average size to which a tortoise grew varied between the islands. He also related that due to the depredations of settlers and sailors, the number of tortoises was unsurprisingly reducing on Charles Island. There might be enough for a further twenty years, but he had already started sending parties north to James Island (Santiago) to kill and salt tortoises to feed the settlement. He told Darwin, "Upwards of 200 lbs of meat have been procured from one . . . In the year 1830 one was caught (which required six men to lift it into the boat) which had various dates carved on its shell. One was 1786." FitzRoy noted that "the quantity of tortoise shells lying about shows what havoc has been made among these helpless animals." He noticed that Lawson was using some of the shells "in an apology for a garden to cover young plants instead of flowerpots."

Although he does not appear to have examined the tortoise shells closely, Darwin, as he wrote in his diary, "industriously collected all the animals, plants, insects and reptiles from this island.—It will be very interesting to find from future comparison to what district or 'centre of creation' the organised beings of this archipelago must be attached." He would continue to tussle with the question of whether there was only a single period or location of creation during the long sea passages to come.

During his explorations, Darwin concluded correctly that, because of its relative smoothness, the lava making up the island had once been beneath the sea, but the land had subsequently been raised up. He also

Charles Island (Floreana), Galápagos, by Philip Gidley King

collected specimens of the islands' plants. Among them was a creeping vine with broad leaves, a rough stem, and tendrils, which it used to spread its way across and over other plants "in great beds injurious to vegetation." Darwin's specimen of this member of the cucumber/squash family is preserved in Henslow's Herbarium at Cambridge University and is unique since no one collected or even noticed its presence afterward or indeed before. It is now presumed extinct, possibly due to the island's wild pigs and goats.

From Charles Island, the *Beagle* voyaged northwest to Albemarle Island (Isabela). Darwin called the island "as it were, the mainland of the Archipelago, it is about 75 miles long and several broad, is composed of 6 or 7 great volcanic mounds from 2 to 3,000 feet high, joined by low land formed of lava and other volcanic substances." As they sailed along the coast, FitzRoy observed "smoke . . . issuing from several places near [one of] the summits but no flame"—apparently the only volcanic activity the *Beagle* men saw in the archipelago.

FitzRoy anchored the *Beagle* in a bay that had once formed part of a volcanic crater. With drinking water aboard running low, FitzRoy had reduced the daily ration per person by a half to "1/2 a gallon for cooking and all purposes." In such a hot part of the world Darwin thought this "a sad drawback to the few comforts which a ship possesses." The crew hoped to find fresh water on Albemarle, but going ashore Darwin was disappointed that some "little pits in the sandstone contained scarcely a gallon and that not good." Even so, it was "sufficient to draw together all the little birds in the country.—Doves and finches swarmed round its margin—I was reminded of the manner in which I saw at Charles Island a boy procuring dinner for his family. Sitting by the side of the well with a long stick in his hand, as the doves came to drink he killed as many as he wanted . . ." This note in his diary seems to be the only explicit reference to the finches of the Galápagos—later so important to his evolutionary researches—that Darwin made while actually in the islands.

Albemarle seemed even more arid and sterile than the other islands. An invitingly "blue and clear" lake at the bottom of a five-hundred-foot-deep crater proved "salt[y] as brine" when Darwin climbed down hoping to relieve his thirst. Exploring the cindery interior he came across "another large reptile in great numbers . . . a great lizard [land iguana], from 10 to 15 lbs in weight and 2 to 4 feet in length, [that] is in structure closely allied to those imps of darkness which frequent the sea-shore.— This one inhabits burrows to which it hurries when frightened with quick and clumsy gait.—They have a ridge and spines along the back; are coloured an orange yellow, with the hinder part of back brick red.— They are hideous animals; but are considered good food." The *Beagle* crew collected forty to take aboard to test this view.

FitzRoy was interested in the small birds he had seen, including some finches, writing, "All the small birds that live on these lava-covered islands have short beaks very thick at the base, like that of a bullfinch. This appears to be one of those admirable provisions of Infinite Wisdom by which each created thing is adapted to the place for which it was intended"—an indication that he appreciated the differences in the birds, even if seeing their adaptations as the product of a Creator.

Leaving Albemarle after just a day, the *Beagle* battled currents for six days to reach James Island to the east, where Darwin, Bynoe, Covington, Henry Fuller—Fitzroy's steward—and another sailor landed in a small cove with provisions for a short stay while the *Beagle* headed

back to Chatham Island to refill its water tanks. FitzRoy is the only person
to mention Covington as one of the party and then only as "Darwin's
servant." Darwin simply refers to "three men" accompanying him and
Bynoe. It was quite typical of him only to refer to Covington, who did
so much to assist him, entirely anonymously. Sometimes he did not even
mention Covington was with him at all.

The five pitched their tents initially in a small valley not far from
the beach whose black sand was so hot that Darwin recalled that even
"in thick boots it was very disagreeable to pass over." In the margin of
his diary, Darwin wrote "freshwater cove of the buccaneers." This was
indeed one of the places where buccaneers careened their ships, caught
tortoises for food, and looked for fresh water, though the only source
near the beach appeared to be "a miserable little spring."

The next day, as Darwin and his companions explored the interior,
they discovered what had attracted the buccaneers. Climbing to over
two thousand feet amid ferns and flowering trees growing in the damp
air they found some springs—one of the island's few watering places—
with "water good and deliciously cold." Many tortoises were drinking
there. Darwin noted "the tortoise when it can procure it, drinks great
quantities of water; Hence these animals swarm in the neighbourhood
of the springs . . . In the pathway many are travelling to the water and
others returning, having drunk their fill.—The effect is very comical
in seeing these huge creatures with outstretched neck so deliberately
pacing onwards.—I think they march at the rate 360 yards in an hour;
perhaps four miles in the 24.—When they arrive at the spring, they bury
their heads above the eyes in the muddy water and greedily suck in great
mouthfuls, *quite regardless* of lookers on." Darwin scrambled onto the back
of one tortoise that continued to move forward, seemingly unperturbed.
Even so, he found it difficult to keep his balance.

That night Darwin and his companions stayed at the camp of some
tortoise hunters who fried tortoise meat for them "in the transparent oil
which is procured from the fat.—The breast-plate with the meat attached
to it is roasted as the Gauchos do the 'carne con cuero.' It is then very
good.—Young tortoises make capital soup—otherwise the meat is but,

to my taste, indifferent food," Darwin wrote. Next day, the hunters showed them a steep-sided crater lake with "beautifully crystallized salt." Fringed with bright green succulent plants, Darwin thought it rather pretty. However, he discovered that "in this quiet spot the crew of a seal-ing vessel murdered their captain. We saw the skull lying in the bushes."

On their return to the beach, they found a big swell had overwhelmed and rendered salty the small spring they had been using for water. With the *Beagle* not expected for a few days, Darwin wrote, "We should have been distressed if an American whaler had not very kindly given us three casks of water (and made us a present of a bucket of onions). Several times during the voyage Americans have showed themselves at least as oblig-ing, if not more so, than any of our countrymen would have been. Their liberality moreover has always been offered in the most hearty manner. If their prejudices against the English are as strong as ours against the Americans, they forget and smother them in an admirable manner." Fitz-Roy also later mentioned his gratitude for the generosity of the American whalers he encountered.

Nine days after their landing on James Island, the *Beagle* returned for them. Before leaving, FitzRoy undertook a further survey of Albemarle Island—Darwin thought it would be difficult to find "in the intertropi-cal latitudes a piece of land 75 miles long so useless to man or the larger animals." Surveying completed, on October 10 the *Beagle* set course for Tahiti, three thousand nautical miles away. The sailors had brought on board just under fifty live tortoises to augment their provisions since tortoises could live up to six months without water. The reason—then of course unknown—was their slow metabolism and ability to metabolize fat into water. Darwin, Covington, and other crew members had also brought aboard small tortoises not to eat but as pets. Darwin had another souvenir—a pipe fashioned from the leg bone of an albatross.

Contrary to the myth that later developed, Darwin had not had a eureka moment. However, he had amassed a vast collection of animal specimens containing a good deal of the evidence on which he would base his subsequent work. During the *Beagle*'s forty days or so in the archi-pelago, he had spent only nineteen full—or more often part—days ashore and visited only four of the islands. Therefore, many of his specimens, including "finch" specimens from other islands, had been collected by other members of the crew and donated to him. However, several crew members, including FitzRoy, Henry Fuller, and Syms Covington, retained some for their own collections, which would later prove useful to Darwin.

As the *Beagle* sailed on, Darwin worked on his ornithological and zoological notes. He confirmed his view that the birds of the Galápagos appeared to be related to those of South America. He had some difficulty identifying the precise species of many of the small birds. At this stage, some seemed to him to be grosbeaks, others finches or blackbirds, and one a wren. As for the mockingbirds, they appeared to differ from known species such as those he had seen in Chile—"I have four specimens from as many islands. These will be found to be 2 or 3 varieties. Each variety is constant in its own island."

In Tierra del Fuego, Darwin had noted with interest the presence of flightless birds (the penguin, steamer duck, and rhea). However, nowhere in his diary or notes on the Galápagos does he mention the Galápagos penguin, a distinct species (the world's third smallest) that lived on some of the islands he visited, or the Galápagos cormorant, the only one of twenty-nine species of cormorant to be flightless. It must be assumed he did not see them. He also did not mention in any detail the three species of booby living on the islands, in particular the dancing of the blue-footed species that so delights modern tourists. This was more understandable since none of the three are endemic to the islands. In his diary he did discuss the magnificent acrobatic frigate bird but was puzzled about its feeding habits.

In his zoological notes Darwin added to his information about the tortoises. "The old ones occasionally meet their death by falling over precipices; but the inhabitants have never found one dead from Natural causes. The males copulate with the female in the manner of a frog. They remain joined for some hours. During this time the male utters a hoarse roar or bellowing . . . The[ir] egg is quite spherical . . . white and hard, the circumference of one was 7 and 3/8 of inch. The young tortoise, during its earliest life frequently falls prey to the Caracara (Galapagos hawk), which is so common in these islands." The overproduction of offspring to allow for losses to predators would feature in Darwin's evolutionary theory. He continued, "The people believe the [tortoises] are perfectly deaf; certainly when passing a tortoise, no notice is taken till it actually sees you; then drawing in its head and legs and uttering a deep hiss, he falls with a heavy sound on the ground as if struck dead."[3]

When a tortoise was caught for food, "a slit is made in the skin near the tail to see if the fat on the dorsal plate is thick; if it is not the animal is liberated and recovers from the wound.—If it is thick it is killed by cutting open the breast plate on each side with an axe and removing from the living animal the serviceable parts of the meat and liver etc etc. . . . The inhabitants when very thirsty sometimes have killed these animals in order to drink the water in the bladder, which is very capacious. I tasted some, which was only slightly bitter."

On the land iguanas, Darwin's additions to his zoological notes included, "They are ugly animals and from their low facial angle have a singularly stupid appearance. Capt. FitzRoy's specimens will give a good idea of their size. Their colors are, whole belly, front legs, head saffron yellow and Dutch orange—upper side of head nearly white. Whole back behind the front legs, upper side of hind legs and whole tail hyacinth red . . . They are torpid slow animals crawling when not frightened with their belly and tail on the ground, frequently they doze on the parched ground . . ." They lived in burrows and "when excavating these holes, the opposite sides of the body work alternately; one front leg scratches the earth for a short time and throws it towards the hind. The latter is well placed [to] heave the soil behind the mouth of hole. The opposite side then takes up the task . . . Those individuals . . . which inhabit the extremely arid land, never drink water during nearly the whole year. These eat much of the succulent Cactus which is in evident high esteem. When a piece is thrown towards them, each will try to seize and carry it away as dogs do with a bone."

---

3. The temperature at which tortoise eggs are incubated, unknown to Darwin, determines the sex of the tortoise, with warmer temperatures tending to produce females, and cooler ones males.

Marine iguanas, Albemarle Island (Isabela), Galápagos, January 2022

On marine iguanas, he added they had sometimes been seen swimming a hundred yards out at sea and their bodies were well fitted for the water. Yet they were reluctant to enter the sea and were always quick to get out of it for reasons Darwin did not understand. He wondered whether it could be because their predators such as sharks lived in the water and not on land. They appeared "to be able to survive a long time without breathing. One was sunk [in the sea] with a weight for nearly an hour and was then very active in its motions . . . I opened the stomach (or rather duodenum) of several, it was largely distended by quantities of minced pieces of seaweed . . . not a trace of any animal matter . . . They have no idea of biting and only sometimes when frightened squirt a drop of fluid from each nostril." (Squirting from the nostrils is now known to be how the marine iguana gets rid of excess salt through a nasal gland which sometimes discharges fluid.)

While Darwin was busy with his notes FitzRoy and his officers were finalizing their charts of the archipelago that proved so accurate they were universally used until well into World War II.

CHAPTER SIXTEEN

Aphrodite's Island

After three and a half weeks at the end of the longest single leg of her voyage, the *Beagle* anchored in Tahiti's Matavai Bay. Everyone aboard was eager to experience the attractions of the Polynesian island fabled for the beauty of its people and of their home, even if, in Darwin's words, it had become "a fallen Paradise." The first Europeans to reach Tahiti were the crew of HMS *Dolphin*, commanded by Samuel Wallis, in June 1767. When the *Dolphin* approached the island and, like the *Beagle*, anchored in Matavai Bay, some Tahitians had thought the ship was "a floating island." Others had recalled a prophecy that, as a result of the chopping down of a sacred tree, newcomers of an unknown kind would arrive and that "this land would be taken by them. The old order will be destroyed and sacred birds of the land and the sea will come and lament what the lopped tree has to dictate. [The newcomers] are coming upon a canoe without an outrigger." After some initial friction, Wallis's crew and the Tahitians became so close that both were sad to see the *Dolphin* leave, not least on the sailors' part because of the friendliness of the people, their willingness to trade, the fertility of their island, and most of all because of the ties between some of the crew and Tahitian women who had uninhibitedly made love to them.

French explorer Louis-Antoine de Bougainville confirmed the impressions of an earthly paradise when he visited the islands soon afterward in his ships *La Boudeuse* and *L'Etoile*. In tribute to Tahiti's "celestial" women, he named the island New Cythera after the island near where the goddess Venus (Aphrodite) reputedly sprang from the sea. On his return to France, de Bougainville referred in his book about the voyage—a copy was in the *Beagle* library—to Tahiti as "the true Utopia." His account focused on the Tahitian islands the discussion of the virtues of natural law based on man's innate and unconscious sense of morality compared to laws dictated by religious leaders or secular rulers for their own benefit, a topic then being debated animatedly in Europe. To philosophers such as Diderot and Rousseau, Tahiti represented a lost golden age, and the Tahitian people epitomized "the noble savage." Darwin's grandfather Erasmus celebrated the Tahitians' free and natural approach to sexuality in his writing, and in one of his poems used a Tahitian marriage ceremony as a metaphor for plant reproduction.[1]

The impression of a fertile pleasure dome was maintained following the first visit of Captain James Cook in HMS *Endeavour* in 1769. Together with naturalist Joseph Banks, who enjoyed many a dalliance with the Tahitian women, including their queen, he spent three months on the island, far longer than either Wallis or de Bougainville. He constructed an observatory at the northeastern tip of Matavai Bay to track the transit of Venus across the face of the sun, important in calculating the distance between Earth, the sun, and other planets. Banks collected and described many of the island's plant species.

By the time of Cook's last visit in 1777, however, western diseases, including sexual ones, and the introduction of alcohol and gunpowder weapons had begun to take their toll. As he left for the last time, Cook wrote: "I cannot avoid expressing it as my real opinion that it would have been far better for these poor people never to have known our superiority in the accommodation and arts that make life comfortable . . . indeed they cannot be restored to that happy mediocrity in which they lived before we discovered them . . ."

When Captain William Bligh visited the island in 1789 in the *Bounty* on the recommendation of Banks to collect breadfruit seedlings to take to the Caribbean as food for slaves, some of the *Bounty* crew's relations

---

1. The English poet and playwright John Dryden coined the phrase "noble savage" in 1672 in his play *The Conquest of Granada*:

   I am free as nature first made man,
   Ere the base laws of servitude began,
   When wild in the woods the noble savage ran.

with local women had, together with Bligh's overbearing nature, contributed to the celebrated mutiny. Some islanders had accompanied Fletcher
Christian and the hard-core mutineers to remote rock-girt Pitcairn Island.

Subsequently conditions had deteriorated further, so that in 1806
Joseph Banks wrote: "Tahiti is said to be at present in the hands of about
one hundred white men, chiefly English convicts [from New South Wales]
who lend their assistance as warriors to the chief whoever he may be,
who offers them the most acceptable wages payable in women, hogs etc;
and we are told that these banditti have by the introduction of diseases,
by devastation, murder and all kinds of European barbarism, reduced the
population of that once interesting island to less than one tenth of what
it was when the Endeavour visited it in 1769."

After this low point, British Protestant missionaries arrived and,
despite some initial setbacks, in 1812 converted the local ruler, Pomare II,
to Christianity. Thereafter, nearly all the Tahitians themselves converted
and mostly accepted the missionaries' somewhat puritanical views on
morality and society. As well as enforcing strict Sabbath observance and
banning alcohol, the missionaries prohibited singing by adults of anything
but hymns and dancing while attempting to inhibit promiscuity and promote modesty in dress. They also understandably proscribed infanticide
formerly practiced by young members of the elite—the *Arioi*—when
their free lovemaking produced unwanted children and the occasional
sacrifice and even consumption of prisoners captured in war. At the time
of the *Beagle*'s arrival the ruler of Tahiti was a woman from the ruling
family known as Queen Pomare IV.

As FitzRoy brought the *Beagle* to its mooring in Matavai Bay, outrigger canoes crammed with friendly Tahitians raced out to greet the new
arrivals and to trade with them. From them the *Beagle*'s crew discovered
that the local day and date differed from their own. On the *Beagle* it was
Sunday, November 15, whereas ashore it was Monday, November 16. The
difference was accounted for by which side of the international dateline
Tahiti was considered to be. Although the line formally lay west of Tahiti
at longitude 180 degrees, the Tahitians calculated their dates as being to
its west. FitzRoy soon adjusted the *Beagle*'s log to reflect the local date.

View of Tahiti by C. Martens

If the *Beagle* had arrived on what the Tahitians considered a Sunday, there would have been no outrushing canoes. The missionaries' strict Sabbath observance laws, agreed upon with the Tahitian ruler, permitted no outriggers to be launched that day. Going ashore Darwin found "crowds of men, women and children were collected on the memorable Point Venus [the site of Cook's and Banks's observatory] ready to receive us with laughing merry faces."

Unlike in so many other places, Darwin was charmed. As a man who throughout his life, and in particular in *The Descent of Man*, often reflected on the divergences between the "civilized" and the "savage" and how the transition between the two occurred, he recognized a nobility in the Tahitians—"a mildness in the expression of their faces, which at once banishes the idea of a savage,—and an intelligence which shows they are advancing in civilization." He admired the physique of the Tahitian men—"the common people . . . have the whole of the upper part of their bodies uncovered; and it is then that a Tahitian is seen to advantage.—In my opinion, they are the finest men I have ever beheld;—very tall, broad-shouldered, athletic, with their limbs well proportioned . . . To see a white man bathing alongside a Tahitian, was like comparing a plant bleached by the gardener's art to the same growing in the open fields.——Most of the men are tattooed, the ornaments so gracefully follow the curvature of the body that they really have a very elegant and pleasing effect . . . The simile is a fanciful one, but I thought the body of a man was thus ornamented like the trunk of a noble tree by a delicate creeper . . ."[2]

Darwin, though, was "much disappointed in the personal appearance of the women; they are far inferior in every respect to the men." They were tattooed similarly to the men but also heavily on their fingers. He disliked "an unbecoming fashion" among them "now almost universal in cutting the[ir] hair or rather shaving it from the upper part of the head in a circular manner so as only to leave an outer ring of hair." The missionaries had tried to dissuade them from this, but "it is the fashion and that is answer enough at Tahiti as well as Paris." Darwin believed the women much needed some becoming clothing to cover their upper bodies. He did, however, admire their custom of putting a red or white hibiscus flower in their hair or behind their ears, together with their wearing of a wreath of coconut leaves on their heads to shade their eyes, both not uncommon in modern-day Tahiti and nearby Polynesian islands.

FitzRoy's view mirrored Darwin's: "The native women had no charm for me. I saw no beauty among them; and either they are not as handsome as they were said to be, or my ideas are fastidious. The men on the other hand exceed every idea formed from the old descriptions . . . [their] personal appearance . . . was to me most remarkable: tall and athletic, with very well-formed heads and a good expression of countenance, they at once made a favourable impression, which their quiet good-humour and tractable disposition afterwards heightened very much."

Returning to the ship together, Darwin and FitzRoy came upon an idyllic scene: "numbers of children were playing on the beach, and had lighted bonfires which illuminated the placid sea and surrounding trees: others in circles were singing Tahitian verses,——we seated ourselves on

---

2. *Tattoo* is a Tahitian word introduced into English by Banks, who had himself ornamented with some.

the sand and joined the circle. The songs were impromptu . . . one little girl sang a line which the rest took up in parts, forming a very pretty chorus,—the air was singular and their voices melodious. The whole scene," Darwin wrote, "made us unequivocally aware that we were seated on the shores of an Island in the South Sea."

Early the next day, perhaps two hundred Tahitians thronged the deck of the *Beagle* to trade. According to Darwin, they now "fully understand the value of money and prefer it to old clothes or other articles." Leaving the thriving market behind, Darwin went ashore and began to climb into the hills behind the bay. As he ascended, the vegetation changed from banana plants and orange, guava, and breadfruit trees—the latter's strong branches reminding an increasingly homesick Darwin of an English oak—through dwarf ferns and coarse grass reminiscent to him of the hills of North Wales. From the hilltop he had a good view of the island of Eimeo (Moorea) opposite Matavai Bay with its jagged cloud-topped green peaks and surf pounding its reef.

Returning through the forest he again mused on the dispersion of flora and fauna between islands and continents: "It must not . . . be supposed that these woods at all equalled the forests of Brazil.—In an island, that vast number of productions which characterise a continent cannot be expected to occur."

The following day, he began a longer expedition into the mountainous interior accompanied by Covington and two Tahitian guides. The latter insisted there was no need to carry food—they could live off the land for the planned two- or three-day trip. As they climbed, following the course of a river that entered the sea near Matavai Bay, the valley became a precipitous ravine "which formed a mountain gorge far more magnificent than anything I had ever beheld." His guides, barefoot like all Tahitians, found tracks along rocky ledges that allowed them to circumvent the frequent waterfalls cascading down the cliffs. Sometimes they scaled cliff faces, using only small hand and toe holds, before lowering ropes to allow Darwin and Covington to climb up after them. Once a guide positioned a tree trunk so they could use it as a ladder to scramble up above a dizzying precipice.

In the late afternoon they made camp amid a mass of fruiting wild banana plants. With "strips of bark for twine, the stems of bamboos and the large leaf of the banana, the Tahitians in a few minutes built an excellent house; and with the withered leaves made a soft bed." They then made a fire by rubbing a stick in a wooden groove, and when it was well alight heaped stones "about the size of a cricket ball" on the burning wood. Once the stones were hot and the wood consumed they placed parcels of fish from the nearby stream and fruit and vegetables wrapped in leaves between and around the stones and piled the whole with earth. When they opened it up a little later Darwin found the cooked food delicious, particularly when washed down with cool water from the stream, drunk from a coconut cup. Among the baked vegetables were "wild yams" and "wild arum," the roots of which "when well baked are good to eat and the young leaves better than spinach." For dessert the guides served the root of "a liliaceous plant called ti," which was "as sweet as treacle."

The missionaries had prohibited the cultivation of the ava plant— the source of the alcoholic cava drunk throughout Polynesia. However, Darwin found the dark, green-leaved plant growing near the stream, chewed some of its leaves, and "found that it had an acrid and unpleasant taste which would induce anyone at once to pronounce it poisonous." He experienced no intoxicating effect, unsurprising since it is the root that is used to produce the drink, not the leaves. Darwin was impressed by how the Tahitians said grace before eating and prayed on their knees before sleeping.

The next morning, after a night in which it rained "but the good thatch of banana leaves kept us dry," how much breakfast the Tahitians consumed amazed Darwin. "I should suppose such capacious stomachs must be the result of a large part of their diet consisting of fruits and vegetables which do not contain in a given bulk very much nutriment." Darwin had taken "a flask of spirits" with him and offered it to the guides to drink. The two men did not refuse, "but as often as they drank a little, they put their fingers before their mouths and uttered the word 'missionary.'"

Throughout that day they climbed farther up toward Tahiti's highest peak at around seven thousand feet. Descending after a second night in the hills, they took another route back to the shore, edging along ridges that were "exceedingly narrow and for considerable lengths steep as the inclination of a ladder . . . When viewing the surrounding country . . . the point of support was so small that the effect was nearly the same as would I imagine, be observed from a balloon." Darwin, clearly conscious of the concept of the noble savage, thought his guides "with their naked tattooed bodies, their heads ornamented with flowers. and seen in the dark shade of the woods, would have formed a fine picture of Man inhabiting some primeval forest." On one of the following days he hired an outrigger canoe—in his view comical because of its extreme narrowness—and its crew to examine Tahiti's reef—the first he had seen—and "its pretty branching coral."

Darwin, perhaps knowing how thorough Banks had been, took few plant or animal specimens from Tahiti, preferring simply to luxuriate in its legendary tropical beauty. On geology, in confirmation of his acceptance of Lyell's view of the rise and fall of land over long periods, he suggested in his diary, "I believe a group of the interior mountains stood as a smaller island in the sea and around their steep flanks streams of Lavas and beds of sediment were accumulated in a conical mass under water. This after having been raised was cut by numerous profound ravines, which all diverge from the common centre; the intervening ridges thus belonging to one slope." In his geological notes he added, "The characteristic feature of this scenery is the depth, narrowness and extreme steepness of the sides of the valleys or rather mountain gorges . . . I believed I saw the effect of running water, continued through so long a succession of ages, as to suffice to wear away several thousand ft in thickness of solid strata."

FitzRoy meanwhile was busy checking his chronometers and validating his observations against the well-known data obtained by Cook and others at Point Venus and supervising some limited survey work. He was also meeting some of the missionaries and the Tahitian chiefs and encountered a larger-than-life European adventurer—"a person who styled himself Baron de Thierry, King of Nuhahiva [Nuka Hiva] and sovereign chief of New Zealand. About the house in which resides this self-called philanthropist, said to be maturing arrangements for civilising Nuhahiva and New Zealand—as well as for cutting a canal across the isthmus of Darien—were a motley group of tattooed New Zealanders, half-clothed natives of Tahiti, and some ill-looking American seamen. I was received in affected state by this grandee, who abruptly began to question me with—'Well, Captain! What news from Panama? Have the Congress settled the manner in which they are to carry my ideas into effect?' I tried to be decently civil to him, as well as to the 'baroness' but could not diminish my suspicions, and soon cut short our conference. In his house was a pile of muskets whose fixed and very long bayonets had not a philanthropic aspect. He . . . was said to be waiting for his ships to arrive and carry him to his sovereignty. Born in England, of French emigrant parents, his own account of himself was that he was secretary of legation to the Marquis of Marialva at the congress of Vienna; and that in 1815 he belonged to the 23rd Light Dragoons (English). In 1816 he was attaché to the French ambassador in London. In 1819 he was studying divinity in Oxford. In 1820–21, he was a student of laws at Cambridge . . . He showed papers to prove these assertions . . . and had succeeded in duping a great many people."

Meeting de Thierry a second time, FitzRoy told him "my suspicions, so plainly, that he said he should appeal to the governor of New South Wales, to the Admiralty, and to the King of England himself, against the unjust suspicions and improper conduct of the captain of the *Beagle*."

FitzRoy was right to suspect de Thierry's colorful, no doubt smoothly and frequently told and embroidered story. It was full of half-truths, omissions, some truths, and some downright falsehoods. His father had assumed the title of baron when fleeing during the revolutionary terror from France, where he had been an equerry at court. He was himself probably born in the Netherlands, not London as he always claimed. He was at the Congress of Vienna briefly on the staff of the Portuguese Marquis of Marialva and equally briefly served in the British dragoons and attended Oxford and Cambridge Universities. He claimed to have purchased land in New Zealand, which he had never visited, for thirty-six axes through the agency of a British missionary from a Maori chief he had met at Cambridge. When investigated by the French and Dutch governments, whom he had attempted to interest in colonizing New Zealand on the basis of his land acquisition, with himself in return to be appointed governor or viceroy, his claim had been found baseless. He had been imprisoned for debt twice, fled bankruptcy in France, married the daughter of an English archdeacon, spent time in the United States and the Caribbean, promoted again briefly the idea of a Panama canal, and, passing through Nuka Hiva in the Marquesas Islands on his way to Tahiti, elected himself its king.

Some more recent visitors to Tahiti, particularly a Russian explorer named Otto von Kotzebue in his book which was in the *Beagle* library, had strongly criticized the British missionaries from the London Missionary Society for removing their natural uninhibited joy from the Tahitians' lives and forcing them to conform to their views of how to behave, thus rendering the Tahitians gloomy and oppressed. However, when they met the missionaries, both FitzRoy and Darwin were impressed by their honesty, the sincerity of their convictions—the senior missionary was about to complete a translation of the Bible into the Tahitian language—and their efforts to help and "civilize" the Tahitians and to remedy some of the evils Europeans had introduced. Darwin was perhaps even more sympathetic to the missionaries than FitzRoy. He was convinced of their "high merit." Their critics, in his view, did not give them credit for what they had achieved and did not compare the current state of morals and society in Tahiti to that which prevailed "twenty years before; not even to that of Europe in this day but to the high standard of Gospel perfection. They expect the missionaries to effect what the very Apostles failed to do.—By as much as things fall short of this high scale, blame is attached to the missionaries, instead of credit for what has been effected."

The missionaries had much reduced "dishonesty, intemperance and licentiousness." The morality of the Tahitian women was much improved in Darwin's view since Cook's and Banks's time. He believed that many critics "disappointed at not finding . . . licentiousness quite so open as formerly . . . . will not give credit to a morality which they do not wish to practice, or to a religion which they undervalue if not despise." He

Missionary chapel, Tahiti, by Philip Gidley King

and FitzRoy attended a service in the chapel, "a large airy framework of wood . . . filled to excess by tidy clean people of all ages and sexes," who paid attention "quite equal to that in a country Church in England."

FitzRoy too acknowledged the benefits the missionaries had brought and the Tahitians' gratitude to them but suggested they might allow "more temporal enjoyments" and "more visible or tangible benefits" to encourage the islanders' adherence, as French Catholic priests were successfully doing in the Gambier Islands. He feared that without such relaxations in the missionaries' strict discipline, the Catholics might gain converts in Tahiti. Only a few years later, the French government would indeed use the refusal by Queen Pomare under the guidance of the British missionaries to permit Catholic priests into Tahiti as the pretext for annexing the island.

Fitzroy himself had official business with Tahiti's queen—to pursue promised but unpaid compensation of some three thousand dollars for the robbery of a British vessel, the *Truro*, and the murder of her master and mate a few years earlier by the inhabitants of the Paamotu (Tuamotu) Islands, which fell under the queen's rule. Pomare arrived from Eimeo for their meeting, according to FitzRoy, "sitting on the gunnel of a whaleboat, loosely dressed in a dark kind of gown, without anything upon her head, hands, or feet and without any kind of girdle or sash to confine her gown, which was fastened only at the throat . . . In her figure, her countenance or her manner there was nothing prepossessing, or at all calculated to command the respect of foreigners." Pomare was apologetic for failing to meet the payment deadline and rapidly agreed to pay in the form mostly of thirty-six tons of pearl oyster shells. Under pressure from FitzRoy she also agreed not to allow any Tahitian to enlist "in a foreign cause," thus depriving, as FitzRoy intended, Baron de Thierry of many of his men.

With good relations established, FitzRoy invited the queen to dine aboard the *Beagle*. Darwin thought her "an awkward large woman, without any beauty, gracefulness or dignity of manners.—She appears to have only one royal attribute, viz a perfect immovability of expression (and that generally rather a sulky one) under all circumstances." She seemed to unbend a little watching "sky rockets" and other fireworks and hearing "seamen's songs," which she noted were certainly not the hymns to which the missionaries restricted the Tahitian adults. She and her party did not leave until past midnight, and "all appeared well contented with their visit." FitzRoy, however, disappointed with the quality and amount of fireworks he had provided, wrote, "Let me repeat a piece of advice given to me, but which from inadvertence I neglected to follow, [to] 'take a large stock of fireworks'" when visiting "distant, especially half-civilized or savage nations."

When they left Tahiti on November 26, Darwin was speaking for himself and the whole crew when he wrote to Henslow, "Tahiti is a most charming spot.—Everything, which former navigators have written is true; 'A new Cythera has risen from the ocean.' Delicious scenery, climate, manners of the people, are all in harmony."

CHAPTER SEVENTEEN

"Not a Pleasant Place"

The *Beagle* nosed its way into the Bay of Islands on the early afternoon of December 21, 1835, and anchored off Pahia, which, when he explored it a little later with FitzRoy, Darwin thought "hardly deserves the title of a village." The *Beagle* would stay in the bay—its only stopping place in New Zealand—for nine days. Even Darwin, ever eager to explore, would only venture within a radius of fifteen miles or so of Pahia. Since Cook had charted New Zealand—skirted and named by Abel Tasman in 1642—and discovered it was two main islands, not one as Tasman thought, the islands had remained under the rule of their local Maori chiefs. However, a motley, ill-disciplined, and shifting collection of whalers—three whale ships were in the bay as the *Beagle* anchored—adventurers of the dubious type FitzRoy believed Baron de Thierry to be and escaped convicts from New South Wales, had settled around the Bay of Islands. Missionaries from the British Church Missionary Society had joined them at Christmas in 1814.

FitzRoy's and Darwin's old shipmate, the artist Augustus Earle, had spent several months in New Zealand in 1827, living among the Maoris with a Maori woman, to the disapproval of the missionaries. In his best-selling book on his travels, published in 1832 and by now in the *Beagle* library, he lambasted the prudery and sanctimonious condescension of the missionaries. In 1833, the British government had appointed a Resident, James Busby, to look after British interests, though he had little power to intervene in local affairs. Busby lived in a newly built residency over which the Union Jack flew at nearby Waitangi. In his well-cultivated garden, vines grew—forerunners of New Zealand's now major wine industry.

During his walk around Pahia, Darwin enjoyed seeing "English flowers . . . roses of several kinds, honeysuckle, jessamine, stocks and whole hedges of sweet briar" in the gardens in front of the missionaries' whitewashed cottages. Next day he set out on foot to explore but found the country impenetrable because of the thick growth of ferns and coarse bush. Walking along the beach instead, he found frequent saltwater creeks and deep brooks blocking his way. He did, however, notice how many hilltops had been turned into *pas*—fortified earthworks from which local Maoris had defended themselves against intruders. The coming of gunpowder weapons had led to most being abandoned in favor of wood and earth stockades built on cleared and leveled ground with zigzagging wooden palisades designed to facilitate enfilading fire. From what he saw and heard, Darwin thought "in no part of the world a more war-like race of inhabitants could be found than the New Zealanders [Maoris] . . . If a New Zealander is struck, although but in joke, the blow must be returned; of this I saw an instance with one of our officers."

With FitzRoy he visited Kororareka (also called Russell) to call upon some of the missionaries. They found this "little village . . . the very stronghold of vice." Its English residents were "of the most worthless character . . . There are many spirit shops and the whole population is addicted to drunkenness." The missionaries said the "only protection . . . on which they rely is from the native Chiefs against Englishmen!" Darwin was not much more impressed by the Maori residents. "Looking at the New Zealander, one naturally compares him with the Tahitian . . . The comparison however tells heavily against the New Zealander. He may perhaps be superior in energy, but in every other respect his character is of a much lower order. One glance at their respective expressions, brings conviction . . . that one is a savage, the other a civilized man . . . No doubt the extraordinary manner in which tattooing is here practiced gives a disagreeable expression to their countenances. The complicated but symmetrical figures covering the whole face, puzzle and mislead an unaccustomed eye."

Darwin suggested, perhaps drawing on his Edinburgh medical studies, "It is probable that the deep incisions, by destroying the play of the superficial muscles would give an air of rigid inflexibility," before continuing, "But besides all this, there is a twinkling in the eye which cannot indicate anything but cunning and ferocity . . . Both their persons

Maori men, sketched by R. FitzRoy

and houses are filthily dirty and offensive; the idea of washing either their persons or clothes never seems to have entered their heads. I saw a chief . . . wearing [a] shirt black and matted with filth; when asked how it came to be so dirty, he replied with surprise, 'Do not you see it is an old one? . . .' the common dress is one or two large blankets generally black with dirt . . . thrown over their shoulders in a very inconvenient and awkward fashion . . . If the state in which the Fuegians live should be fixed as zero in the scale of governments, I am afraid that the New Zealand would rank but a few degrees higher, while Tahiti, even as when first discovered, would occupy a respectable position."

The British Resident James Busby helped Darwin accept a missionary invitation to visit their agricultural settlement at Waimate fifteen miles away. The journey convinced Darwin that New Zealand had "one great natural advantage . . . the inhabitants can never perish from famine. The whole country abounds with fern, and the roots of this, if not very palatable, yet contain much nutriment;—A native can always subsist on them and on the shellfish, which is very abundant." He took part in the ceremony of "nose rubbing," which, in his view, should "more properly [be] called pressing noses. The women . . . began uttering something in a most dolorous plaintive voice, they then squatted themselves down and held up their faces; [we] standing over them placed the bridges of [our] own noses at right angles to theirs, and commenced pressing; this lasted rather longer than a cordial shake of the hand would with us; as we vary the force of the grasp of the hand in shaking, so do they in pressing. During the process they utter comfortable little grunts, very much . . . as two pigs do when rubbing against each other."

As they approached Waimate, "the sudden appearance of an English farmhouse and its well dressed fields, placed there as if by an enchanter's wand, was exceedingly pleasing . . . there were large gardens, with every fruit and vegetable which England produces and many belonging to a warmer clime . . . asparagus, kidney beans, cucumbers, rhubarb, apples and pears, figs, peaches, apricots, grapes, olives, gooseberries, currants, hops, gorse for fences and English oaks! And many different kinds of flowers." There was "that happy mixture of pigs and poultry which may be seen so comfortably lying together in every English farm-yard . . . where the water of a little rill has been dammed up . . . a large and substantial watermill had been erected." The whole scene reminded Darwin of England, but he thought neither this nor "the triumphant feeling at seeing what Englishmen could effect" on land cleared only five years previously was as important as "the object for which this labour had been bestowed,—the moral effect on the native inhabitant of New Zealand."

The missionary system differed from that in Tahiti. In the latter, "much more attention is there paid to religious instruction . . . here more to the arts of civilization." Darwin thought this well suited to local circumstances. In the evening he was pleased to see young Maoris from the settlement "very merry and good humoured . . . playing cricket; when I thought of the austerity of which the missionaries have been accused, I was amused at seeing one of their sons taking an active part in the game." The young Maori women who acted as household servants had a "clean tidy and healthy appearance, like that of dairymaids in England . . . a wonderful contrast with [those in] the filthy hovels in Kororareka." The missionaries' wives had tried to persuade the women not to be tattooed, but they had resisted, saying, "We really must just have a few lines on our lips; else when we grow old our lips will shrivel and we shall be so very ugly."

During an overnight stay at the missionaries' house, Darwin became thoroughly convinced of the benefits the missionaries were bringing "in improving the moral character" of those who not so long ago had been cannibals and teaching them "the arts of civilization. It is something to boast of, that Europeans may here, amongst men who, so lately were the most ferocious savages probably on the face of the earth, walk with as much safety as in England." He thought "it would be difficult to find a body of men better adapted [than the missionaries] for the high office which they fulfil" and was "indignant" about Earle's "unjust" criticisms—"I know without doubt [the missionaries] always treated him with far more civility, than his open licentiousness could have given reason to expect." While in New Zealand, both he and FitzRoy contributed to the building fund for Christ Church in Kororareka, New Zealand's oldest surviving church, in which the first service took place a few days after the *Beagle* sailed.[1]

The next morning, Christmas Eve, two missionaries took Darwin into the forest "to show me the famous Kauri pine. I measured one . . . and found it to be thirty-one feet in circumference . . . The trunks are . . . remarkable by their smoothness, cylindrical figure, absence of branches, and having nearly the same girth for a length from sixty even to ninety feet. The crown . . . where it is irregularly branched is small and out of proportion to the trunk; and the foliage is again diminutive as compared to the branches. The forest . . . was almost composed [completely] of the Kauri . . . the great ones from the parallelism of their sides stood up like gigantic columns . . . The timber of this tree is the most valuable product of the island."

As he returned to the *Beagle*, Darwin noticed great quantities of New Zealand hemp (flax) plants. Hemp was the second most valuable export after timber and used for ships' ropes and rigging and also for clothing. While he journeyed on, he pondered the ecology of New Zealand. He saw few birds in the woods and was surprised that so large a place as New Zealand with a good climate and varied habitats "should not possess one indigenous animal with the exception of a small rat." Already conscious of competition among species, with the best suited to the conditions

---

1. The missionaries' house is New Zealand's second oldest surviving European building. The first is the mission house at nearby Kerikeri.

prospering at the expense of others, he added, "It is moreover said that the introduction of the common Norway [rat] has entirely annihilated the New Zealand species . . . from the northern extremity of the island. In many places, I noticed several sorts of weeds, which like the rats I was forced to own as countrymen. A leek, however, which has overrun whole districts and will be very troublesome, was imported lately as a favour by a French vessel . . ."

Back on the *Beagle*, Christmas Day was passed, according to Covington, seemingly as homesick as everyone else, "pretty merrily, considering the place." FitzRoy, Darwin, and many of the crew attended the Christmas service in the mission church in Pahia, conducted partly in English, partly in the local language. FitzRoy noted, "The two entire services were mixed, and the whole extended to such a length that had even the most eloquent divine occupied the pulpit, his hearers could scarcely have helped feeling fatigued."

In his remaining time in the bay, Darwin accompanied Sulivan and Busby to the rocks and caves at Waiomio. On the way they witnessed funeral rites: "The daughter of the chief of this place, who yet followed heathen customs, had died five days before; the hovel in which she had expired was burnt to the ground; her body being enclosed between two small canoes, was placed upright in the ground and protected by an enclosure bearing wooden images of their gods, and the whole was painted bright red, so as to be conspicuous from afar. Her gown was fastened to the coffin, and her hair being cut off was cast at its foot. The relatives of the family had torn the flesh of their arms, bodies and faces, so as to be covered with clotted blood and the old women looked most filthy, disgusting objects."

Continuing, the party found the caves and limestone rocks at Waiomio were sacred to the Maoris and used as burial places and so did not enter. The local people, following custom, gave each of them a basket of roasted sweet potatoes for their journey back to the ship. Darwin noticed that among the women doing the cooking was "a man-slave" captured in warfare and thought "it must be humiliating to a man thus to be employed in what is only considered as woman's work."

Throughout his stay, Darwin collected insect and plant specimens and made geological notes. He recorded, "the soil is volcanic . . . we passed over slaggy lavas and craters could clearly be distinguished on several of the neighbouring hills" and also his belief that, as in South America, parts of New Zealand had been elevated from the sea.

FitzRoy, with only longitudinal measurements and reprovisioning to oversee, spent some time with Busby. They discussed the measures Busby was taking, with the help of local Maori chiefs, to thwart Baron de Thierry if he arrived, and the paucity of powers Busby had to intervene in local disputes.

In his published narrative, FitzRoy recommended, "New Zealand much requires assistance from the strong but humane arm of a powerful European government. Sensible treaties should be entered into by the head of an over-awing European force and maintained by the show, not physical action, of that force until the natives see the wonderful effect of a changed system." Although he did not say so, FitzRoy very clearly had Britain in mind as the European force since he also recommended the stationing of a naval frigate in the area and warned against the possibility of existing French incursions into the islands' affairs increasing.

He noticed that the Maoris recognized the seeming inevitability of western domination and "say frequently, 'the country is not for us; it is for the white men!' and they often remark upon their lessening numbers. Change of habits, European diseases, spirits, and the employment of many of their finest young men in whale-ships (an occupation which unhappily tends to their injury), combine to cause this diminution. Wearing more clothes (especially thick blankets) exposes them to sudden colds, which often end fatally." FitzRoy was interested in the physiognomy of the Maoris, some of whose heads he sketched, and in their tattoos, observing that "the lines upon their faces are not, however, arbitrary marks, invented or increased at the caprice of individuals . . . they are heraldic ornaments far more intelligible to the natives of New Zealand than our own armorial bearings are to many of us in these unchivalric days."

Like Darwin, FitzRoy recognized the Maori were a warlike race and commented on their war dance—the haka: "What exaggerated distortions of human features could be contrived more horrible than those they then display? What approach to demons could human beings make nearer than that which is made by the Zealanders when infuriating, maddening themselves for battle by their dance of death!"

FitzRoy examined a Maori war canoe: "Seventy feet in length, from three to four in width, and about three in depth . . . Her lower body formed out of a single tree—the New Zealand kauri . . . the upper works by planks of the same wood; the stem and the stern, raised and project-ing, like those of the galleys of old, were carved and hideously disfigured, rather than ornamented, by red, distorted faces with protruding tongues and glaring mother-of-pearl eyes. Much carving of an entirely different and rather tasteful design decorated the sides . . . From forty to eighty men can embark in such canoes. But their day is gone! In a few years scarcely a war canoe will be found . . ."

He also described the Maoris' thatched huts: "These roofs slope downwards, lengthways as well as sideways; so that the front of the hut is the highest part . . . Besides the door, through which a man cannot pass excepting upon his hands and knees, there is neither window nor aperture of any kind . . . The eaves of the roof project two or three feet beyond the front so likewise do the sidewalls." The cooking was done outside as in Polynesia by placing food wrapped in leaves on hot stones in a pit and covering it with earth. In front of the houses were often "large planks generally painted red" set in the ground, whose meaning he could not discover.

FitzRoy became as convinced as Darwin of the merits of the mission-aries' work. When he described to them his efforts to bring Christianity and "civilization" to the Fuegians, he found it "very gratifying . . . to mark [their] lively interest . . . in every detail connected with the Fuegians, and our attempt to establish Richard Matthews in Tierra del Fuego. Again and again they recurred to the subject, and asked for more information; they could not hear of my calling the attempt 'a failure.' 'It was the first step,' said they, 'and similar in its result to our first step in New Zealand. We failed at first; but, by God's blessing upon human exertions, we have at last succeeded far beyond our anticipations.'" He arranged with the missionaries that Matthews, who had remained on the *Beagle* following his removal from his intended mission in Woollya Cove, should join his brother, a well-established missionary at Waimate.[2]

As the *Beagle* departed from New Zealand on December 30, 1835, an unenthusiastic Darwin wrote, "I believe we were all glad to leave New Zealand; it is not a pleasant place; amongst the natives there is absent that charming simplicity which is found at Tahiti; and of the English the greater part are the very refuse of Society. Neither is the country itself attractive.—I look back but to one bright spot and that is Waimate with its Christian inhabitants."

---

2. Richard Matthews did not succeed as a missionary. After becoming involved in disputes with the local Maori he left Waimate and was later embroiled in a scandal about the misappropriation of funds at another mission station.

CHAPTER EIGHTEEN

"A Rising Infant"

Early in the morning of January 12, 1836, the *Beagle* passed the lighthouse designed by convict architect Francis Greenway at the entrance to the harbor of Australia's Port Jackson. As the *Beagle* sailed farther in, Darwin noticed the bare and horizontal strata of the sandstone cliffs and shoreline. "Large stone houses, two or three stories high, and windmills" began to appear, although the ship did not come to anchor in Sydney Cove beneath lowering skies and with lightning flashing until around 1:30 P.M. Almost immediately there was a disappointment for the whole crew—they had arrived earlier than anticipated, hence no letters from home. Darwin wrote to his sister Susan lamenting that this would probably also be the case when they reached Cape Town, so that by the time he reached England, "I shall not have received a letter dated within the last eighteen months . . . I feel much inclined to sit down and have a good cry."

Only forty-eight years after Captain Philip had led the First Fleet crammed with male and female convicts transported from Britain into Port Jackson, Darwin's first impressions after taking a walk ashore were favorable. He "returned full of admiration at the whole scene.—It is a most magnificent testimony to the power of the British nation: here, in a less promising country, scores of years have effected many times over more than centuries in South America.—My first feeling was to congratulate myself, that I was born an Englishman . . . Sydney has a population of twenty-three thousand and is . . . rapidly increasing; it must contain much wealth; it appears a man of business can hardly fail to make a large fortune . . . [A] convict . . . who is always driving about in his carriage, has an income so large that scarcely anybody ventures to guess at it."

Darwin quickly hired a guide and two horses for an expedition over the Blue Mountains to the pastoral center of Bathurst, 130 miles to the west. On the first leg of his journey, toward Paramatta, the great number of "ale-houses" even compared to England surprised him—at that time over half of the income of the colony's government came from alcohol taxes and public house licenses. Traveling on, he noticed "extreme uniformity" in the vegetation. There was much thin pasture and open woodland containing monotonous stands of eucalyptus trees with leaves "of a peculiar light green tint" and pale peeling bark. Darwin believed the trees' evergreen nature deprived the Australians of the joys of seeing spring burst into leaf—"one of the most glorious . . . spectacles in the world." Always eager to compare, he could not imagine "a more complete contrast in every respect" between what he called the "arid sterility" of this landscape and the "forest of Valdivia and Chiloe."

Sometimes he passed work gangs of convicts "dressed in yellow and grey clothes, and . . . working in irons" guarded by armed sentries. Convicts were still being transported from Britain—the last convict ship reached New South Wales in 1850 and the last to Australia as a whole in 1868.

Toward sunset he came across "a score of the Aboriginal blacks . . . each carrying in their accustomed manner a bundle of spears and other weapons.—By giving a leading young man a shilling, they were easily detained and threw their spears for my amusement.—They were all partly clothed and several could speak a little English; their countenances were good-humoured and pleasant and they appeared far from such utterly degraded beings as usually represented.—In their own arts they are admirable; a cap being fixed at thirty yards distance, they transfixed it with the spear delivered by the throwing stick, with the rapidity of an arrow from the bow of a practiced archer; in tracking animals and men they show wonderful sagacity and I heard many of their remarks, which manifested considerable acuteness.—They will not however cultivate the ground, or even take the trouble of keeping flocks of sheep which have been offered them; or build houses and remain stationary.—Nevertheless, they appear to me to stand some few degrees higher in civilisation, or more correctly a few lower in barbarism, than the Fuegians."

Like a modern tourist, Darwin bought himself a boomerang from them and another for Henslow. He realized the Aboriginal population was decreasing due to European encroachment, alcohol, and diseases such as measles—an irreversible process that he thought similar to what he had already observed in Tahiti and New Zealand. Darwin would not read the works of Thomas Malthus, who argued that in a world of limited resources only the strong would prosper, for another two years and, when he did, Malthus's views would have a major impact on his evolutionary thinking. Yet, intriguingly, Darwin's comments on the seemingly inevitable decline in Aboriginal numbers had Malthusian echoes: "as the difficulty of procuring food increases, so must their wandering habits; and hence the population, without any apparent deaths from famine, is repressed in a manner extremely sudden compared to what happens in civilized countries, where the father may add to his labour, without destroying his offspring."

Crossing the Nepean River by ferry, Darwin ascended the Blue Mountains. Reaching the plateau at around three thousand feet in the vicinity of Wentworth Falls and what is now Katumba, he was suddenly confronted by "the most stupendous cliffs I have ever seen," plunging "1500 ft beneath ones feet" and forming a great amphitheater in which the cliffs stretching away for miles resembled a "bold sea coast." He later took a walk along the base of the cliffs to view the waterfalls descending from them. The next morning he woke early and, walking to the cliff edge, saw in the valley "the thin blue haze," which gave the mountains their blue effect and hence their name.[1]

Darwin could not make up his mind how the large, deep valleys had been cut out from the sandstone. His original view that it was by surface water and weather erosion, rather than a later one that the cliffs

---

1. This part of the Blue Mountains is now a UNESCO World Heritage site and touristic, with a scenic railway and cable car, but hike along what is now called the Charles Darwin Walk and you will find fewer people and view the valley and cliffs above as they were in Darwin's time.

had been submerged and the erosion produced by currents and marine action, proved correct.

Riding on, Darwin reached one of the area's first sheep stations, where shearing of fifteen thousand sheep had just finished. Although the Scottish manager welcomed him, he did not find the accommodation overly comfortable and felt uneasy about being surrounded at night by "more than forty hardened, profligate convicts . . . quite impossible to reform" who worked on the farm. Unlike the slaves he had encountered in South America, Darwin thought they had "no just claim for compassion."

The next day he joined another supervisor——the only other free man on the station——on a kangaroo hunt, but they failed to find a single kangaroo or even a dingo, although they did finally capture a kangaroo rat hiding in a hollow tree trunk. This was the first marsupial Darwin had a chance to examine, having failed to come across any of the small marsupials in Chile. He noticed how Europeans and their hunting dogs had reduced the wildlife: "A few years since this country abounded with wild animals; now the Emu is banished to a long distance and the kangaroo is become scarce . . . it may be long before these animals are altogether exterminated, but their doom is fixed."[2]

During the hunt, Darwin encountered large flocks of "white cockatoos" (sulfur-crested cockatoos). Relaxing in the evening, he walked with the manager the short distance to the nearby Cox's River that the drought had dried into a series of pools. In one, several duckbilled platypuses were "diving and playing about the surface of the water" like "water rats." The manager shot one. Darwin wrote, "I consider it a great feat, to be in at the death of so wonderful an animal" and examined the corpse to find "it a most extraordinary animal; the stuffed specimens do not give at all a good idea of the . . . head and beak; the latter becoming hard and contracted," whereas the beak was in fact soft. Only first seen by them four decades previously, Europeans still did not realize the animal

---

2. Nearly two hundred years later, more than a dozen larger species of Australian animals have already become extinct, and more are under threat.

First known sketch of duck-billed platypus, 1799

laid eggs, refusing to believe those Aboriginal people who told them so. Nearly a further half century passed before their egg-laying was accepted.

That evening Darwin made a significant entry in his diary, writing that earlier, "I had been lying on a sunny bank and was reflecting on the strange character of the animals of this country as compared to the rest of the world. An unbeliever in everything beyond his own reason, might exclaim, 'Surely two distinct Creators must have been [at] work; their object however has been the same and certainly the end in each case is complete.'" Although he went on to muse that possible explanations might be that the creations were distinct in time in Europe and Australia, or that the Creator "rested in his labour," he was clearly again contemplating other reasons why totally different species lived in similar environments in widely varying locations and yet made similar adaptations to their environment, and how to explain or reconcile such thoughts with the concept of a single, unchanging Creation and a single Creator.

Riding on and crossing the watershed of the Blue Mountains, on one side of which the waters ran into the ocean and on the other into the interior, Darwin learned—presumably from his guide—the meanings of some terms used in New South Wales—"A 'squatter' is a freed or 'ticket of leave' man, who builds a hut with bark in unoccupied ground, buys or steals a few animals, sells spirits without a licence, buys stolen goods and so at last becomes rich and turns farmer: he is the horror of all his honest neighbours . . . The 'Bush Ranger' is an open villain, who lives by highway robbery and plunder; generally he is desperate and will sooner be killed than taken alive . . ."

Eventually Darwin and his guide reached Bathurst, the end point of their journey. Darwin had been "half roasted with the intense heat" and experienced "the Sirocco-like wind of Australia . . . While riding, I was not fully aware . . . how exceedingly high the temperature was.—Clouds of dust were travelling in every part, and the wind felt like that which has passed over a fire.—I afterwards heard the thermometer . . . stood at 119 degrees [F] [48°C] . . ."

Dusty Bathurst—the first town founded beyond the Blue Mountains— was suffering from drought and did not impress Darwin. During his outward journey, he had observed much burnt countryside strewn with blackened tree stumps. On the first evening of his return he passed fires that, by morning, were "raging," and in his diary for January 23, noted laconically, "We passed through large tracts of country in flames; volumes of smoke sweeping across the road."[3]

On his way back to Sydney, Darwin visited Captain Philip Parker King, overall commander of the first *Beagle* voyage, at his Australian home. Darwin had last seen him on December 27, 1831, waving farewell from the Plymouth mole to the *Beagle* and his son, midshipman Philip Gidley King. Young King recalled how after disembarking from the *Beagle* he met his parents again. "I thought [my father] at first much aged but this impression wore off after a day or two. My mother had waited for me just outside the house. I had not seen her since 25 May 1826. My feelings at finding myself with a mother after such a separation were as strange as the situation was novel." His mother's low-key diary entry conceals what must have been massive joy: "My dear Philip Gidley after an absence of ten years is restored to me in health and strength, and apparently very amiable." His father had decided that young King should leave the navy

---

3. What Darwin witnessed is a reminder that even then bushfires were common in the area, though not on the scale of the 2019–20 fires.

and remain in Australia. Although both FitzRoy, and indeed Darwin, had nothing but good to say of him, his father had concluded for no known reason, "He has no chance of succeeding in the service and here he will do very well."

Sometime during Darwin's expedition to the Blue Mountains, Fitz-Roy, who seems again to have been in poor health, whether mental or physical is unclear, quarreled with his old commanding officer. According to young King, there was "some unpleasantness about the surveys and charts." Beyond this there is little surviving evidence of the quarrel other than that FitzRoy appeared to be contrite, even lobbying Philip Parker King's mother on his behalf and writing to young midshipman King, "Could your good father be induced to forgive my late offences?" and inviting the elder King to visit the *Beagle*. No more is known other than a note from midshipman King: "The misunderstanding was happily made up." Probably FitzRoy's "hot coffee" temperament had got the better of him over some relatively trivial issue. After the *Beagle* departed, Philip Parker King wrote to Beaufort about FitzRoy, "I regret to say he has suffered very much and is yet suffering from ill health—he has had a very severe shake to his constitution which a little <u>rest</u> in England will I hope restore for he is an excellent fellow and will I am satisfied yet be a shining ornament to our service."

The Kings took Darwin to a convivial lunch party of twenty people at the house of Hannibal McArthur, a member of one of the colony's leading families. Darwin recorded in his diary, "It sounded strange to my ears to hear very nice looking young ladies exclaim, 'Oh we are Australian, and know nothing about England.'" Some of these young ladies were McArthur's daughters, two of whom would in due course marry Darwin's shipmates, Lieutenant Wickham and young Philip Gidley King.

Returned to Sydney, Darwin wrote the only two letters he dispatched from there—one to Henslow and the other to his sister Susan. The latter contained pleas to her to explain to his father that he needed yet more money. Sydney was "a most villainously dear place; and I stood in need of many articles." Consequently he had already drawn a bill for one hundred pounds. Half had been to pay FitzRoy for his mess bills—"The remaining fifty is for current expenses; or rather I grieve to say it was for such expenses; for all is nearly gone." He continued, "Tell my father I really am afraid I shall be obliged to draw a small bill at Hobart"——the *Beagle*'s next port of call. He confessed that "I have been extravagant and bought two watercolour sketches, one of the Santa Cruz river and another of Tierra del Fuego; 3 guineas each, from Martens who is established as an artist at this place. I would not have bought them if I could have guessed how expensive my ride to Bathurst turned out."

Darwin told Susan he was homesick but that the *Beagle* would be unlikely to return until at least September, "but, thank God the Captain is as home sick as I am, and I trust he will rather grow worse than better. He is busy in getting his account of the voyage in a forward state for publication. From those parts, which I have seen of it, I think it will be well written, but to my taste is rather deficient in energy or vividness of description. I have been for the last 12 months on very cordial terms with him.——He is an extraordinary, but noble character." Perhaps thinking of FitzRoy's quarrel with Philip Parker King, Darwin continued, "Unfortunately however [he is] affected with strong peculiarities of temper. Of this, no man is more aware than himself, as he shows by his attempts to conquer them. I often doubt what will be his end, under many circumstances I am sure, it would be a brilliant one, under others I fear a very unhappy one."

Although he wrote to Susan, "This is really a wonderful colony; ancient Rome, in her imperial grandeur, would not have been ashamed of such an offspring," Darwin's views of New South Wales were in fact becoming more mixed, causing him to abandon some "Utopian ideas" he had entertained about it. In his diary he identified an "open profligacy" in society, even among those who should have known better. There was "much jealousy between the children of the rich emancipist and the free settlers; the former being pleased to consider honest men as interlopers." He went on: "The whole population poor and rich are bent on acquiring wealth; the subject of wool and sheep-grazing amongst the higher orders is of preponderant interest. The very low ebb of literature is strongly marked by the emptiness of the booksellers shops . . .

"There are some very serious drawbacks to the comforts of families, the chief of these is perhaps being surrounded by convict servants. How disgusting to be waited on by a man, who the day before was by your representation flogged for some trifling misdemeanour? The female servants are of course much worse; hence children acquire, the use of the vilest expressions.

"On the other hand, the capital of a person will without trouble produce him treble interest as compared to England; and with care he is sure to grow rich. The luxuries of life are in abundance, and very little dearer, as most articles of food are cheaper, than in England. The climate is splendid and most healthy, but to my mind its charms are lost by the uninviting aspect of the country . . . I am not aware that the tone of Society has yet assumed any peculiar character; but with such habits and without intellectual pursuits, it can hardly fail to deteriorate and become like that of the people of the United States." He had never visited America, and his view of the deficiencies of American society, which he later crossed out in his diary, may well have come from his abhorrence of slavery and the society surrounding it.

Darwin did not believe agriculture could ever succeed widely in New South Wales given the relatively poor soil and paucity of water. Sydney's future would "depend upon being the centre of commerce for the Southern Hemisphere; and perhaps on her future manufactories . . . possessing coal, she always has the moving power at hand—I formerly imagined that Australia would rise into as grand and powerful a country as N. America, now it appears to me . . . that such future power and grandeur is very problematical." Transportation did not really reform convicts, but "as a means of making men outwardly honest,—of converting vagabonds most useless in one hemisphere into active citizens of another . . . it has succeeded to a degree perhaps unparalleled in history."

FitzRoy's views of the colony were remarkably similar to Darwin's, even down to the lack of books in the bookshops.

Six days' passage from Sydney, the latter part, according to Darwin, "very cold and squally," brought the Beagle to Tasmania. As the ship entered the Derwent River on which Hobart stood, Syms Covington, despite the foul weather, sketched the lighthouse at the harbor entrance. The country was much greener than the parched landscape around Sydney. At the base of the mountains and along the bay "the bright yellow fields of corn and dark green ones of potato crops appear very luxuriant," Darwin noted.

The next morning Darwin went for a walk. To him Hobart was only a town compared to the more citylike Sydney. However, "the streets are fine and broad . . . the shops appeared good. The town stands at the base of Mount Wellington, a mountain [of] 3,100 feet, but of very little picturesque beauty; from this it receives a good supply of water, a thing which is much wanted in Sydney." Impressive warehouses lined the shore. A small fort, insignificant compared to those Darwin had seen in South America, guarded the settlement.

In his diary Darwin described the population: "The inhabitants [of Hobart] for this year are 13,826: in the whole of Tasmania, 36,505.—The Aboriginal blacks are all removed and kept (in reality as prisoners) in a Promontory, the neck of which is guarded. I believe it was not possible to avoid this cruel step; although without doubt, the misconduct of the Whites first led to the Necessity." The last sentence concealed a massive tragedy for which Darwin was entirely correct to blame the misconduct of the British, who had first come to the island to establish a penal colony in 1803 and afterward also encouraged free settlers to emigrate there. Their continuing seizure of Aboriginal land—a million acres being allocated to free settlers between 1824 and 1831—led to justified complaints by the Tasmanian Aboriginal people, the Palawa, whose numbers are estimated to have been around five thousand in 1803, about loss of hunting grounds and the killing of one of their main food sources, the kangaroo.

With repression continuing, the Aboriginals mounted a guerrilla campaign in the 1820s and early 1830s against the intruders in what became known as the "Black War." When the settlers were inevitably victorious, as Darwin wrote, they first corralled the Aboriginal people into a peninsula. However, by the time of Darwin's visit, the few—less than one hundred—who had survived the fighting and European disease were isolated on Flinders Island in the Bass Strait. There, and after a move back to the mainland subsequent to the *Beagle*'s visit, numbers declined further due to harsh conditions and disease.[4]

With this stain on British history out of sight, Darwin and FitzRoy shared similar views of Tasmania. Darwin wrote to his sister Catherine, "All on board like this place better than Sydney—the uncultivated parts here have the same aspect as there; but from the climate being damper, the gardens, full of luxuriant vegetables and fine corn fields, delightfully resemble England . . . it is a most admirable place of emigration. With care and a very small capital [a man] is sure soon to gain a competence, and may, if he likes, die wealthy . . . There is a better class of society [than in Sydney]. Here, there are no convicts driving in their carriages, and revelling in wealth . . ." He had attended "an excellent concert of first rate Italian Music" and heard of a costume ball just before his arrival attended by more than one hundred people.

In his narrative, FitzRoy emphasized the "less profligate and more reclaimable class" of convict than those in Sydney and the "excellent local government," which "restrained the licentious, and encouraged the moral to a far greater extent" than in Sydney. Corn of "excellent quality" ripened well. Altogether, "natural advantages are greater and likely to increase as the country is cleared and inhabited because rain is now almost too plentiful."

Darwin's activities in Tasmania, as in so many places, focused more on geology than zoology, but even his geological work was limited because of what he called the "flying visits" nature of his Antipodean landings, designed mainly for FitzRoy to make "chronometrical measurements," compared to his much more detailed surveying work in South America. He found evidence of the rise and fall of the land, with layers of seashells elevated above sea level. In a small quarry near Hobart he discovered fossil impressions of leaves and plants no longer growing in Tasmania.

---

4. When in 1876 a woman named Truganini died, she was widely believed to the last survivor of pure Tasmanian Aboriginal stock and totemic of their extermination. However, descendants of Aboriginal women's relationships, often forced, with sealers and others have preserved much of their culture.

He was also the first to identify evidence of an extinct volcano. He made two attempts to climb Mount Wellington, failing the first time because of the density of the vegetation, but on the second when he hired a guide succeeding after "five and a half hours of hard climbing." The summit was broad and flat and "composed of huge angular masses of naked greenstone [dolerite]." As he climbed he noticed large tree ferns over twenty-five feet high, with trunks six feet in circumference, whose foliage resembled "most elegant parasols," creating "a shade gloomy like that of the first hour of night."

Bad weather delayed the departure of the *Beagle*, "so full of home-sick heroes," until February 17, 1836. Just before they sailed, Darwin reported proudly to his sister Catherine, for onward transmission to his father, that he had not needed to draw money in Hobart as he had prophesied he might when writing from Sydney. He was not looking forward to the voyage across the Great Australian Bight to King George's Sound on the western tip of the Australian mainland: "1800 miles of most stormy sea.—Heaven protect and fortify my poor stomach."

On March 6, Darwin recorded the partial fulfillment of his forebodings—"came to an anchor in . . . King George's Sound. Our passage has been a tolerable one; and what is surprising, we had not a single encounter with a gale of wind.—Yet to me, from the long Westerly swell, the time has passed with no little misery."

Again, Darwin and FitzRoy had similar views on King George's Sound and the embryonic city of Albany only settled by Britons in 1827 and at the time of the *Beagle*'s visit consisting of "30 to 40 small white-washed cottages." Darwin wrote, "We stayed there eight days and I do not remember since leaving England having passed a more dull, uninteresting time." FitzRoy's reaction was, "Had inclination been our guide instead of duty I certainly should have felt much disposed to 'put the helm up' and make all sail away from such an uninviting place."

The only subject of any great interest to either man was the local Aboriginal people. Darwin described them as being "good-natured and good-humoured . . . Moreover they are quite willing to work and to make themselves very useful; in this respect they are very different from those in the other Australian colonies . . . they are very remarkable by the extreme slightness of their limbs, especially their legs; yet, without, as it would appear, muscles to move their legs, they will carry a burthen for a longer time than most white men.—Their faces are very ugly, the beard is curly and not at all deficient, the skin of the whole body is very hairy and their persons most abominably filthy. Although true Savages, it is impossible not to feel an inclination to like such quiet good-natured men." FitzRoy, phrenologist as he was, sketched one of their heads and recorded they had "features smaller and less marked than are usual among savages; but their foreheads are higher and more full." He too noticed the "lathy thinness of their persons . . . totally destitute of fat" and also that some had pieces of bone stuck through the cartilage of their noses.

A tribe of Aboriginal people "from a distance"—known as the white cockatoo men—were visiting the local Aboriginal people and, according to FitzRoy, "as the residents wished to conciliate them, a 'corobbery' [a celebratory gathering] was proposed and Mr. Darwin ensured the compliance of all the savages by providing an immense mess of boiled rice, with sugar, for their entertainment." After painting spots and lines on their dark bodies with white pigment and with fires blazing in the darkness, Darwin described how the white cockatoo men and the local tribesmen formed separate lines, "the whole set running either sideways or in Indian file into an open space and stamping the ground as they marched all altogether and with great force.—Their heavy footsteps . . . accompanied by a kind of grunt, and by beating their clubs and weapons, and various other gesticulations, such as extending their arms and wriggling their bodies. It was a most rude barbarous scene, and to our ideas without any sort of meaning . . . Perhaps these dances originally represented . . . scenes such as wars and victories; . . . One [was] called the Emu dance in which each man extended his arm in a bent manner so as to imitate the movement of the neck of one of those birds. In another . . . one man took off all the motions of a Kangaroo grazing in the woods, whilst a second crawled up and pretended to spear him . . . the group of nearly naked figures viewed by the light of the blazing fires, all moving in hideous harmony, formed a perfect representation of a festival amongst the lowest barbarians.——In T. del Fuego we have beheld many curious scenes in savage life, but I think never one where the natives were in such high spirits and so perfectly at their ease."

To FitzRoy some of the dancing resembled "persons working in a treadmill; but their imitation of snakes, and kangaroos in a kind of hunting dance, was exceedingly good and interesting . . ." While watching, "I could not but think of our imprudence in putting ourselves so completely into their power: about thirty unarmed men being intermixed with a hundred armed natives." Darwin recalled that once the dancing was over, "the whole party formed a great circle on the ground and the boiled rice and sugar was distributed to the delight of all."

As the *Beagle* left Australia on March 14, 1836, Darwin addressed a portentous and somewhat jaundiced goodbye to the continent: "Farewell Australia, you are a rising infant and doubtless some day will reign a great princess in the South; but you are too great and ambitious for affection, yet not great enough for respect; I leave your shores without sorrow or regret."

CHAPTER NINETEEN

"Myriads of Tiny Architects"

The *Beagle*'s next destination was the Keeling (Cocos) Islands in the Indian Ocean, some 1,250 nautical miles off the western Australian coast. There FitzRoy's orders from Beaufort were to take chronometer readings but also, if time permitted, to study and chart the reefs—reefs and their formation were at the time an often debated but imperfectly understood topic.

After a seventeen-day passage, its latter part hindered by gales, Fitz-Roy described their approach: "A long but broken line of cocoa-palm trees, and a heavy surf breaking upon a low white beach, nowhere rising many feet above the foaming water, was all we could discern till within five miles of the larger Keeling (there are two distinct groups), and then we made out a number of low islets, nowhere more than thirty feet above the sea, covered with palm trees, and encircling a large shallow lagoon . . ." Passing "cautiously between patches of coral rock clearly visible to an eye at the mast-head," they anchored.

Captain William Keeling, in the East India Company's service, discovered the lonely, uninhabited islands in 1608–9. Little or no notice was taken of them from then until 1823, when Briton Alexander Hare established himself and a small party of Malays upon the Southern Keeling Island, which he thought favorable for commerce and for maintaining his seraglio of Malay women. Two or three years later, an English merchant captain named Ross settled on one of the islands, together with his ship's mate named Liesk and their families. Hare quickly departed after the Malays, whom he treated as slaves, deserted to Ross. Now, in FitzRoy's opinion "still slaves and only less ill-used" than previously, the Malays worked for Ross harvesting coconuts, being paid a pittance and unable to leave. Ross was currently absent on a visit to Singapore, recently founded by Stamford Raffles, to trade coconut oil and nuts for provisions, leaving

Liesk in charge. With no clergyman having visited the island or likely
to do so, Liesk prevailed upon FitzRoy, as ship's captain, despite "some
scruples" to baptize his and his English wife's six children.

Syms Covington described the islands in his diary as "complete forests
of cocoa nut trees . . . Can wade from one island to another when the
tide is low . . . The water . . . always being clear, the beautiful branches
of Coral can be seen from the ship's side, the fish constantly passing and
repassing amongst the Coral has a most beautiful effect . . . immense
numbers of small fish of different species of the most brilliant colours I
ever saw or fancy could paint . . . On Sunday the 3rd April was caught
a shark 8 feet long which put a stop to our bathing which before was at
every evening by moonlight."

FitzRoy quickly fabricated a device for measuring the movement and
height of the tides and made charts of the islands so accurate they became
the standard reference for more than one hundred years.

Darwin's longing for home perhaps showed in his comparison of the
colonies of gannets, frigate birds, terns, and noddies on the Keeling Islands
with "a rookery in the fresh budding woods of England! The gannets,
sitting on their rude nests look at an intruder with a stupid yet angry
air. The noddies, as their names expresses, are silly little creatures . . ."

The giant clams fascinated Covington. They possessed "one of the
largest shells in the world. Would take a very strong man to lift one with
the animal in." He may have exaggerated when he estimated the biggest
at "nine feet long." Darwin and FitzRoy spent a long time examining "the
gigantic shells of the clams . . . into which," according to Darwin, "if a
man were to put his hand, he would not, as long as the animal lived, be
able to withdraw it."

Darwin and FitzRoy referred in similar terms to the "monstrous"
coconut or robber crabs.[1] Darwin described how they inhabited "the low
strips of dry coral land; they live entirely on the fruit of the cocoa nut
tree . . . tearing fibre by fibre with their strong forceps the husk of the

---

1. Coconut crabs can grow to three feet from leg tip to leg tip, and the largest of them
can exert a force of some 3,300 Newtons (740-pound force) with their claws.

Coconut (robber) crab

nut. This process they always perform at the extremity, where [the nuts']
three eyes are situated. By constant hammering the shell in that soft part
is broken and then by the aid of their narrow posterior pincers the food
is extracted. I think this as curious a piece of adaptation and instinct as
I ever heard of. The crabs . . . live in burrows which frequently lie at
the foot of the trees. Within the cavity they collect a pile, sometimes as
much as a large bagful of the packed fibre and husks and on this they rest.
Their flesh is very good food; in the tail of a large one there is a lump of
fat, which when melted gives a bottle full of oil."

Darwin pondered how the plants and trees on the islands, which had
originally only been coral, got there. When he was writing up his diary
for publication in 1837, he told Henslow, "I shall want to know number
[of] species of plants at Galapagos and Keeling, and at the latter whether
seeds could probably endure floating on salt water." He would later conduct experiments suspending seeds in brine to see whether they floated
and subsequently testing whether they germinated.

Darwin's greatest interest, however, was in the geology of the coral
reefs. While on the west coast of South America, he had debated whether
and where subsidence might have taken place to compensate for the elevation of the land in the Andes and elsewhere. Even though he had only
had limited opportunity to examine coral reefs while in the Pacific—
the Indian Ocean Keeling Islands were the only occasion he set foot on

an atoll—before reaching them he already believed that the subsidence might occur on the ocean floor and that might be key to how coral reefs formed. Contrary to Lyell's view that reefs rose from the crater lips of submerged volcanoes, he postulated that they might originate as shore or fringing reefs surrounding volcanic peaks. As the peaks slowly subsided, the coral, which he believed could only live relatively close to the surface, had to grow higher to survive. When the mountain disappeared fully, the coral reef surrounding it formed the lagoon and atoll as a marker of the previous presence of land.

Exploring the Keeling Islands, Darwin was delighted to find confirmatory evidence of his theory, which he would use in all his future publications on the topic. He found that any coral always above the waterline died and identified various types of coral polyps just beneath the water's surface, some four to eight feet across. He also described the stinging effect of coral. When he touched his face with some branching coral, the pain was "instantaneous." The sensation continued strong for a few minutes "but was perceptible half an hour afterwards." On touching the skin of his arm, "red spots" appeared that looked as if they would produce "watery pustules."

Going out some two thousand feet beyond the main reef with FitzRoy in a boat, Darwin watched the captain drop a lead line to 1,200 fathoms—7,200 feet, an immense distance for the time—without reaching the bottom. The two men found it difficult to take soundings nearer the reef to determine how deep the living coral polyps descended because of the movement of the breakers. In FitzRoy's words, "Judging however, from impressions made upon a large lead, the end of which was widened, and covered with tallow hardened with lime, and from such small fragments as we could raise I concluded that the coral was not alive at a depth exceeding 7 fathoms [42 feet] below low water." Darwin concluded, "The submarine slope of this coral formation is steeper than that of any volcanic cone," confirming his view that "we must look at a lagoon island as a monument raised by myriads of tiny architects [the corals] to mark the spot where a former land lies buried in the depths of the ocean." Onboard the *Beagle* Darwin sketched cross-sections to show what he thought occurred as the peaks subsided and the living coral rose.

The formation and development of coral as well as of coral islands remained of great interest to Darwin. Just over a year later, even before he envisaged a tree of life as a metaphor for his developing evolutionary theory, he saw the branching of coral from its base as such an analogy. As the base of the coral died, new life grew above and branched out in all directions with no clear pattern. Similarly, evolutionary developments, branching out randomly, could not be traced back to their now disappeared source, meaning the origin of life was cloaked in antiquity. Examination of the coral polyps also seems to have helped him to a realization that there was no clear boundary between plants and animals.

During their stay on the Keeling Islands the *Beagle*'s crew brought on board for food some of the fat coconut-fed pigs bred there, together with coconuts, poultry, and some turtles. Darwin found the method of catching the latter curious—"the water is so clear and shallow that although at first the turtle dives away with much rapidity, yet a canoe or a boat under sail will after no very long chase overtake it; a man standing ready in the bows at this moment dashes through the water [onto] its back. Then clinging with both hands by the shell of the neck, he is carried away till the turtle becomes exhausted and is secured." FitzRoy added that the turtle was secured by being "turned on its back."

With Darwin delighted with a great deal of newly acquired and valuable information, and with its new provisions, on April 12, 1836, the *Beagle* set sail for Mauritius. The island was the site of one of the most notorious and earliest extinctions by arriving Europeans of an indigenous, endemic creature—the dodo, extinct by the end of the seventeenth century. Mauritius had been a British possession since 1810, when British forces seized it from the French during the Napoleonic Wars, retaining it after the peace as a valuable strategic staging post on the way to India.[2]

Walking around the "clean and regular streets" of Port Louis, the capital, Darwin noticed that despite the time in British hands "the general

2. Mauritius was named in honor of Prince Maurice van Nassau, Dutch ruler during a previous period of Dutch occupation.

character . . . is quite French. Englishmen speak to their servants in French, and the shops are all French; indeed I should think that Calais or Boulogne were much more Anglified." In contrast to Sydney there were many and well-stocked bookstores. There were several theaters where operas as well as "common plays" were performed. It bespoke "our approach to the old world of civilisation, for in truth both Australia and America may be considered as New Worlds." Though the British had brought greater prosperity to the island, massively increasing sugar production and "macadamising" (asphalting) the roads, benefiting the French residents, "yet the English government is far from popular."[3]

Since their arrival, the British had brought both convicts and indentured laborers from India to the island. Darwin "had no idea that the inhabitants of India were such noble looking men . . . many of the older men had large moustachios and beards of a snow white color; this together with the fire of their expressions, gave to them an aspect quite imposing." Some convicts had not been transported for "moral faults" but "for not obeying, from superstitious motives, the English government and laws." Darwin saw one such man "of high cast [sic] who had been banished because he would not bear witness against his neighbour . . . [he] was also remarkable as being a confirmed opium eater, of which fact his emaciated body and strange drowsy expression bore witness." Because of "their outward conduct, their cleanliness, and faithful observance of their strange religious enactments, it was impossible to look at these men with the same eyes as at our wretched convicts in New South Wales."

Darwin climbed the 2,600-foot-high La Pouce mountain—"so called from a thumb like projection"—working out that the island was volcanic and had risen over the years, the sea having formerly lapped around the bottom of the mountains and, on its receding, stranding on land chunks of coral and seashell beds.

Darwin and Stokes stayed with the colony's surveyor general, Captain John Lloyd, for two nights at his estate outside Port Louis. "Capt. Lloyd

---

3. Even today, though English is the official language, most Mauritians still speak a French Creole.

took us to the Riviere Noire which is several miles to the southward, in order that I might examine some rocks of elevated coral," Darwin recounted. "We passed through pleasant gardens and fine fields of sugar cane growing amidst huge blocks of lava." Lloyd had the only elephant on the island, on which Darwin rode part of the way. He found its "perfectly noiseless step" surprising but felt "the motion must be fatiguing for a long journey."

From Mauritius Darwin posted letters to his cousin William Darwin Fox and his sister Caroline. In both he made clear his desire to be home. He told Fox, "I thank my good stars I was not born a sailor—I will take good care no one shall ever persuade me again to volunteer as 'Philosopher' (my accustomed title) even to a line of Battle Ship.—Not but what I am very glad I have come on the expedition; but only that I am still gladder it is drawing to a close." He now understood why sailors celebrated arriving home by "light[ing] their tobacco pipes with Pound notes." To Caroline he wrote, "There is no country which has now any attractions for us, without it is seen right astern and the more distant and indistinct the better. We are all utterly homesick." Formerly he would have traded a year of his ordinary life for an hour in the tropics, "but now one glimpse of my dear home, would be better than the united kingdoms of all the glorious tropics."

He wrote to both Fox and Caroline about his enthusiasm for geology, telling his sister, "I am in high spirits about my geology.—and even aspire to the hope that, my observations will be considered of some utility by real geologists." He told both that to pursue his work he would probably need to move to Cambridge or London. He described to Caroline how, while at sea, he was starting to work up his notes: "I am just now beginning to discover the difficulty of expressing one's ideas on paper. As long as it consists solely of description it is pretty easy; but where reasoning comes into play, to make a proper connection, a clearness and a moderate fluency, is to me . . . a difficulty of which I had no idea."

Darwin related to Caroline that FitzRoy was "daily becoming a happier man" and still busying himself with writing up the voyage—"I sometimes fear his 'book' will be rather diffuse, but in most other respects it certainly will be good; his style is very simple and excellent." What's more, he had asked Darwin if he could have access to his diary to use in his book: "Of course I have said I am perfectly willing, if he wants materials; or thinks the chit-chat details of my journal are any ways worth publishing. He has read over the part, I have on board and likes it."

FitzRoy's reading of Darwin's diary may explain why, at this stage of the voyage, they recorded similar reactions to what they saw. If, as some experts believe, a written comment—"Good, but first point not quite clear"—against Darwin's theory on coral reefs is in FitzRoy's handwriting, it underlines not only FitzRoy's continuing interest in natural history but the closeness with which the two men were working and the latter's openness at this stage to thoughts that geological formations at least might be subject to considerable change since Creation.

As the *Beagle* departed from Port Louis on May 9 for the Cape of Good Hope, Darwin summed up his visit to Mauritius—"Since leaving England I have not spent so idle and dissipated a time. I dined out almost every day in the week: all would have been very delightful, if it had been possible to have banished the remembrance of England. Pleasant as the society appeared to us, it was manifest even during our short visit that no small portion of jealousy, envy and hatred was common here, as in most other small societies.—Alas, there does not exist a terrestrial paradise where such feelings have not found an entrance!"

After the Dutch under Jan van Reebeck established the Cape Colony in 1652, it quickly became an important entrepot. Darwin called it "a great inn, on the great highway to the east." Britain occupied the Cape Colony during the Napoleonic Wars, and the subsequent peace treaty confirmed their possession. Britain had abolished slavery in the Cape, as elsewhere, in 1833. Two years later some Dutch (Boer) settlers, eager both for new land and to put themselves beyond British reach, had begun their great trek north across the Orange River. Darwin quickly discovered that many of the remaining Dutch spoke some English, producing "a wide difference between this colony and that of Mauritius. This however does not arise from the popularity of the English, for the Dutch, as the French at

Mauritius, although having profited to an immense degree by the English government, yet thoroughly dislike our whole nation. In the country, universally there is one price for a Dutchman, and another and much higher one, for an Englishman."

Journeying from Simonstown, where the *Beagle* had berthed, toward Cape Town, Darwin saw great numbers of the large Boer bullock carts pulled by oxen, of the type used for the great trek. They had "eighteen oxen all being yoked together in one team . . . the line looks as long as if all the cows in a field had been caught and tied together for sport." He admired three-thousand-five-hundred-foot-high Table Mountain behind Cape Town, a "great mass of horizontally stratified sandstone . . . the upper part forms an absolute wall, often reaching into the region of the clouds . . . so high a mountain, not forming part of a platform and yet being composed of horizontal strata, must be a rare phenomenon."

Darwin described the Hottentots, or "Hodmadods as old Dampier calls them," as "the ill-treated aboriginals of the country."[4] Darwin described them as small and looking like "partially bleached negroes . . . [with] most singularly formed heads and faces. The temple and cheekbones project so much, that the whole face is hidden from a person standing in the same side position, in which he would be enabled to see part of the features of a European." The more numerous Malays, descended from slaves imported by the Dutch, appeared "a fine set of men . . . always . . . distinguished by conical hats, like the roof of a circular thatched cottage, or by a red handkerchief on their heads." Darwin estimated the total population of Cape Town as around fifteen thousand people and the colony as a whole some two hundred thousand.

FitzRoy seems to have found so little of interest in Cape Colony that in his narrative he passed over it without any comment other than that the *Beagle* visited it. Darwin set out on a short journey of exploration, hiring two horses and a young Hottentot groom. The latter "spoke English very well, and was most tidily drest [*sic*]; he wore a long coat, beaver

---

4. The name Hottentot means "stuttering" in Dutch, a disparaging reference to the Hottentots' click language.

hat and white gloves." Darwin spent the first night at Paarl, which he noted was set amid a considerable number of vineyards—Jane Austen's favorite wine came from the Cape. The next day he crossed some bare mountains. With his mind alert as usual to comparisons, they reminded him "of Northern Chili [sic], but the rocks there possess at least a brilliant coloring." He reached the town of "French Hoeck," settled by French Protestants (Huguenots), which he found "one of the prettiest places I saw in the colony." Returning to Cape Town, Darwin noted seeing only a few roebuck and some vultures, and there is no evidence that he ever saw any African big game in the wild.

Darwin and FitzRoy dined in Cape Town with the famous astronomer and polymath John Herschel. He had come out to South Africa in 1834 with his family to map the stars of the Southern Hemisphere, just as his equally famous father William Herschel had previously done for the Northern Hemisphere. By the time he returned to Britain in 1838 to be knighted for his efforts, he had recorded the location of nearly sixty-nine thousand stars, cataloged many nebula and double stars, done pioneering work on sunspots, and observed Halley's Comet.

Darwin had long admired this eccentric figure after reading at Cambridge his book advising on the philosophy and practice of the study of natural history. He found meeting Herschel a "most memorable event," although neither man made a record of their conversation. Darwin simply wrote to Henslow that Herschel was "exceedingly good natured," even if at first acquaintance his manners appeared "rather awful." Much later, in his autobiography, Darwin expanded a little. Herschel "never talked much but every word which he uttered was worth listening to. He was very shy and he often had a distressed expression. Lady Caroline Bell at whose house I dined at the Cape of Good Hope, admired Herschel much, but said that he always came into a room as if he knew that his hands were dirty, and that he knew that his wife knew that they were dirty."

CHAPTER TWENTY

"A Great Name among the Naturalists of Europe"

On June 18, the *Beagle* set sail in a gale for St. Helena. While in the Cape, both Darwin and FitzRoy had believed the missionaries there were being unfairly blamed for fomenting disruption and dissent among the indigenous people and supporting them against commercial interests—very similar to criticisms of the missionaries they had heard in Tahiti and New Zealand. To attempt to defend the Cape missionaries by analogy, during the voyage to St. Helena the two men jointly penned a paper to an evangelical monthly, *The South African Christian Recorder*, giving their favorable views of the missionaries' "civilising" activity in the Pacific. This, the first venture into print by either man, again underlined the similarity of their views at the time.

Twenty days after leaving the Cape, the *Beagle* arrived at St. Helena, more than one thousand nautical miles west of the Angolan coast of southwest Africa and more than two thousand nautical miles east of Brazil. Darwin described the island's "forbidding aspect," rising "like a huge castle from the ocean. A great wall, built of successive streams of black lava, forms around its whole circuit, a bold coast.——Near to the town [Jamestown], as if in aid of the natural defence, small forts and guns are everywhere built up and mingled with the rugged rocks. The town extends up a flat and very narrow valley . . . When approaching the anchorage, there is one striking view; an irregular castle perched on . . . a lofty hill and surrounded by a few scattered fir trees, boldly projects against the sky."

A Portuguese expedition discovered the uninhabited island, only forty-seven square miles in area, in 1502 and named it St. Helena after the mother of the Roman emperor Constantine. The British arrived in 1659. St. Helena's greatest claim to fame was not its function as a staging post on the South Atlantic route between Europe and the Cape of Good Hope but as Emperor Napoleon's place of exile. After his final defeat at the Battle of Waterloo in 1815, rather than surrender to the Prussian authorities who had threatened to execute him, Napoleon threw himself on "the hospitality of the British nation," expecting to be housed in comfortable exile on an estate in southern England. Instead he and his thirty-strong entourage were immediately dispatched to St. Helena. The fortifications that Darwin noticed on his arrival and others throughout the island had been constructed to prevent any attempt by French forces to free the emperor.

Longwood House on the plateau in the center of the island had been improved to serve as Napoleon's residence. There he lived in considerable splendor. Dinner was lit by the imperial candelabras and served by menservants in the Tuileries livery of green coats with silver lace. He had paths in the garden lowered, short though he was, so he could walk without being observed over the wall by sightseers. He spent much of his day pacing about dictating his memoirs. When he died after six years on the island, the autopsy diagnosed bowel and stomach cancer, which had afflicted many of his family. Other causes have been advanced, in particular poisoning by arsenic either administered, as Covington recorded hearing, by one of his officers in the pay of the restored French royal family or, as more recently suggested, seeping from the dye in the wallpaper in his damp bedroom.

THE TOMB OF NAPOLEON, AT ST. HELENA.

Napoleon's Tomb, St. Helena

Darwin found lodgings "within stone's throw of Napoleon's tomb." The tomb was "close by cottages and a frequented road [and did not] create feelings in unison with the imagined resting place of so great a spirit." The state of Longwood

House was "scandalous, to see the filthy and deserted rooms scored with the names of visitors, to my mind was like beholding some ancient ruin wantonly disfigured." Darwin wrote to Henslow that on one of the nights he spent in his lodgings, it was "blowing a gale of wind, with heavy rain, and wretchedly cold: if Napoleon's ghost haunts his dreary place of confinement, this would be a most excellent night for such wandering spirits."

Covington sketched Napoleon's tomb and described it in more detail, correctly noting its lack of an inscription. There had been a dispute about how to refer to Napoleon. To Covington the grave was "simple for so great a man," being no more than "a large oblong stone." Iron railings surrounded it and, in turn, these were enclosed within wooden palings. An old soldier was stationed by the tomb to prevent anyone damaging it. When he visited Longwood House, Covington wrote that it was badly decayed and that visitors were allowed to play in Napoleon's billiard room, where they could also purchase wine.

Over the next few days Darwin explored the island, where the scenery reminded him of Wales. His guide, an elderly former goatherd, "was of a race many times mixed, and although with a dusky skin, he had not the disagreeable expression of a mulatto . . . It was strange to my ears to hear a man nearly white, and respectably dressed, talking with indifference of the times when he was a slave." Proto-ecologist as he was, Darwin particularly noted the impact of imported species on indigenous ones. As he knew from his reading, of the 746 plants, only 52 were indigenous, the rest being imports mainly from Britain, such as yellow-flowered gorse bushes. Many flourished better than in their home country, to the detriment of the indigenous flora, causing Darwin, again conscious of competition among species for life-sustaining resources, to remark, "The many imported species must have destroyed some of the native kind; and it is only on the highest and steepest ridges, that the indigenous flora is now predominant."

Similarly, imported goats and hogs had destroyed much vegetation and many young trees, leading to woodland being replaced, once any surviving old trees had died, by pasture. "There can be little doubt," Darwin wrote, "that this great change in the vegetation affected not only the land-shells [snails] causing," as he had discovered, "eight species to become extinct, but likewise a multitude of insects."

Darwin found evidence "to the geologist . . . of high interest" that volcanic St. Helena had existed as an island "from a very remote epoch." Nevertheless, "obscure proofs . . . of the elevation of the land are still extant . . . the central and highest peaks form parts of the rim of a great crater, the southern half of which has been entirely removed by the waves of the sea; there is, moreover, an external wall of black basaltic rocks, like the coast-mountains of Mauritius, which are older than the central volcanic streams." The latter, in his view, correctly suggested that since the island had been raised from the sea, there had been further successive volcanic eruptions.

Leaving St. Helena on July 14, the *Beagle* arrived five days later at Ascension Island, some seven hundred nautical miles to the northwest and, at thirty-four square miles, even smaller than St. Helena. First discovered by a Portuguese sailor on Ascension Day in 1501 (May 21 that year), uninhabited Ascension Island thereafter provided an occasional anchorage for ships in the Atlantic. William Dampier's HMS *Roebuck* had sunk there in 1701 while returning from his second voyage to Australia. After two months he and his crew were rescued and returned to Britain.[1]

The British government formally established a military garrison on the island in 1815 to prevent French forces occupying it in any attempt to free Napoleon from St. Helena. Only British marines, their wives, children, and support staff, including some former slaves—a total of around 260 people—lived on the island, which, according to Darwin, resembled "a huge ship kept in first rate order." He believed many marines thought "it better to serve their one-and-twenty years on shore, let it be what it may, than in a ship." He would have agreed had he been a marine.

Darwin described the island as "volcanic" and "arid" with "smooth conical hills of a bright red colour, with their summits generally truncated

---

1. Three hundred years later marine archaeologists from the Western Australia Maritime Museum in Fremantle found the wreck of the *Roebuck* and recovered some large natural history specimens, such as shells from Papua New Guinea, which Dampier was unable to recover before his ship sank.

rising distinct out of a level surface of black horrid lava. A principal mound in the centre of the island, seems the father of the lesser cones." The coast was "lashed by a wild and turbulent sea." What Darwin called "volcanic bombs . . . masses of lava . . . shot through the air whilst fluid" as the volcano erupted lay scattered around the landscape. From examining them and other volcanic material such as tuff, he concluded that they had subsequently been subject to action by water, probably in a large lake now dried up by the presently prevailing hot climate. As he had also felt in South America, natural climate change appeared indisputable. "At some former epoch the climate and productions of Ascension were very different from . . . now. Where on the face of the earth can we find a spot on which close investigation will not discover signs of that endless cycle of change to which this earth has been, is, and will be subjected?"

He killed and included among his specimens two varieties of rats from the island. Their ancestors probably originally escaped from ships, but he noticed they were a third smaller than those in Europe and further differentiated by "the color and character of their fur." One variety lived on the summits of the hills and the other on the coast. Darwin commented tellingly that "as at the Galapagos, [they] have varied from the effect of the new conditions to which they have been exposed: hence the variety on the summit of the island differs from that on the coast."

Darwin had been considering the impact of the environment on the adaptation of species ever since leaving the Galápagos. Some time since then, perhaps during the long days crossing the South Atlantic, he had composed five pages of ornithological notes about the Galápagos. They seem to have been written with a new pen and have few crossings out or changes, suggesting that they were the product of considerable thought and may have been preceded by a rough draft. The pages include the following crucial sentences:

I have specimens [of Galapagos mockingbirds] from four of the larger islands . . . In each Is. each kind is *exclusively* found; habits of all are indistinguishable . . . When I recollect the fact that . . . the Spanish can at once pronounce, from which Island any tortoise may have been brought;——when I see these Islands in sight of each other, and possessed by but a scanty stock of animals, tenanted by these birds, but slightly differing in structure and filling the same place in Nature, I must suspect they are only varieties. The only fact of a similar kind of which I am aware, is the constant asserted difference—between the wolf-like Fox of East and West Falkland Islds.

If there is the slightest foundation for these remarks the Zoology of Archipelagos will be well worth examining; for such facts undermine the stability of Species.

These are the first incontrovertible, if nuanced and guarded, statements of his emerging belief in evolution, and they have echoes in his thoughts about the Ascension rats.

Some letters from home finally caught up with Darwin at Ascension and, to his delight, they contained good news. Henslow had assembled a paper to the Cambridge Philosophical Society in November 1835, bringing together extracts from the information Darwin had sent him from South America. His sister Susan relayed that the society had received the paper well. Sedgwick had praised Darwin to Dr. Butler, Darwin's old headmaster and a college friend of Sedgwick. In turn, Butler had written to Darwin's father with an extract from Sedgwick's letter: "[Darwin] is doing admirably in S. America, and has already sent home a collection above all praise—It was the best thing in the world for him that he went out on the Voyage of Discovery—There was some risk of his turning out an idle man; but his character will now be fixed, and if God spare his life, he will have a great name among the Naturalists of Europe."

In another letter, his sister Caroline told Darwin that when his father received a copy of the society's paper, he "did not move from his seat till he had read every word of your book and was very much gratified . . . Your frank unhackneyed mode of writing was to him particularly agreeable." Darwin recalled in his autobiography that "after receiving the letters, I clambered over the mountains of Ascension with a bounding step and made the volcanic rocks resound under my geological hammer!"

In addition to delight at the reception of his findings, and what that meant for his future, Darwin must have felt great pleasure that his father, of whom he remained nervous, still using his sisters as intermediaries to approach him on difficult matters and who probably shared Sedgwick's worries about his potential idleness as well as doubts about the worth of the *Beagle* voyage, was gratified by his efforts. He would find it easier to persuade Dr. Darwin that rather than go into the church he should pursue a scientific life—a decision left unsaid explicitly, even if implicit in Darwin's later correspondence from the voyage and unchallenged by his father and family. His last positive reference to a clerical career had been in early spring 1833 in a letter to his cousin Fox.

Much to the dismay of Darwin and the rest of his homesick crew, the meticulous FitzRoy decided that, to be certain of his longitudinal measurements, he had to visit Salvador de Bahia in Brazil for a third time. Safely arrived, Darwin was "glad to find my enjoyment of tropical scenery, from the loss of novelty, had not decreased even in the slightest degree." He believed attempting to convey this beauty to others who had not seen it "an hopeless endeavour . . . Who else from seeing a plant in an herbarium can imagine its appearance when growing in its native soil? Who, from seeing choice plants in a hothouse, can multiply some into the dimensions of forest trees, or crowd others into an entangled mass? Who, when examining in a cabinet the gay butterflies, or singular Cicadas, will associate with these objects the ceaseless harsh music of the latter, or the lazy flight of the former—the sure accompaniments of the still glowing noon day of the Tropics . . . the land is one great wild, untidy, luxuriant hothouse."

Darwin tried, in one last walk, to fix in his mind as well as in his notes, an impression of the tropics, which even then he knew "must sooner or later fade away. The forms of the Orange tree, the Cocoa nut, the Palms, the Mango, the Banana will remain clear and separate, but the thousand beauties which unite them all into one perfect scene, must perish: yet they will leave, like a tale heard in childhood, a picture full of indistinct, but most beautiful figures."

Due to bad weather after leaving Salvador de Bahia, the *Beagle* was forced to put in to Recife in Pernambuco state, five hundred miles north on the Brazilian coast. During this short stop, the only item of scientific interest to Darwin was "the reef that forms the harbour. It runs . . . several miles in a perfectly straight line, parallel to and not far distant from the shore; it varies in width from thirty to sixty yards; it is quite dry at low water . . . and is composed of obscurely stratified hard sandstone." He could not decide how it was formed—whether before the land was elevated to its present level the reef had been a submerged sandbar that had consolidated as it rose, or the currents had eroded a previous large above-water sandbar to its hard central core.

Leaving Recife, the *Beagle* called once more at St. Jago in the Cape Verde islands, to which Darwin devoted only three brief paragraphs in his diary—since it was "the beginning of the unhealthy season, I confined my walks to short distances . . . Our old friend the great Baobab tree was clothed with a thick green foliage, which much altered its appearance. As might be expected, I was not so much delighted with St. Jago as during our former visit; but even this time I found much in its Natural History very interesting." He also recorded the presence of some "slaving vessels" in the harbor. His voyage had only increased his abhorrence of slavery—a topic frequently referred to in his diary—and confirmed his belief that, as the abolitionists' pamphlets proclaimed and his grandfather Josiah Wedgwood had famously inscribed on his pottery plaque, all men were brothers and a single species. It is noteworthy that at a time when skulls of Aboriginals, Hottentots, and other indigenous people, and even their skins, were being brought back to Europe, Darwin did not include in his collection of specimens any human remains.

By September 20 the *Beagle* was anchored at the town of Angra on the island of Terceira in the central group of the nine island Azores archipelago in the North Atlantic, some thousand miles west of mainland Portugal, to which they had belonged since the mid-fifteenth century. During a ride into the interior, Darwin met a number of peasants. He did not remember ever seeing "handsomer young men with more good-humoured pleasant expressions . . . The men and boys are all dressed in a plain jacket and trowsers, without shoes or stockings, their heads are barely covered by a little blue cloth cap with two ears and a border of red; this they lift in the most courteous manner to each passing stranger. Their clothes although very ragged, appeared singularly clean as well as their persons . . . Each man carries in his hand a walking staff about six feet high; by fixing a large knife at each extremity, they can make this into a formidable weapon.—Their ruddy complexions, bright eyes and erect gait made them a picture of a fine peasantry; how different from the Portuguese of Brazil!" He much regretted that poverty caused so many Azoreans to emigrate to Brazil, where their indentured labor contracts differed "but little from slavery."

As he crossed the volcanic landscape, Darwin noticed that "from the long traffic of the bullock wagons, the solid lava, which formed in parts the road, was worn into ruts of the depth of twelve inches," something which had only been remarked in "the ancient pavement of Pompei" but not elsewhere in Italy. Always curious and concerned to ascribe reasons for similar features in different locations, he suggested that this was because in the Azores the wheels had "a tire surmounted by singularly large iron nobs, perhaps the old Roman wheels were thus furnished." He also noticed how the islanders made low walls from volcanic rocks to delineate their fields and protect their vineyards, "thus covering the country with a network of black lines."

The only specimens the tired naturalist collected in the Azores were six rocks, and geology was clearly his greatest interest there. In particular he visited the fumaroles, or steam vents, in the center of the island at Furno do

Vineyard lava stone protective walls, Terceira, Azores, October 2021

Enxofre—a phenomenon he had little opportunity to observe elsewhere. In a depression in the land he compared to an ancient stone quarry, he found "the bottom was traversed by several large fissures, out of which, in nearly a dozen places, small jets of steam issued as from the cracks in the boiler of a steam engine. The steam close to the irregular orifices is far too hot for the hand to endure." It had little smell, but since it blackened iron and produced "a peculiar rough sensation" on the skin, he assumed, correctly, that the steam must contain some acid.

As the *Beagle* left the Azores and headed home for Britain, Darwin set down in his diary his views on "the advantages and disadvantages, the pain and pleasure of our five years wandering." He believed such voyages were valuable if the traveler had a decided interest in a particular branch of knowledge that could be acquired from it but perhaps not simply for sightseeing. In the latter case, disadvantages such as being away from home and friends and after a while shipboard issues such as "the want of room, of seclusion, of rest—the jading feeling of constant hurry—the privation of small luxuries, the comforts of civilization . . . and lastly even of music and the other pleasures of imagination" became a factor. Then there was seasickness to take into account: "I speak from experience, it is no trifling evil cured in a week." He was not sure even most sailors actually enjoyed being at sea.

However, as he acknowledged, such relatively trivial complaints showed that life at sea, ship design, and navigation had improved markedly in the six decades since Cook's voyages. Now "a yacht with every luxury of life might . . . circumnavigate the globe . . . the whole Western shores of America are thrown open; and Australia is become . . . a rising continent . . . a hemisphere has been added to the civilized world." Darwin looked back with a kind of "extreme delight" at his land and small boat journeys that "no scenes of civilisation could create." He had retained a patriotic chauvinism and indeed an ability to stereotype other nationalities and groups throughout the voyage. British colonies had raised within him "high pride and satisfaction. To hoist the British flag seems to draw as a certain consequence wealth, prosperity and civilisation"—something only confirmed by the lesser progress achieved by other nations over greater time, as exemplified in South America. Even though all men were brothers, to be British was best.

In his reflections, Darwin again praised the civilizing influence of the British missionaries in the Pacific. His first sight "in his native haunt of a real barbarian—of man in his lowest and most savage state" caused his mind to go back over past centuries and ask: "Could our progenitors be such as these?" As an indication of his own thought processes of comparing, contrasting, and reasoning why, everywhere he went, he wrote, "As a number of isolated facts soon become uninteresting, the habit of comparison leads to generalisation; on the other hand as the traveller stays but a short space of time in each place, his description must generally consist of mere sketches instead of detailed observation. Hence arises, as I have found to my cost," he commented modestly, "a constant tendency to fill up the wide gaps of knowledge by inaccurate and superficial hypotheses."

Darwin concluded, "I have too deeply enjoyed the voyage not to recommend to any naturalist to take all chances . . . In a moral point of view, the effect ought to be to teach him good humoured patience, and selflessness, the habit of acting for himself, and of making the best of everything . . . in short, he should partake of the characteristic qualities of the greater number of sailors."

PART THREE

After the *Beagle*

"A Peacock Admiring His Tail"

The *Beagle* anchored in Falmouth on October 2, 1836, after a voyage lasting four years, nine months, and six days. Darwin, who had spent some three-fifths of the time ashore, disembarked immediately. Perhaps surprisingly, he would never go abroad again.

The long months of the voyage that he had begun as an inexperienced, if promising, amateur natural historian had matured him into someone certain that he wished to take a place among the leading scientific men, as Adam Sedgwick had recently suggested—and not in a country parsonage. He was now a young man in a hurry in so many ways—in a hurry to be rid of the debilitating seasickness that had reoccurred as the *Beagle* crossed the Bay of Biscay, in a hurry to see his fond family once more, in a hurry to visit his Cambridge mentors and discuss with them his collection of more than 5,400 specimens of all kinds from the voyage, in a hurry to establish himself in the scientific community, and, more privately, in a hurry to work on the ideas of how species might change over time—"transmutation," he called it—that he had begun to develop, particularly in his ornithological notes, in the last months of the voyage. As he later recalled, the next months would be "the most active ones which I ever spent." They would define his future.

Darwin immediately boarded a mail coach to take him home to Shrewsbury, and as he jolted onward, was surprised that "the stupid people on the coach did not seem to think the fields one bit greener than usual but I'm sure, we should have thoroughly agreed, that the wide world does not contain so happy a prospect as the rich, cultivated land of England." He reached The Mount late on October 4 after two days of exhausting travel, so late that his family had already gone to bed. He surprised them by suddenly appearing at breakfast the next morning. There was much delight as he and his sisters embraced. His father, although not a great believer in phrenology, thought "the shape of his head is quite altered," a view Darwin himself did not discount in the light of his own mental development. Darwin was pleased when an old family dog immediately responded to his whistle and ran up to him in the yard, tail wagging, ready to accompany him on a walk. Some of his family's farmhands used the celebration of his arrival to get drunk.

Darwin swiftly penned a letter to FitzRoy assuring him "the five years voyage has certainly raised me a hundred percent [in the eyes of my family] . . . I do indeed hope all your vexations and trouble with respect to our voyage . . . have come to a close.—If you do not receive much satisfaction for all the mental and bodily energy you have expended in His Majesty's Service, you will be most hardily treated.—I put my radical sisters into an uproar at some of the *prudent* (if they were not *honest* Whigs, I *would* say shabby), proceedings of our Government . . . But I am no renegade, and by the time we meet, my politics will be as firmly fixed and as wisely founded as ever they were. I thought when I began this letter I would convince you what a steady and sober frame of mind I was in. But I find I am writing most precious nonsense . . . God bless you—I hope you are as happy, but much wiser than your most sincere but unworthy Philos. Chas. Darwin."

Before embarking on his period of frantic activity he spent ten slightly uneasy days at The Mount with his family, all trying to readjust themselves to one another after the nearly fifth of his young life he had spent abroad. He fled first to Cambridge to discuss with John Henslow where to deposit his specimens, then he was off to London, where his brother Erasmus now lived as a leisured man of letters with a wide circle of intellectual acquaintances. There he sounded out various museums and societies as repositories for his collections. As Henslow had forewarned him, he found the Zoological Society unenthusiastic—"the Zoologists seem to think a number of undescribed creatures rather a nuisance"—and he described the behavior of the speakers at one of their meetings he attended as "snarling at each other, in a manner anything but like that of gentlemen."

A more fruitful encounter came at the end of October—a first meeting with his geological hero Charles Lyell. Darwin's 1,383 pages of *Beagle* notes on geology, compared to his 368 on zoology, marked him out as at least as much a geologist as a zoologist at this stage in his career. Darwin found Lyell "*most* good-natured," interested in his specimens and description of the Chilean earthquake and its consequences, receptive to his ideas—particularly since they confirmed his own gradualist theories—and more than willing to introduce him into scientific society. Lyell immediately had him elected a member of the Geological Society—a not inexpensive privilege, Darwin complained. He later recalled, Lyell's "delight in science was ardent, and he felt the keenest interest in the future progress of mankind. He was very kind-hearted and thoroughly liberal in his religious beliefs or rather disbeliefs; but he was a strong theist."

Lyell thought Darwin a "glorious" addition to his circle. At a tea party at his house he introduced him to Richard Owen, the newly appointed Hunterian Professor at the Royal College of Surgeons (RCS). Owen, who five years later would give the dinosaur its name, was one of Britain's foremost anatomists and a confirmed Tory. As such, he was both politically and professionally opposed to Darwin's old mentor from his Edinburgh days, Robert Grant, now at University College London, whose ostracism by the Zoological Society he had recently helped orchestrate.

Only five years older than Darwin, Owen had a deep interest in fossils as well as, like Grant, in invertebrates and anatomy. Darwin had several discussions with him, some in his dissecting room at the RCS. Grant, around this time, offered to help Darwin with his specimens, but Darwin refused him. He perhaps feared that too close an association with this socialist and outspoken critic of the Church, of the establishment—and in particular of the Cambridge scientific establishment—might prove counterproductive in his search for a place in science, especially at a time when fears of revolution stalked the drawing rooms of the well-heeled. Besides, Darwin, although a liberal, as a wealthy and privileged man had no love for the revolutionary or radical that might disturb the social hierarchy in which he had such a comfortable place. Instead of involving himself with Grant, he decided to pass his fossils to Owen for professional identification.

During this period a letter from FitzRoy reached Darwin in response to the one Darwin had dispatched from The Mount. Part of it read:

"The account of your family—and the joy tipsy style of the whole letter were *very* pleasing. Indeed Charles Darwin I have *also* been *very* happy—even at that horrid place Plymouth—for that horrid place contains a *treasure* to *me* which even *you* were ignorant of! Now guess—and think and guess again. Believe it, or not,—the news is *true*—I am going to be *married*!!!!!! to Mary O'Brien.—*Now* you may know that I had decided on this step, long *very long* ago.—All is settled and we shall be married in December [They indeed married on 27 December.] . . . Pray call on my sister in Stratton Street—she longs to see you . . . Money matters are better than *you* think."

Darwin's reaction to FitzRoy's engagement reflected his own determination to concentrate on his work. FitzRoy had chosen "a most inconvenient time to marry." Darwin showed no interest in rumors picked up by his father in an apothecary's and eagerly relayed to him by his sister Caroline that FitzRoy had a child elsewhere and on his engagement had revised some commitments made to support it. (There is no surviving evidence.) Around this time Sulivan too became engaged. His fiancée was Sophia Young, the admiral's daughter he had failed to take leave of in Plymouth just before the *Beagle* sailed because he fell asleep and whom again he had not mentioned to his fellows. That two officers could keep secret from Darwin and their fellows attachments strong enough to survive years away says much about the buttoned-up nature of shipboard conversation on emotions.

Darwin himself visited the *Beagle* one last time before she was paid off. She had arrived at Greenwich on October 28. There the eleven chronometers that remained in working order on the *Beagle*—some had broken and others had been loaned to other surveys during the voyage—were checked against Greenwich Mean Time. They were only thirty-three seconds off after almost five years! The number loaned or broken justified FitzRoy's personal expenditure on extra chronometers. FitzRoy was busy putting in order the eighty-two coastal charts and eighty plans of harbors that were a testament to his fulfillment of his orders to make detailed and comprehensive surveys of the coastlines he visited. Francis Beaufort would later praise FitzRoy and his work highly to Parliament.

Amid all the bustle and confusion on the *Beagle*, as he collected his crates of specimens all carefully packed by Covington, Darwin was disconcerted "how to begin," writing, "All I know is, that I must work far harder, than [these] poor shoulders have ever been accustomed to do." He was soon on the road again back to Shropshire where, in mid-November, he at last found time to visit Maer and his uncle Jos Wedgwood, who had played such a part in persuading Dr. Darwin to allow him to join the *Beagle*, and his cousins. A letter from one of them, Elizabeth Wedgwood, suggests that he again found it not entirely easy to reassume his old place in the family. She began with comments about his appearance: "He looks very much thinner, but it has improved his looks, and his countenance is so pleasant that his plainness does not signify . . . it is impossible not to be quite fond of him—it is very pleasant to see him so ready to enjoy himself," but then continued, "We had begun to be low at dinner for he was shy and we could not get on."

In mid-December, Darwin rented a small house in Cambridge, where he was rejoined as an amanuensis cum servant by Syms Covington and by a small tortoise from the Galápagos that had survived the *Beagle* voyage. He immediately began working on the allocation to experts and further cataloging of his specimens, frequently consulting Henslow and Sedgwick for advice and the use of their scientific contacts. On a visit to London he was introduced to a botanist at the Linnaean Society and recalled, "I felt very foolish, when [he] remarked on the beautiful appearance of some plant with an astoundingly long name, and asked me about its habitation. Someone else seem[ed] quite surprised that I knew nothing about a carex from [I] do not know where, I was at last forced to plead most entire innocence, and that I knew no more about the plants, which I had collected, than the Man in the Moon."

Darwin did eventually find good homes for nearly all of his specimens. His birds went to John Gould, an expert ornithologist and taxidermist at the Zoological Society. The son of a gardener, he augmented his small one-hundred-pound salary as the society's animal preserver by producing illustrated books on birds and animals. More of Darwin's animal specimens, carefully preserved aboard the *Beagle*, some in spirits, went to specific members of the Zoological Society with stern instructions that they must be studied, mounted, and described properly rather than being left to rot or gather dust. His marine iguanas from the Galápagos went to a professor at Oxford University. He dispatched his fish specimens to his Cambridge friend and Henslow's brother-in-law, the clergyman and ichthyologist Leonard Jenyns to whom Henslow had offered the position of *Beagle* naturalist before Darwin.

Owen soon had news for Darwin about the first tranche of fossils Darwin had deposited with him. Darwin told his sister Caroline that "some of them are turning out great treasures. One animal, of which I have nearly all the bones is very closely allied to the Ant Eaters, but of the extraordinary size of a small horse. There is another head, as large as a Rhinoceros, which as far as they can guess, must have been a gnawing animal. Conceive a Rat or a Hare of such a size—What famous cats they ought to have had in those days!" Owen soon had more news. What Darwin had thought was a Megatherium was a giant extinct armadillo belonging to a group Owen named *Glyptodon,* from the Greek for "grooved or carved tooth." The gnawing animal appeared to be a kind of giant capybara, now known as a *Toxodon.* There were fossils of several ground sloths and of a giant camel-like animal seemingly related to a guanaco or perhaps a llama that Owen named *Macrauchenia patachonica* and that Darwin had first thought was a mastodon. Taken together, these identifications suggested to Darwin some kind of continuity or succession, in form if not in size, between extinct and existing regional species. The only real exception to this apparent process of regional succession between types that were distinct from those found elsewhere were the "horse teeth." Owen confirmed these indeed belonged to some kind of ancient form of horse that had become extinct in South America until reintroduced by the conquistadores. "A curious fact," Owen called their presence.

On January 4, 1837, Darwin presented his first paper to the Geological Society. It was titled "Observations of Proofs of Recent Elevations on the Coast of Chili [sic]." He was applauded by Lyell as president and felt like "a peacock admiring his tail." Soon afterward he was elected to the council of the society. By then, Gould had almost completed his study of Darwin's birds. Some of those from the Galápagos had posed him a quandary. He had spent considerable time working on the collection called by Darwin "finches, wrens, blackbirds, and grossbeaks [sic]" and concluded they were in fact "a series of ground Finches which are so peculiar [that they might form] an entirely new group, containing 12 species." They had different types of beaks. To Gould there seemed a distinct possibility that the finches each came from different islands, but he could not be sure because Darwin had packed some of them together and had labeled them insufficiently. Discomfited, Darwin somewhat sheepishly asked FitzRoy to loan him the birds he had taken from the Galápagos to help resolve the question of their origins.

The two men had met little since the *Beagle* was paid off, although Darwin had gone to tea with FitzRoy to meet his new wife, "so very beautiful and religious a lady," Darwin observed. Their relationship seems quickly to have become increasingly distant and strained. Perhaps buoyed by the reception of his geological work, Darwin now felt less need either to accommodate himself to FitzRoy's moods or to refrain from questioning his views too closely. Perhaps FitzRoy's opinions were changing under the influence of his marriage. Whatever the case, they still got on sufficiently well for FitzRoy to agree to loan Darwin his Galápagos birds, which he had labeled much better than Darwin.

Together with another four birds belonging to Syms Covington and two from FitzRoy's steward Henry Fuller, they allowed Gould to come with a degree of certainty to the accurate conclusion that they were indeed finches unique to individual islands. Their different types of beaks have subsequently been found to be adaptations to their individual eating habits. Birds with heavier beaks are adapted to crack seeds. Those with more delicate, tweezer-like beaks are suited to probing for insects.

Beaks of Galápagos finches by J. Gould

Others eat cacti or—highly unusual in birds—green leaves. Even more exotically, the vampire finch lands on the back of boobies (gannets) and pecks to draw blood, which it then drinks. The woodpecker finch has adapted to fill the niche occupied by the woodpecker elsewhere. Lacking the woodpecker's long, barbed tongue, it has developed the use of a tool—a twig or cactus spine—to probe into the wood for grubs.

Darwin's Galápagos mockingbirds were better labeled, and Gould stated unequivocally that they were three separate species and that each came from a different island—a conclusion that Darwin, while still on the *Beagle*, had asserted in his ornithological notes "would undermine the stability of species." News came too from Oxford that the Galápagos marine iguanas appeared to differ from island to island and from elsewhere that the same was true of the Galápagos tortoises—as Nicholas Lawson had told Darwin when in the islands—and also of the islands' rodents. In March 1837, Gould told the Zoological Society that Darwin's two types of South American rheas were definitely separate species, not simply varieties, adding further weight to Darwin's emerging thinking about how species might evolve. Gould gave the name *Rhea darwinii* to the southern version, the half-eaten carcass of which Darwin had rescued from the cooking pot. (It is now known as the *Pterocnemia pennata*.)

By this time, Darwin was beginning to work on his contribution to the publication of information on the two voyages of the *Beagle*, for which FitzRoy had agreed a contract with his publisher Henry Colburn. Rather than Darwin's diary augmenting FitzRoy's account, there would now be three separate volumes. One would be by Captain King on the first voyage of the *Beagle*, one on the second voyage by FitzRoy, and the third by Darwin on the second voyage's research. Darwin struggled to combine elements from his *Beagle* geological and zoological notebooks with the accounts from his diary into a comprehensive narrative, even if, as he admitted, it might nevertheless seem a bit of "a hodge-podge."

By March, Darwin had decided that to pursue his researches and further his contacts he had to move to "dirty odious London." Bolstered by a four-hundred-pound annual allowance from his father, he took lodgings at 36 Great Marlborough Street near his brother Erasmus, or Eras. Like Erasmus, his father's generosity freed Darwin from any need to seek a salaried appointment or an academic post. Financial independence allowed him independence of thought and the freedom to use his time as he wished. Darwin's letters home to his family during the *Beagle* voyage had revealed his own growing inclination for a life in science. This preference, reinforced by Adam Sedgwick's prediction that Darwin would "have a great name among the Naturalists of Europe," had clearly convinced Dr. Darwin that a scientific life alone should be his and not one in the church, as originally intended when Darwin embarked on the *Beagle*.[1]

Both Eras and Lyell were keen to introduce him into their overlapping London social circles. Central to both was Charles Babbage, the mathematician, computer pioneer as the builder of the Difference Engine calculating machine, social and political liberal, and author of a

---

1. Darwin's voyage on the *Beagle* had cost his father £1,800—£600 to equip him, and £1,200 in expenses during the expedition, a not dissimilar annual expenditure to his new allowance.

treatise in which he stated controversially that his God introduced laws by which Creation occurred rather than undertaking it directly himself—a concept that he believed reconciled religion and new developments in science. Babbage's Saturday evening gatherings were a must for both the fashionable and intellectual and, as Lyell pointed out, were remarkable not only for the conversation and contacts to be made but also the number of pretty women present. At this time Lyell's wife appears to have encouraged her unmarried sisters to set their caps at Darwin there and elsewhere, but to no avail.

The Darwin family and Dr. Darwin in particular were themselves less concerned with Charles's marriage prospects than the fear that Eras, whom historian and essayist Thomas Carlyle, a friend of both Babbage and Eras, called idle, might be so indolent as to be unable to escape the snares of Harriet Martineau. A radical activist, proto-feminist, and political author, she was an enthusiastic and knowledgeable proponent and popularizer of the views of the clergyman Thomas Malthus, who had died during the *Beagle*'s voyage. Malthus's pessimistic thesis, published as long ago as 1798 in his book *An Essay on the Principles of Population*, was that populations, if left unchecked, would quickly outstrip the production of food for them to eat since population increased in a geometric ratio (i.e., 1, 2, 4, 8, 16, 32, etc.) while food supply increased only by an arithmetic ratio (i.e., 1, 2, 3, 4, 5, 6, etc.). A savage competition would soon begin for scarce resources. Vice, poverty, famine, disease, and then war would quickly follow, with their greatest impact on the poor, weak, and vulnerable. Malthus believed such was God's will and that undue charity to the poor and weak would only exacerbate overpopulation and lead to more trouble in the long run. The poor must struggle to survive by their own efforts. His still controversial theories had given rise to the 1834 Poor Law and the subsequent introduction of workhouses.

When he actually met her, Martineau's intellect and breadth of knowledge impressed Darwin. Nevertheless, he consoled his sisters, "I was astonished to find how ugly she is," adding "our only protection from so admirable a sister-in-law is in her working him [Eras] too hard. He begins to perceive, (to use his own expression), he shall be not much better than her 'nigger' [*sic*] . . . She already takes him to task about his idleness.——She is going some day to explain to him her notions about marriage——Perfect equality of rights is part of her doctrine. I much doubt whether it will be equality in practice." Later Darwin couldn't help teasing his family that Harriet Martineau had been "noon, morning, and night" with Eras and had "been as frisky lately as the rhinoceros" (recently imported to London Zoo). Martineau herself, unaware of his detailed misogynistic scrutiny, summed up Charles Darwin at the time as "simple, childlike, painstaking, and effective."

At the end of May 1837, Darwin presented another paper to the Geological Society, laying out his opinion——based on his researches during the *Beagle* voyage and in particular on the Keeling Islands——that coral reefs were the last remnants of disappearing mountains. He had already explained it to Lyell, with whose views it conflicted, who told Darwin, "I could think of nothing for days after your lessons on coral reefs, but of the topic of submerged continents." He quickly accepted the validity of Darwin's opinion, which became generally accepted during his lifetime but was not confirmed until the 1950s when deep borings were made on Bikini Atoll in the Marshall Islands in the Pacific at the time of the United States' nuclear bomb tests there.

Darwin found life in London a mixed blessing. "I thought Cambridge a bad place from good dinners and other interruptions, but I find London no better, and I fear it may grow worse . . . I miss a walk in the country very much; this London is a vile smoky place, where a man loses a great part of the best enjoyments of life. But, I see no chance of escaping even for a week from this prison, for a long time to come."

In July 1837, as he later wrote, Darwin "opened first notebook on 'transmutation of species.' Had been greatly struck from about month of previous March on character of S. American fossils, and species of Galapagos Archipelago. These facts origin . . . of all my views." This brown notebook (now known as the B notebook) had the word "Zoonomia" inscribed alone on its first page. He was soon drawing on page 36 a rudimentary tree of life, which showed species as the branches of an ancient mother tree, some of which became extinct while some flourished, reflecting beneath it: "Heaven knows whether this agrees with Nature." It was certainly

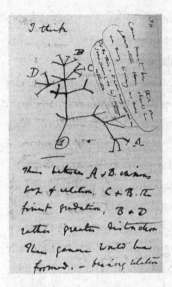

Charles Darwin's "Tree of Life"

at variance with the commonly accepted view of a strict, ordered, and immutable hierarchy of living things with humans at the apex and uniquely possessed of a soul.

Over this period Darwin scribbled in his notebooks what was a series of stream-of-consciousness jottings—"mental riot-ing," he called such thinking—sometimes contradictory but pushing toward the con-clusion that transmutation was undeniable. "It is absurd to talk of one animal being higher than another—We consider those where the cerebral structure/intellectual faculties most developed as highest—A bee doubtless would when the instincts were [highest] . . ." "If all men were dead then monkeys make men . . . men angels." His strong anti-slavery beliefs coalesced with his thinking when he wrote, "Animals—whom we have made our slaves we do not like to consider our equals—Do not slaveholders wish to make the black man other kind."

He continued: "Propagation explains why modern animals same type as extinct which is law almost proved"; "animals, on separate islands, ought to become different if kept long enough—apart, with slightly dif-ferent circumstances.—Now Galapagos Tortoises, Mocking Birds, Falk-land Fox—Chilloe Fox"; "Animals differ in different countries in exact proportion to the time they have been separated . . . Countries longest separated, greatest differences—if separated from immense ages pos-sibly two distinct types, but each having its own representatives—as in Australia. This presupposes time when no mammalia existed; Australian Mamm. were produced from propagation from different set . . . This view supposes that in course of ages . . . every animal has tendency to change"; "Tree of life should perhaps be called Coral of life, base of branches dead so that passages cannot be seen"—in other words, the analogy of how coral branched while its progenitors at its base died—as he had seen in the Keeling Islands—might explain why the early stages of life, and some of the transitions ("passages") by which present forms developed, could not be identified.

Darwin emphasized that sexual reproduction was essential to transmutation since it gave the possibility of variation, "to adapt and alter race to changing world," whereas vegetative propagation, by cuttings, for example, did not. Species died out when conditions changed adversely too fast for them to adapt. As part of his researches, Darwin began assembling as much information as he could on selective domestic breeding to allow him to probe analogies with breeding in nature. He asked questions of all with an interest, from his father's coachman and his own barber—both keen dog breeders—to other natural scientists, plundering his friends' and family's address books for contacts. He also began reading around the subject, particularly those volumes relating to philosophical speculations about life and the role of a deity.

A conventional man with an unconventional, not to say heretical, idea, he kept his thoughts on transmutation entirely private as he sought a government grant to fund the publication of a series of books on the zoological findings of the *Beagle* voyage, believing it would be beneficial to have "the gleanings of my hands, after having passed through the brains of abler naturalists collected together in one work." He envisaged individual volumes written by those experts who had or were examining specimens in their field, such as Gould on the birds. He himself would edit them. He followed his custom of using trusted intermediaries, one of whom was Henslow, to smooth an approach to the Chancellor of the Exchequer, Thomas Spring-Rice, while himself writing to Beaufort that the costs of amassing these collections "creditable to the country . . . even to the purchase of materials for the preservation of the specimens, together with a salary for an assistant, has been willingly defrayed by myself." After a delay occasioned by the death of William IV and accession of his eighteen-year-old niece Victoria as queen on June 20 and a subsequent general election won again by the Whigs, Spring-Rice—reappointed as chancellor—wrote to Darwin that the treasury would award him one thousand pounds to help fund the work, in particular the colored illustrations. Emboldened by the size of the grant, Darwin decided to try to make it stretch to fund further volumes on *Beagle* geology, which he would himself write—an indication that at this time he still considered his main expertise was in geology, not zoology. His first book would be on coral reefs, followed by one on the geology of volcanic islands and another on the geology of South America.[2]

By this time, only ten months after leaving the *Beagle*, Darwin had already completed a draft of his volume on the voyage researches and submitted it to Colburn, the publisher. In it, the Galápagos finches assumed greater prominence. In a teasing first public hint about his theory, he highlighted the differences between the beaks of finches on different islands: "In the thirteen species of ground-finches, a nearly perfect gradation may be traced from a beak extraordinarily thick, to one so fine, that it may be compared to that of a warbler. I very much suspect, that certain members of the series are confined to different islands; therefore if the collection had been made on any *one* island, it would not have presented so perfect a gradation. It is clear, that if several islands have each their peculiar species of the same genera, when these are placed together, they will have a wide range of character." However, as quickly as he had broken cover, he retreated again, concluding, "But there is not space in this work, to enter on this curious subject." He would be slightly bolder in the second edition of 1845, speculating, "Seeing this gradation and diversity of structure in one small intimately related group of birds, one might really fancy that from an original paucity of birds in this archipelago, one species had been taken and modified for different ends."

He had found "writing is most tedious and difficult work," but when the first proofs arrived in September 1837, he "sat . . . gazing in silent admiration at the first page of my own volume." In November, he wrote FitzRoy a now missing letter containing a draft of his preface and his acknowledgments, probably intended both for this book and for the first volume of the zoology series now nearing publication.

FitzRoy was appalled at the perfunctory nature of his draft and wrote back: ". . . I will now tell you frankly my ideas upon the subject of prefaces to *any* of yours [*sic*] works immediately resulting from the

---

2. One thousand pounds is roughly equivalent to 110,000 pounds today.

Beagle's voyage. He continued, "Most people (who know anything of the subject) are aware that your going in the *Beagle* was a consequence of my original idea and suggestion—and of my offer to give up part of my own accommodations—small as they were—to a scientific gentleman who would do justice to the opportunities so afforded.—Those persons also know how much the Officers furthered your views—and gave you the preference upon all occasions—(especially Sulivan—Usborne—Bynoe and Stokes)—and think—with me—that a plain acknowledgement— without a word of flattery—or fulsome praise—is a slight return due from you to those who held the ladder by which you mounted to a position where your industry—enterprise—and talent could be thoroughly demonstrated—and become useful to our countrymen—and—I may truly say—to the world."

FitzRoy regretted the lack of acknowledgment of their own friendship and reaffirmed his astonishment "at the total omission of any notice of the officers—either particular—or general.—My memory is rather tenacious respecting a variety of transactions in which you were concerned with them; and others in the *Beagle*. Perhaps you are not aware that the ship which carried us safely was the first employed in exploring and surveying whose Officers were not ordered to collect—and were therefore at liberty to keep the best of all—nay, all their specimens for themselves. To their honour—they gave you the preference."

FitzRoy concluded, "Believe me Darwin—I esteem *you* far too highly to break off from you willingly—I shall always be glad to see you—and if there is any question to be discussed let us *talk* it over here—or in your room—before referring it to the partial views and perhaps selfish feelings of persons who neither know, nor feel for, you—or for me—as your Father would feel for either of us. Pray believe me Very sincerely yours Robt. FitzRoy."

Darwin did not take up the offer of a meeting as far as is known, but hastily added some emollient compliments, including thanks to the officers. This smoothed matters over, superficially at least, but marked a further distancing and deterioration in his relationship with FitzRoy. Only a few months later Darwin was writing to an inquirer, "The Captain is going on very well,—that is for a man, who has the most consummate skill in look-ing at everything and everybody in a perverted manner." The exchange on Darwin's preface did not reflect well on either man but least well on Darwin, who was at times more inclined to seek favors than to acknowledge or return them—an accusation FitzRoy had leveled at him when they quarreled in Valparaiso. The apparently perfunctory thanks he initially expressed sug-gest that a somewhat swollen-headed Darwin felt he had outgrown FitzRoy and indeed lacked any recognition that without FitzRoy he would never have made the voyage and acquired the information on which, as FitzRoy pointed out, he was now rising to fame. FitzRoy was basically correct in his complaints on behalf not only of himself but of his officers, even if the expression of his remonstrances was redolent of one of those "hot coffee" moments, of which his officers had complained aboard the *Beagle*.

Perhaps a less likely but alternative reason for Darwin's lack of atten-tion to the feelings of his companions may be that he was feeling unwell. In September he had told Henslow, "I have not been very well of late with an uncomfortable palpitation of the heart," and would soon be writing to his sister Caroline, "I find the noddle [brain] and the stomach are antagonistic powers, and that it is a great deal more easy to think too much in a day, than to think too little—What thought has to do with digesting roast beef,—I cannot say, but they are brother faculties. I am living very quietly, and have given up all society." These illnesses, even if intermittent, would only get worse as the years passed.

Despite Darwin's last comment to his sister, his ill health did not hamper his social life entirely. On June 21, 1838, he was elected a mem-ber of the Athenaeum Club; another new member that day was Charles Dickens. Darwin would often sit reading in the club's extensive library and wrote that he felt "like a gentleman, or rather like a Lord . . . I enjoy it the more, because I expected to detest it . . . one meets so many people there, that one likes to see."

During the spring of 1838, Darwin visited the Zoological Society Gar-dens to observe the orangutan Jenny, only recently brought to Britain from Borneo by a returning sailor and sold to the zoo for £150 in late

1837. Ironically, at a time when any suggestion of a relationship between human and ape was considered anathema, both she and a chimpanzee named Tommy, a previous brief resident of the zoo, were usually dressed in children's clothes and their behavior almost universally interpreted anthropomorphically.[3] Darwin's own first impressions also emphasized the similarity with humans. When the keeper showed Jenny an apple and would not hand it to her, he noted, "She threw herself on her back, kicked and cried, precisely like a naughty child.—She then looked very

Jenny the Orangutan

sulky and after two or three fits of passion, the keeper said 'Jenny if you will stop bawling and be a good girl, I will give you the apple.'—She certainly understood every word of this, and, though like a child, she had great work to stop whining, she at last succeeded, and then got the apple, with which she jumped into an arm chair and began eating it, with the most contented countenance imaginable."

Further visits to Jenny confirmed Darwin's wish to study in more detail the links between human and ape. His initial reaction, written in two of his notebooks, was significant: "Compare the Fuegian and Ourang outang and dare to say difference so great . . ."; "Man in his arrogance thinks himself a great work, worthy the interposition of a deity, more humble and I believe truer to consider him created from animals." The last thought would not be published until more than thirty years later in his *The Descent of Man*, which followed on from *On the Origin of Species*.

During the summer of 1838, Darwin traveled to Scotland, journeying by steam packet. He wrote smugly that he "enjoyed the spectacle, wretch that I am, of two ladies and some small children quite seasick, I

3. London Zoo only stopped chimpanzees' tea parties in 1972, and a leading British tea brand, Brooke Bond, ran anthropomorphic chimpanzee advertisements until 2002.

being well." During his visit he studied what were known as the "parallel roads"—distinct geological bands along the sides of Glen Roy. Basing his analysis on his examination of apparently similar strata in Chile, he decided they were remnants of successive sea beaches moved up to higher levels by the effect of Lyell's gradual movement of the Earth—a conclusion he published in a paper to the Royal Society. This was, perhaps, the only occasion he was entirely and publicly wrong—"one long gigantic blunder," he himself called it just a few years later when the parallel roads were proved to be margins of a disappeared lake built up by a glacier damming the valley entrance. Perhaps when reaching his erroneous conclusion Darwin's mind was full of too much else to pay careful attention. Perhaps for once, rather than study the evidence to reach a theory, he had brought preconceived ideas from the *Beagle* voyage, praised by the Geological Society, and assumed that what he saw fitted them.

On his return to London, Darwin had a rare meeting with FitzRoy, who "made a kind of growl" as he criticized Lyell's latest work, *Elements of Geology*, causing Darwin to write to Lyell, "I never cease wondering at his character, so full of good and generous traits but spoiled by such an unlucky temper.—Some part of . . . his brain wants mending: nothing else will account for his manner of viewing things."

Meanwhile, Darwin continued his extensive reading in the library of the Athenaeum Club and elsewhere, with the result that, as he recorded in his autobiography, "In October 1838, that is, fifteen months after I had begun my systematic enquiry, I happened to read for amusement Malthus on *Population,* and being well prepared to appreciate the struggle for existence which everywhere goes on from long-continued observation of the habits of animals and plants, it at once struck me that under these circumstances favourable variations would tend to be preserved, and unfavourable ones to be destroyed. The result of this would be the formation of new species. Here, then, I had at last got a theory by which to work."

It is perhaps surprising that given his acquaintance with Harriet Martineau, a great proponent of Malthus's theory, Darwin had taken so long to read him and did so only for amusement. However, reading Malthus seems to have been much closer to a eureka moment than other occasions such as his Galápagos visit. Malthus gave him an explanation for how the transmutation he had speculated upon toward the end of the *Beagle* voyage, and whose presence had been to a great extent established by the expert study of his *Beagle* specimens, had occurred. Even though variation appeared randomly in nature, those variations that were best suited to their environment, as his grandfather Erasmus Darwin had theorized and his own detailed observations aboard the *Beagle* suggested, allowed those possessing them to survive and to outbreed others. Less than two years after his return on the *Beagle* and when he was still under thirty years of age, the structure of what he now called "my theory" was essentially complete, even if he felt the need for further evidence and data to test and to confirm it.

CHAPTER TWENTY-TWO

"It Is Like Confessing a Murder"

At the same time as secretly pondering transmutation and Malthus's views on the perils of excessive reproduction, Darwin was contemplating marriage—at first simply as a concept with no particular prospective partner in mind. Compulsive lister, note-maker, and analyzer as he was, he set down on a sheet of blue paper two columns, *Marry* on the left-hand side and *Not Marry* on the right:

This is the question:

Marry

children——(if it Please God)
—Constant companion, (and friend in old age) who will feel interested in one,—object to be beloved and played with.——better than a dog anyhow.——Home, and someone to take care of house—Charms of music and female chit-chat.—These things good for one's health.——*but terrible loss of time.*——
My god, it is intolerable to think of spending one's whole life, like a neuter bee, working, working, and nothing after all.—— No, no won't do.——Imagine living all one's day solitarily in smoky dirty London House.——

Not Marry

Freedom to go where one liked—choice of Society *and little of it.*——Conversation of clever men at clubs—Not forced to visit relatives, and to bend in every trifle.——to have the expense and anxiety of children —perhaps quarrelling—Loss of time.——cannot read in the Evenings—fatness and idleness— Anxiety and responsibility—less Money for books and etc—if many children forced to gain one's bread.——(But then it is very bad for one's health to work too much)
Perhaps my wife won't like London; then the sentence is

Only picture to yourself a nice
soft wife on a sofa with good
fire, and books and music perhaps
—Compare this vision with the
dingy reality of Grt. Marlbro' St.
Marry—Marry—Marry Q. E. D.

banishment and degradation into
indolent, idle fool.

On the back of the paper, he added to the analysis, "It being proved necessary to marry—When? Soon or Late. The Governor says soon for otherwise bad if one has children—one's character is more flexible— one's feelings more lively, and if one does not marry soon, one misses so much good pure happiness.—

"But then if I married tomorrow: there would be an infinity of trouble and expense in getting and furnishing a house,—fighting about no Society—morning calls—awkwardness—loss of time every day— (without one's wife was an angel and made one keep industrious)— Then how should I manage all my business if I were obliged to go every day walking with my wife.—Eheu!! I never should know French,—or see the Continent,—or go to America, or go up in a Balloon, or take solitary trip in Wales—poor slave, you will be worse than a negro— And then horrid poverty (without one's wife was better than an angel and had money)—Never mind my boy—Cheer up—One cannot live this solitary life, with groggy old age, friendless and cold and childless staring in one's face, already beginning to wrinkle. Never mind, trust to chance—keep a sharp look out.—There is many a happy slave—"

Given that Darwin exhibited a sense of humor in his private letters, one cannot believe that there was not at least a trace of a smile on his face as he wrote, however patriarchal he sounded. Nevertheless, his conclusion to marry was in earnest. The question was whom? Prior to the *Beagle* voyage, his sisters had thought he might marry his Maer cousin Fanny Wedgwood, but she had died during the voyage. His own youthful enthusiasms had centered on Fanny Owen, to whom even after his return he had sent flowers and found great difficulty in referring to as Mrs. Biddulph. (Her marriage was rumored to be unhappy.) His eldest sister

Charles Darwin, 1840

Caroline had recently married her Maer cousin, Josiah Wedgwood III.[1] Darwin's mind seems quickly to have turned to Emma, the youngest of his Maer cousins. Still unmarried at the advanced age, for a woman of the time, of thirty, she was intelligent, educated, spoke Italian, German, and French, and was said to have read *Paradise Lost*, Darwin's favorite book on the *Beagle*, when she was four. She already seems to have had feelings for Darwin, and when he proposed one Sunday in November 1838 in the library at Maer, she readily accepted before, despite the excitement, going off to teach at the local Sunday school.

Their families were delighted. Her father, Darwin's "Uncle Jos," placed five thousand pounds in a trust fund for Emma and gave her a personal allowance of four hundred pounds per annum. The use of a trust fund circumvented the then legal position that all married

---

1. Their grandson would be the composer Ralph Vaughn Williams.

women's property was their husbands'—something not changed until 1882. Dr. Darwin gave the happy couple ten thousand pounds in stocks in canals, railways, and property, designed to produce an income of six hundred pounds per annum.

Emma recorded her views of her fiancé: "He is the most open, transparent man I ever saw, and every word expresses his real thoughts. He is . . . perfectly sweet tempered, and possesses some minor qualities that add particularly to one's happiness, such as not being fastidious, and being humane to animals."

Darwin's openness, albeit also his selfishness, showed in a letter he wrote to Emma. His happiness rested "on quietness and a good deal of solitude; but I believe the explanation is very simple, and I mention it because it will give you hopes, that I shall gradually grow less of a *brute*,—it is that during the five years of my voyage (and indeed I may add these two last) which from the active manner in which they have been passed, may be said to be the commencement of my real life, the whole of my pleasure was derived, from what passed in my mind, whilst admiring views by myself, travelling across the wild deserts or glorious forests, or pacing the deck of the poor little *Beagle* at night.—Excuse this much egotism,—I give it you, because I think you will humanize me, and soon teach me there is greater happiness than holding theories, and accumulating facts in silence and solitude."

His openness also showed in his talking, despite warnings from his father, to Emma about his by then dwindling religious beliefs and very probably about his new theory. A genuine and wholehearted Christian believer, as exemplified by her Sunday school teaching, Emma clearly feared the consequences of his apostasy becoming total—that they might not share the afterlife. She wrote to her fiancé, "Since you are gone some sad [thoughts] have forced themselves [into my mind] of fear that our opinions on the most important subject should differ widely. My reason tells me that honest and conscientious doubts cannot be a sin, but I feel it would be a painful void between us. I thank you from my heart for your openness with me and I should dread the feeling that you were concealing your opinions from the fear of giving me pain. It is perhaps foolish

Emma Darwin, 1840

of me to say this much but my own dear Charley we now do belong to each other and I cannot help being open with you. Will you do me a favour? . . . it is to read our Saviour's farewell discourse to his disciples which begins at the end of the 13th Chap. of John. It is so full of love to them and devotion and every beautiful feeling. It is the part of the New Testament I love best . . . though I can hardly tell why I don't wish you to give me your opinion about it." In addition to the devotional and beautiful feelings and wise and altruistic morality, the passage also warned, "If a man abide not in me, he is cast forth as a branch and is withered; and men gather them . . . and they are burned."

What physical contact the two had before marriage is obviously unknown, but after the engagement, Darwin, ever a lister and detached observer, wrote, "November 27th sexual desire makes saliva to flow, yes *certainly*—curious association; I have seen Nina [the dog] licking her chops,—someone has described slovering teethless-jaws, as picture of disgusting lewd old man, ones tendency to kiss, and almost bite, that which one sexually loves is probably connected with flow of saliva and hence the action of *mouth* and jaws—Lascivious women are described as biting so do stallions always." He went on to speculate that blushing too was sexual, driving "blood to surface exposed, face of man . . . bosom in woman; like erection."

In preparation for their marriage, which took place on January 29, 1839, they had rented a house in London's Upper Gower Street. Despite its yellow curtains, bright blue walls, and red upholstered furniture, which caused Darwin to name it Macaw Cottage after the South American parrot, and the initial presence of a dead dog in the garden, they both liked it. Darwin did not look forward to the prospect of the wedding ceremony and celebration, dreading "the awful day," but it passed quickly. Mr. and Mrs. Charles Darwin were soon on the train to London's Euston Station.[2]

As the newlyweds adjusted to married life, Darwin, who had been elected a Fellow of the Royal Society five days before his wedding—not a cheap honor, the membership admission fee was seventy pounds—continued his scientific work. He did so without Syms Covington, who left his employment at the end of 1838 with a leaving present of only one pound, and a few months later emigrated to Australia. Before their marriage Emma had written to Darwin, "I am afraid poor Covington will hate the sight of me." However, it is not clear why Covington left or was let go. Darwin remained in friendly contact with him and occasionally asked him for information on and specimens of Australia's nature.

In early summer 1839, Colburn published four volumes on the two voyages of the *Beagle*—the three originally envisaged and a fourth containing 350 pages of appendices. At the end of his volume on the second

---

2. The railway system was the great innovation of the 1830s. In September 1830, the world's first passenger railway—the Liverpool and Manchester—had proved a success despite a local Member of Parliament being run over and killed by a locomotive on the first day of operation. By the end of the 1830s, rail lines were snaking across the country. By 1845, there were 2,240 miles of railway track laid, and thirty million passengers were being carried each year.

voyage of the *Beagle,* FitzRoy included two final chapters—numbers 27 and 28—almost certainly written in response to what he had read in Darwin's volume, available to him in proof since autumn 1837, and what he had heard about Darwin's views. Chapter 27 contained speculations about how races diverged following the Creation, the fortunate becoming West Europeans, and the most fortunate becoming Englishmen, whereas others degenerated to the condition of the Fuegians. He attacked any thought that humans were created in a juvenile or primitive state: "That man could have been first created in an infant, or a savage, state appears to my apprehension impossible . . . because—if an infant—who nursed, who fed, who protected him till able to subsist alone? and if a savage, he would have been utterly helpless . . . and after a few hours of apathetic existence he must have perished. The only idea I can reconcile to reason is that man was created perfect in body, perfect in mind . . . Have we a shadow of ground for thinking that wild animals or plants have improved since their creation?"

FitzRoy's last chapter, 28, was a refutation of Lyell's geological theories, about which he had "growled" a year earlier, and a recantation of any doubts about the literal truth of the Old Testament he had expressed during the *Beagle* voyage. He wrote, "I suffered much anxiety in former years from a disposition to doubt, if not disbelieve, the inspired History written by Moses. I knew so little of that record, or of the intimate manner in which the Old Testament is connected with the New, that I fancied some events there related might be mythological or fabulous . . . Much of my own uneasiness was caused by reading works written by . . . geologists who contradict, by implication, if not in plain terms, the authenticity of the Scriptures; before I had any acquaintance with the volume they so incautiously impugn." He referred in particular to his experience in 1834 when, on the expedition along the Santa Cruz river in Patagonia, he had said to Darwin that the vast plains with diluvial detritus some hundred feet in depth "could never have been effected by a forty days' flood." He now recanted this view, affirming he believed thoroughly in the flood and Noah's ark. He suggested that the ark (which Sir Isaac Newton had calculated from Genesis was 537 feet long, 80 feet wide, 51 feet high, and weighed 18,231 tons) was too small for some animals—such as the giant rhinoceros, whose fossils had been found a few years before, when London's Trafalgar Square was being built—to enter and that some men "in their moral blindness" had refused to go aboard. "As the creatures approached the ark, might it not have been easy to admit some, perhaps the young and small, while the old and large were excluded?"

Darwin, who now privately believed "the Old Testament was no more to be trusted than the sacred books of the Hindoos [sic]"—after all, his transmutation theory would obliterate the first chapters of Genesis and with them Adam and Eve—wrote to his sister Caroline, "You will be amused with FitzRoy's Deluge Chapter.—Lyell, who was here today, has just read it, and he says it beats all the other nonsense he has ever read on the subject.—I was delighted a few days since by hearing the news that the FitzRoys are going to move to a house, given them by their uncle Col. Wood, 15 miles from London.—I am right glad of it—Although I owe very much to FitzR. I for many reasons, am anxious to avoid seeing much of him." One surprising reason he admitted for avoiding FitzRoy was that he found his wife "rather too patronizing."

Darwin had sent a copy of his own volume to the elderly Alexander von Humboldt in Potsdam and was delighted when his hero wrote back fulsomely: "You told me in your friendly letter that, when you were young, my manner of studying and depicting nature in the torrid zones contributed to exciting in you the ardour and desire to travel in far-off lands. Considering the importance of your work, Sir, this may be the greatest success that my humble work could achieve."

Reviews of the four volumes were generally favorable, although Fitz-Roy's chapter on the flood was criticized. One reviewer wrote, "On this subject the gallant captain has got quite beyond his depth." *The Edinburgh Review* praised FitzRoy for expending "considerable sums from his private funds to complete the survey of the Peruvian coast." *The United Service Journal* called Darwin's volume a "fund of entertaining instruction." *The Quarterly* thought Darwin "a first-rate landscape-painter with a pen," giving

Joseph Hooker

"ample materials for deep thinking." Darwin's natural history colleagues also praised his work. Owen said, "It is as full of good original wholesome food as an egg."

Darwin received an unrelated and spontaneous compliment when on September 9, 1839, his old ship-mates John Wickham and John Lort Stokes, now commander and first lieutenant of the *Beagle* on a mission to survey the Australian coast, took her into a harbor in northwestern Australia, which was unknown to Europeans. Noting some unusual geological features, they believed "this afforded us an opportunity of convincing an old shipmate and friend that he still lived in our memory," and so they named the harbor Darwin. The town that grew up around it was renamed Darwin in 1911 and is now the capital of Australia's Northern Territory.[3]

Also in 1839, Darwin met Joseph Hooker, who would become his good and lifelong friend and great supporter. The twenty-two-year-old Hooker was the son of the director of Kew Gardens and would in due course succeed his father in that post. Already a keen botanist but unable to fund independent travel, he was about to embark on James Clark Ross's expedition to the Antarctic in the ships HMS *Erebus* and HMS *Terror* as the assistant surgeon on the *Erebus*. Darwin bumped into Hooker while Hooker was walking with Robert McCormick, the former surgeon of the *Beagle* who was to be the *Terror*'s surgeon. Hooker was impressed by

3. In 1845, two years after the *Beagle*'s return, the navy decommissioned her, stripping her of her masts and spars. The hulk was moored in the River Roach near Paglesham in the Essex marshes in southeast England and used by the coastguard service as a watch station. In 1870 she was sold for scrap for £525. In 2019 marine biologists using drones with infrared cameras began searching the mudbanks near Paglesham, where they hope remains of the *Beagle* still lie.

Darwin's "animated expression, heavy beetle brow, mellow voice, and delightfully frank and cordial greeting to his former shipmate."

On December 27, 1839, the Darwins became parents with the birth of William Erasmus Darwin. As the birth approached, Darwin's headaches and stomach problems increased. "What an awful affair a confinement is: it knocked me up, almost as much as it did Emma herself," Darwin confessed. He was proud of his new son, enjoyed playing with him unlike many a Victorian father, and from his very first days began assiduously making jottings in a notebook reserved for the purpose about his behavior and progress to add to his material on human development. On one occasion he gave William a mirror to compare his responses to seeing himself to those of Jenny the orangutan. On another, he tested William's reaction to loud noises, sneezing and shouting behind the baby to assess his sense of fear.

Throughout 1840 Darwin was frequently ill. In a letter to FitzRoy he used his illness as a reason for not having encountered his old captain in London: "My health has been very indifferent during the two last months . . . My stomach as usual has been my enemy . . . I am not surprised at our never having met in the street for I do not go to the west end of the town more than once a week." He continued, "However others may look back to the *Beagle*'s voyage, now that the small disagreeable parts are well nigh forgotten, I think it far the *most fortunate circumstance in my life* that the chance afforded by your offer of taking a naturalist fell on me—I often have the most vivid and delightful pictures of what I saw on board the *Beagle* pass before my eyes.—These recollections and what I learned in Natural History I would not exchange for twice ten thousand a year."

Due to his "long continued illness" Darwin did much less work on editing the *Beagle*'s zoological volumes and his own *Beagle* geology books than he would have hoped during the year, spending much of it at The Mount and Maer. Weighing him as usual on his visits, as he did all his children, his father found he had lost ten pounds in a year and eighteen since he returned from the *Beagle*. He now weighed ten stone eight pounds (148 pounds), quite light for a man six feet tall. Even so, Darwin complained that his father was not entirely sympathetic to his symptoms. A second child was born to the Darwins on March 2, 1841—a girl, Annie. Darwin again immediately began noting her behavior and development. His health improved a little after the birth, and he was able to do more work but nevertheless spent much of the summer at Maer and The Mount a semi-invalid. At The Mount, he persuaded his father that for many reasons, including his health, the need for more space for his growing family and staff, and for "pure air" and a garden, he should move out of London and its "dirt, noise, vice and misery." Dr. Darwin agreed to advance him up to three thousand pounds at 4 percent interest against his inheritance to fund a suitable purchase.

Darwin was buoyed by the prospect of leaving London, where increasing unrest from unruly elements among supporters of the People's Charter was an additional incentive to move. Chartism was a working-class movement that, not content with the changes brought by the 1832 Reform Act, with living standards reducing through an economic depression, and the introduction of the 1834 Malthus-inspired Poor Law, made six demands: votes for all men; equal electoral districts; abolition of the requirement that Members of Parliament be property owners; payment for MPs; annual general elections; and the secret ballot.

Darwin's health improved further, and in early 1842 he sent to the publishers Smith, Elder & Co. the completed draft of the first of his books on the geology of the *Beagle* voyage. The subject was coral reefs, the topic he had researched in the Pacific and then in the Keeling Islands and Mauritius in the Indian Ocean. One of the work's key features was the integration of detailed charts and cross-sections into the text and the inclusion of an unprecedented map of the reefs of the Pacific and Indian Oceans. Darwin rightly believed that such images spoke "more plainly to the eye, than any description could do to the ear." His second geological volume—on the geology of volcanic islands—would appear in 1844, and the third and final one on the geology of South America in 1846.

In May 1842 at Maer, as strikes and serious Chartist rioting broke out around the Midlands potteries, which included those of the Wedgwoods, Darwin produced a thirty-five-page sketch of his transmutation theory. Although written in pencil in a scrawling hand and full of crossings out and

additions, it laid out the same basic structure and arguments as would his
On the Origin of Species many years later. Darwin began with a description
of selective breeding by humans of domestic animals, followed by one of
varieties and transmutation in nature as he had observed on the *Beagle*. He
then described the theory of natural selection based on his understanding
of Malthus's work as the mechanism by which the changes in the natural
world came about. He also postulated a role for sexual selection whereby the
strongest animals best suited to their environments had the greatest success
in attracting mates and thereby produced the most offspring. He went on
to make another important statement: "As all Mammals have descended
from one stock, we ought to expect that every continent has been sometime connected"—a concept that he came to doubt and that was not fully
established until Alfred Wegener's *Origins of Continents and Oceans* in 1912.

Darwin pointed out the impact of geographical isolation, time,
and changing environmental conditions. He then listed the difficulties
of his theory, including where evidence was thin and what questions
remained. He concluded, "The graduations by which each individual
organ has arrived at its present state, and each individual animal with
its aggregate of organs, probably never could be known, and all present
great difficulties. I merely wish to show that the proposition is not so
monstrous as it first appears." He did not address the role of any Creator
or divine being in any of this.

Back in London, with Chartists rioting in the street outside and
around nearby Euston Station and troops called in to confront them, he
and Emma, pregnant again for the third time in three years, turned to
house-hunting in earnest. Their attention soon alighted on Down House,
a former parsonage in the Kentish Weald village of Downe, some sixteen
miles from London but that could be reached by a combination of horse
and carriage—Darwin quickly bought a new carriage for one hundred
pounds—and train in around two hours. Down House was not a particularly beautiful or grand building, but set in eighteen acres overlooking
the southern downs, it seemed to serve their purpose well.

They quickly completed the purchase for 2,200 pounds and in mid-
September 1842 moved in. Shortly afterward, Emma gave birth to a

daughter, Mary, who only lived three weeks. As the grieving parents recovered from their terrible loss, they embarked on improvement works. Since Darwin found "the publicity of the place at present is intolerable," one of their first acts was to have the lane that ran alongside and close to the house, allowing people to stare into Darwin's study, moved farther away. They had a bank raised between the lane's new position and the house and topped it with a flint wall, achieving the privacy Darwin required. Later Darwin would place a mirror strategically by the front door that he could see from his study, forewarning him of any visitors intruding on the quiet and seclusion he increasingly craved, whether ill or not. The house had no running hot water or bathrooms, though this was not unusual for the times, even for the wealthy. His frequent sickness led Darwin to install a curtained-off alcove containing a makeshift lavatory in a corner of his study, where he could vomit when illness struck. Darwin kept his snuff jar just outside the study door so that he would have to make an effort to get it, in an attempt to avoid using quite so much of it.

Over the next years, Darwin and Emma developed a daily routine at Down House—"clockwork" Darwin called it—that did not vary greatly, except when interrupted by illness, throughout the rest of his life. He got up and went for a walk at around 7:00 A.M. After breakfast, which he took alone at 7:45, he worked in his study until 9:30, when he moved to the drawing room to read his professional post and then lay on the sofa while Emma read to him family letters and perhaps something from a novel, until around 10:30, when he went back to his study to work until 12:00 or 12:30. He then went for a walk with one of his dogs along "the sand walk," "a narrow strip of land one and a half acres in extent with a gravel path around it not far from the house," which Darwin had had laid out soon after buying the property. After lunch, at which, like at other meals, he had "a boy-like love of sweets," he read his newspaper and wrote his letters and the manuscripts of his books, sitting "in a huge horse-hair chair by the fire, his papers supported on a board resting on the arms of the chair."

At around 3:00 P.M. Darwin retreated to his bedroom to relax by smoking a cigarette—he became a keen smoker as well as snuff

1. The Exterior from the Garden.—2. Mr. Darwin's Study.
THE HOME OF THE LATE CHARLES DARWIN, DOWN, KENT

Charles Darwin's study, Down House

taker—while Emma again—or in later life, one of his children—read to him from a novel. According to his son Francis, the only nonscientific work Darwin read himself was the newspaper; everything else, such as novels or travel books, was read aloud to him: "He would not enjoy any story with a tragical end, and for this reason he did not keenly appreciate George Elliott . . . Walter Scott, Miss Austen and Mrs Gaskell were read and reread until they could be read no more . . . He did not read out-of-the-way books but generally kept to the books of the day obtained from a circulating library. His literary tastes and opinions were not on a level with the rest of his mind."

Darwin went for another short walk at 4:00, then returned to his study to work until around 5:30. After dinner, he usually played two games of backgammon with Emma. They were both competitive and kept their scores over a long period. Darwin boasted to a friend, "Now the tally with my wife in backgammon stands thus; she, poor creature, has won only 2,490 games, whilst I have won, hurrah, hurrah, 2,795 games!" Also sometimes in the evenings, Emma—a talented pianist who had in her youth had a few lessons from Frédéric Chopin in Paris—played the piano for him. Darwin would then usually read some scientific books or periodicals before retiring to bed at 10:30.

Further children were born to Emma and Charles Darwin in quick succession—Henrietta (1843), George (1845), Elizabeth (1847), Francis (1848), Leonard (1850), Horace (1851), and finally Charles Waring (1856). Emma was not a particularly fastidious or tidy household manager. One of her daughters wrote that her mother's childhood nickname, Little Miss Slip-Slop, was revealing: "She was never tidy or orderly as to little things . . . My father said after he married that he made up his mind to give up all his natural tastes for tidiness, and that he would not allow himself to feel annoyed by her calm disregard for such details." Their relations, particularly Darwin's sisters, sometimes criticized both Emma's housekeeping and her decision to allow her children to run free rather than to discipline them to conform to social norms. One particular criticism was of her allowing them to develop cockney accents acquired from their servants, of whom they had a considerable number. These included a butler, Joseph Parslow, who would remain with the Darwins for the rest of his working life, a cook, a children's nurse who had previously looked after the novelist William Thackeray's daughters, a footman, maids, and gardeners.

The expanding post office system allowed Darwin to keep in touch with scientific contacts in London, elsewhere in Britain, and overseas. In 1840 Rowland Hill had introduced the penny post with the world's first postage stamp—the penny black. Soon twenty-five thousand postmen were carrying six hundred million letters a year across Britain. Though many have disappeared, more than eight thousand letters written by Darwin survive as well as nearly seven thousand written to him—a boon to historians.

In 1843, Darwin was once more in contact with Robert FitzRoy, this time writing to him to introduce a friend of his, the secretary of the Geological Society, whose son wanted a position with FitzRoy. In mid-1841

THE EVOLUTION OF CHARLES DARWIN

the Whig government led by Lord Melbourne, the guide and favorite of Queen Victoria in her first three years on the British throne, had fallen leading to an election. Lord Londonderry (Castlereagh's brother), Fitz-Roy's uncle and a good friend of the prospective Tory prime minister Sir Robert Peel, offered his nephew the chance to stand as Tory candidate for the Durham constituency, where he owned large estates and thus exerted considerable influence. FitzRoy had no hesitation in accepting. In the previous Parliament the constituency, which returned two members, had elected one Tory and one Whig. Now, each party initially put forward only a single candidate, leaving the election an apparent formality. However, another faction within the Tory party nominated a second Tory candidate—twenty-six-year-old William Shepherd—turning the election into a real contest.

FitzRoy and Shepherd collaborated in the hope of returning two Tory MPs until—Shepherd claimed—a voter disclosed to him that Lord Londonderry, his landlord, had insisted he should vote for FitzRoy rather than for Shepherd. Shepherd publicly and violently lambasted Londonderry before resigning his candidacy. Although the withdrawal made FitzRoy's election a certainty, he loyally defended his uncle, publicly attacking Shepherd's "disgraceful desertion of the Tory party . . . unparalleled in history." In response Shepherd labeled FitzRoy "a liar, a slanderer and—a coward and a knave."

The hostility between the two men continued after the formality of FitzRoy's election to the House of Commons as a member of the victorious Tories.

On August 25, 1841, Shepherd, whirling a horsewhip above his head, accosted FitzRoy in London's Mall, shouting, "Captain FitzRoy! I will not strike you but consider yourself horsewhipped." FitzRoy did actually strike Shepherd, hitting him with his umbrella and then, discarding it, grappled his opponent to the ground, whereupon a friend of Shepherd materialized, shouting, "Don't strike him, Captain FitzRoy, now he's down!" FitzRoy withdrew with as much dignity as he could muster.

The incident provided some sensational copy for the public press but was soon forgotten. FitzRoy began his parliamentary duties speaking, on the basis of his experiences on the *Beagle*, about colonization and on maritime subjects. He proposed that to improve maritime safety all masters or chief mates of merchant ships should be required to have formal qualifications—a measure gradually adopted over the next decade, leading to a system of certification or "tickets."

By that time, however, FitzRoy had long ceased to be a Member of Parliament. In spring 1843 he resigned his seat to become the second Governor of New Zealand, and it was in that capacity that Darwin wrote to him on behalf of his friend's son and received a sympathetic but noncommittal reply.

Darwin had all this time been working on his transmutation—evolutionary—theory. In early 1842 he had raised it with Lyell only to receive a skeptical response. Now, somewhat surprisingly, he mentioned it to young botanist and surgeon Joseph Hooker. He had returned a few months earlier from James Clark Ross's Antarctic expedition, which had penetrated to 78 degrees, 10 minutes south, through the sea now named after him to the great ice barrier (Ross Ice Shelf)—a record that stood for nearly sixty years. Hooker had told his father how much he had benefited from Darwin's diary during his voyage, and Darwin had offered to lend him his Patagonian botanical specimens to help him with writing up his own work—*Flora Antarctica*.

Darwin wrote to Hooker, "I have been now ever since my return [from the *Beagle*] engaged in a very presumptuous work and which I know no one individual who would not say a very foolish one.—I was so struck with distribution of Galapagos organisms etc etc and with the character of the American fossil mammifers, etc etc that I determined to collect blindly every sort of fact, which could bear any way on what are species . . . At last gleams of light have come, and I am almost convinced . . . that species are not (it is like confessing a murder) immutable." Hooker responded with interest, sparking a growing friendship, "I shall be delighted to hear how you think that this change may have taken place, as no presently conceived opinions satisfy me on the subject."

While corresponding with Hooker, Darwin was working up a longer, better constructed abstract of his theory. When complete it ran to some 189 pages of his own handwriting but to 231 when Darwin asked a local schoolmaster to transcribe a clean copy. Darwin then wrote a letter to Emma into which he had clearly put much thought: "My dear Emma, I have just finished my sketch of my species theory. If, as I believe that my theory is true and if it be accepted even by one competent judge, it will be a considerable step in science. I therefore write this, in case of my sudden death, as my most solemn and last request, which I am sure you will consider the same as if legally entered in my will, that you will devote £400 to its publication and further will yourself, or through Hensleigh [Wedgwood], take trouble in promoting it.——I wish that my sketch be given to some competent person, with this sum to induce him to take trouble in its improvement and enlargement.——I give to him all my books on Natural History, which are either scored or have references at end to the pages, begging him carefully to look over and consider such passages, as actually bearing or by possibility bearing on this subject . . . I also request that you (or some amanuensis) will aid in deciphering any of the scraps [of paper] which the editor may think possibly of use." He went on to suggest Lyell would be the best editor. Other possibilities might include Henslow or Hooker. He concluded, "If there should be any difficulty in getting an editor who would go thoroughly into the subject and think of the bearing of the passages marked in the books and copied out on scraps of paper, then let my sketch be published as it is . . ."

Then, in October 1844, a book by an anonymous author titled *Vestiges of the Natural History of Creation* appeared and quickly became a bestseller. Page-turningly well written and designed for the "ordinary reader," not the natural history experts or "dogs of clergy" as the author dubbed them, it contained an impressionistic theory of the development of the universe, and of life on Earth from the smallest, simplest creatures such as mites through to human beings in a process of continuous improvement. The work lacked detailed evidence or data or any clear mechanism for the transmutation it postulated. Some of its theorizing was clearly incorrect, such as suggesting that insect life was spontaneously generated by electricity or that individual embryos passed through several evolutionary changes in the womb and even, if that process were interrupted, one species could give birth to another, for example a goose to a rat. Nevertheless, despite the detail being poor and the geology striking Darwin as "bad, and his zoology far worse," after he read *Vestiges* in the British Museum reading room, he was downcast that in some sense his own thinking was not unique since, like himself, the author believed that "species are shown not to be immutable."

The storm of outrage from scientific and clerical sources that broke over *Vestiges* did absolutely nothing to encourage Darwin to hasten the publication of his own, more evidence-based abstract. In a time of Chartist unrest in Britain and incipient revolutions across Europe, *Vestiges* challenged conventional ideas not only of religion, but of the structure of society exemplified in the popular hymn "All Things Bright and Beauteous," written four years later. The hymn begins:

All things bright and beauteous,
All creatures great and small,
All things wise and wondrous,
The Lord God made them all.

. . .

Each little flower that opens,
Each little bird that sings,
He made their glowing colours,
He made their tiny wings.

. . .

The rich man in his castle,
The poor man at his gate,
God made them, high or lowly,
And ordered their estate.

The *North British Review* denounced *Vestiges* as "poisoning the fountain of science and sapping the foundations of religion." Adam Sedgwick, in a review, considered it lacking any science and any reason—immoral, irrelevant, irreligious, and so bad he thought that it could have been written by a woman, even if no self-respecting woman would stoop so low. Privately Sedgwick believed that Ada Lovelace, the poet Lord Byron's daughter and the pioneer of computer software, was the degenerate author.

Among several other rumored authors of *Vestiges* was Darwin himself—a suggestion about which he wrote, "I ought to be much flattered and unflattered." The author was in fact Robert Chambers, a successful Edinburgh publisher. He never publicly confirmed his authorship, although it slowly became known in intellectual circles. Once asked why he never claimed the work as his own, Chambers "pointed to his house, in which he had eleven children, and then slowly added 'I have eleven reasons.'"

Hating controversy and confrontation and concerned about his reputation, Darwin plowed on privately, accumulating the data he knew he would need to avoid the castigation of the interlinked religious, academic, and scientific communities if and when he published his own theory. As he wrote in his autobiography, during his copious reading, "I followed a golden rule, namely . . . whenever [I encountered] a published fact, a new observation or thought . . . which was opposed to my general results, to make a memorandum of it without fail and at once; for I had found by experience that such facts and thoughts were far more apt to escape from the memory than favourable ones."

Darwin plunged into what turned out to be an eight-year study of barnacles, both living and fossils, sparked as he wrote by "a most curious form," lacking a shell, he had found on the coast of Chile while on the *Beagle* expedition. He initially examined some preserved specimens from that voyage, which he had not studied since his return home, but later sought specimens from far and wide, including from Syms Covington, now settled and married in New South Wales in Australia. In some ways this protracted study may have been a retreat, conscious or otherwise, from the maelstrom he feared if he went public with his evolutionary theory. Nevertheless, it added substantially to the depth of evidence supporting the theory in regard to not only the gradual nature of variations but also to the great frequency of their occurrence—species were constantly evolving. As he wrote later, it was "of considerable use" in his description of natural classification in *On the Origin of Species*. His discovery that not all barnacles were hermaphrodite, possessing both male and female sexual organs—as was thought to be the case—but that in a few species rudimentary males lived within the much larger females' shells, strengthened his belief that humans and indeed all animals had distant hermaphrodite ancestors, which would explain vestigial or redundant organs, such as male nipples.

As he propelled himself backward and forward from his desk to a specimen table under the window of his study on a specially modified chair with brass legs and strong castors, his immersion in his barnacles was so deep that one of his sons thought all fathers were so preoccupied, asking a friend where his father "did his barnacles." In later life Darwin doubted "whether the work was worth the consumption of so much time," but nevertheless his published monographs on his barnacle work, together with that on coral reefs, would be cited in the award to him of the Royal Society's Royal Medal in late 1853.

CHAPTER TWENTY-THREE

"Most Hasty and Extraordinary Things"

In 1845, Darwin was back in contact with Robert FitzRoy, writing him a consoling letter on his return to Britain following his dismissal as Governor of New Zealand. On his arrival, FitzRoy had found much changed in that country since the *Beagle* departed in early 1836, but it remained in deep internal crisis, with continuing disputes between Maoris, missionaries, and would-be settlers over the governance of the country and over land rights in particular.

Shortly after the *Beagle* left, Edward Gibbon Wakefield had formed the New Zealand Company to import settlers to farm the land. When he heard Wakefield's story, FitzRoy might well have recalled Baron de Thierry, the self-styled sovereign chief of New Zealand whom he had encountered in Tahiti on the *Beagle* voyage and denounced as an impostor. Although from a well-established family in Britain, Wakefield's early life had been scandalous. He had been imprisoned for persuading a fifteen-year-old girl to run away from school with him, lying to her that her father's business had failed and that if she wished to save any of the family fortune she must marry him so that it could be put in his name to protect it from creditors.[1]

Plausible and silver-tongued, Wakefield rehabilitated himself following his release with his plans for the colonization of New Zealand. Partly in response to his pressure, partly to that from the Church of England missionaries in New Zealand who wanted freedom from the prevailing chaos to allow them to get on with their work of "civilizing" and converting the Maoris, and partly to rumors of French and American ambitions

---

1. De Thierry had indeed arrived in New Zealand, only for Maori chiefs to repudiate his land claims.

in the islands, the British government had appointed William Hobson, a naval officer, to bring some order to the country.

On February 6, 1840, only five months after his arrival, Hobson and forty-five North Island chiefs signed the Treaty of Waitangi, under which the chiefs recognized British sovereignty while they and their followers acquired "all the rights and privileges of British citizens." Four months later, after obtaining the signatures of further chiefs, Hobson proclaimed British sovereignty over both main islands of New Zealand and was confirmed as the first governor.

Sovereignty only exacerbated disputes about land that, under the terms of Hobson's appointment, only the underfinanced administration could buy from the Maoris and sell on to settlers. In September 1842 Hobson died with the new colony nearly bankrupt. When the news reached London, FitzRoy appeared an obvious candidate for the post. In 1838 he had told a parliamentary committee that "the missionaries have opened the way . . . and now a more efficient and secure power ought to step in." In 1840, the secretary of the Church Missionary Society had proposed FitzRoy for the governorship, being fully aware of FitzRoy's sympathy for the missionary cause and writing, "He is deeply interested in New Zealand . . . He combines many valuable qualifications for an office which is certainly one of much difficulty and delicacy." Events would show that delicacy in difficult situations would never be one of FitzRoy's attributes, any more than in his *Beagle* days.

FitzRoy arrived in Auckland in late December 1843 and immediately quarreled with his deputy, who resigned and left on the same ship on which FitzRoy had arrived. The first major problem with which Fitz-Roy had to deal was the massacre at Wairu, near Nelson, by Maoris of nine white settlers. Wakefield's New Zealand Company claimed to have bought some parcels of land there from Maori chiefs. However, the two local chiefs, Rauparaha and Rangihaeta, insisted they owned the land and had made no such sale. Governor Hobson had ordered a land claims commission to investigate, but progress was slow.

Several members of the Wakefield family were in New Zealand supporting settlement efforts. One, Edward's brother Arthur, an ex-army officer, anticipated the commission's decision by sending surveyors onto the disputed land to split it into individual plots for his newly arrived and impatient would-be settlers. The Maoris harassed the surveyors and torched one of their huts. Arthur Wakefield persuaded the local magistrate to issue a warrant against them and set out with a motley band to enforce it. When they foolishly confronted a better-disciplined and equally well-armed (with muskets) band of Maoris defending their chiefs and their land, shots were exchanged, and most of Wakefield's men fled. He and eight others waved a white flag, inviting a truce to talk. The Maoris took it as a sign of surrender, and following—as they later claimed—their usual custom, "tomahawked" all their prisoners to death.

The killings brought outraged cries for vengeance from the settlers, led by Jerningham Wakefield, Edward's son and Arthur's nephew, who suggested that the Maoris must be "crushed like a wasp in the iron gauntlet of armed civilisation." FitzRoy's first action was to visit Nelson and lecture the settlers about the importance of "a good understanding between the settlers and the natives" before continuing that the settlers must do "all in their power to conciliate the natives, to forgive them, and to make allowances for them because they were natives, even if they were in the wrong."

FitzRoy dealt with the two Maori chiefs in one of their own *pas* or fortified villages. They again insisted the lands were theirs and reiterated that killing prisoners was tribal custom. After half an hour's reflection, FitzRoy gave his verdict: "In the first place the white men were in the wrong. They had no right to survey the land which you had not sold until [the land commissioner] had finished his inquiry; they had no right to build the houses they did on that land. As they were, then, first in the wrong I will not avenge their deaths," before continuing, "I have to tell you that you have committed a horrible crime in murdering men who had surrendered themselves . . . White men never kill their prisoners. For the future let us live peaceably and amicably—the Paheka [white man] with the native and the Maori with the Paheka." FitzRoy's solution satisfied neither party. Both perceived him as weak.

FitzRoy had next to struggle with the country's financial problems. Disobeying his explicit instructions, he issued promissory notes on behalf of his administration, which did not have the funds to back them up. To encourage trade he abolished customs duties but introduced property taxes and then reversed these decisions in the face of public outcry and further declines in revenue. He did all this on his own initiative without even informing his superiors in London. In this he followed his practice on the *Beagle* of doing what he thought best, as when he bought the *Adventure* without permission.

More landownership problems quickly followed. Some Maori chiefs had become adept at selling land actually possessed by their fellows to the government, prompting confrontation when the administration tried to take possession. In a rare letter to his friend Beaufort, FitzRoy confessed, "The task I have to grapple with here is beyond anything I could have anticipated in extreme difficulty . . . I have been obliged to . . . do many rash things (as they may be deemed) but—at whatever responsibility— the course I think best for our Queen and country—that I will follow."

Although some Maori chiefs supported FitzRoy, others became increasingly disenchanted with the behavior of both the settlers and— as they saw it—the pusillanimous governor. In July 1844, Hone Heki, a local chief, cut down the flagpole from which the Union Jack flew at Kororareka in the Bay of Islands, which the *Beagle* had visited, to demonstrate his contempt for the British. FitzRoy reerected the flagpole and told senior local chiefs he would take no action against Hone Heki if they kept him under control.

The still rebellious and mischievous Hone Heki cut down the flagpole again in January 1845 and then again that March after it had been reerected once more. On that occasion, the two hundred Maoris he had with him looted Kororareka and, possibly accidentally, caused fires that destroyed much of the settlement. FitzRoy pursued Hone Heki and his companions but to little success. In October 1845, FitzRoy received news from London of his recall. Settlers celebrated by burning his effigy.[2]

---

2. Christ Church, to the building of which FitzRoy and Darwin contributed, still bears the marks of bullets.

The minister at the Colonial Office, Lord Stanley, gave FitzRoy as cause for his dismissal his failure to consult or even inform London rather than disagreement with most of his policies, leaving Stanley unable to respond to petitions on behalf of the New Zealand Company, which had powerful supporters in Parliament. Acting without authority, allied with his quick temper, had again cost FitzRoy dear. His greater willingness to conciliate the Maoris rather than to act against them—now admired by many historians in New Zealand—and his obvious preference for the missionaries' spiritual "civilizing" aims over the temporal commercial objectives of the New Zealand Company would have surprised no one who knew him well from the *Beagle*. As much the victim of circumstances as of his temperament and political naïvete, FitzRoy returned to Britain sunk in gloom, believing his reputation "deeply and irreparably injured."

Bartholomew Sulivan, FitzRoy's lieutenant on the *Beagle*, happened to be visiting Down House when Darwin received a letter telling him of FitzRoy's dismissal. After the *Beagle*, Sulivan had undertaken further surveys of the Falkland Islands and elsewhere, sending back geological specimens and fossils to Darwin. While recognizing the general good sense of FitzRoy's policies, Sulivan ascribed FitzRoy's failure to "*Beagle* [all] over again, *temper violent*, saying *any thing* to any body, doing most hasty and extraordinary things . . ."

In his letter of consolation to FitzRoy, Darwin wrote, "I fear you must have undergone much trouble and vexation and been ill repaid except by the consciousness of your own motives, for the sacrifices which I am aware you made in accepting the Governorship." Despite his own illness, his usual desire for seclusion, and his previous avoidance of FitzRoy, Darwin invited FitzRoy and his family to stay for a few days at Down House, something that showed a strong remaining degree of affection for his old captain in his troubles. There is no record FitzRoy took up the invitation.

In 1843, Emma's father and Darwin's uncle and great supporter, Josiah Wedgwood II, had died, and by 1848 Darwin's own father Robert was declining fast. Traveling for once without Emma, he visited his father at The Mount that May. Depressed and oppressed by his own and his father's health he wrote to Emma, "Oh Mammy I do long to be with you and under your protection for then I feel safe." This single sentence says so much about their relationship—how in his illness and anxiety he saw Emma as a mother figure as much as a wife—someone who provided both stability and protection from the uncertainties of the world.

Darwin returned to Down House after two weeks. His father lingered until November. Although Darwin had plenty of time to get to Shrewsbury for the funeral, he delayed for a few days for unknown reasons and did not arrive at The Mount until the cortege had left for the church. He did not follow it but remained at the family home. The death of their fathers left Emma and Charles Darwin with assets of some eighty thousand pounds and an annual income of some three thousand to four thousand pounds, at a time when a farm worker might receive around fifty pounds a year in pay.[3]

Over the months following his father's death, Darwin became even more obsessive about his own worsening health—he sometimes felt too unwell to work more than an hour or two a day on his barnacles. From July 1849 to January 1856 he kept a daily health diary, initially using a notation based on a meteorological recording system he remembered FitzRoy used on the *Beagle*, employing pluses and minuses and underlinings to record the ups and downs of his condition but later using simple words. He divided the pages with mornings on one side and nights on the other, just as he had when writing the pros and cons of marriage, and produced monthly summaries. He began reading medical textbooks and journals, which only seemed to remind him of and agitate him about the many dreadful diseases from which he might be suffering.

After trying well-known London doctors without success, and on the recommendation of his old *Beagle* shipmate Bartholomew Sulivan, Darwin moved his growing family and servants temporarily to Malvern in March 1849, to try the newfangled water cure at Dr. Gully's much-hyped,

---

3. Eighty thousand pounds is roughly equivalent to 10 million pounds in today's prices, while their income was equivalent to between four hundred thousand and five hundred thousand pounds.

celebrated, and expensive establishment. There, over the next three months, Dr. Gully subjected him—or he subjected himself—to a vigorous and invigorating, not to say exhausting regime. It included a special bland diet, excluding many of the sweet desserts Darwin loved, homeopathic remedies that Darwin took "obediently without an atom of faith," and brisk walks in the local hills. A few months previously Darwin had told Syms Covington, to whom he had sent an ear trumpet to alleviate his increasing deafness, "I have not been able to walk a mile for some years." He was also trying to restrict his snuff taking to six pinches a day. Darwin described to Hooker the water cure itself: "I am heated by Spirit lamp till I steam with perspiration, and am then suddenly rubbed violently with towels dripping with cold water; have two cold feet-baths and wear a wet compress all day on my stomach . . . I feel certain that the Water cure is no quackery."

Darwin believed the treatment was doing him good and felt so much better that after his return to Down House he had a hut built in the garden. There, whatever the weather, over the next few years, his butler Parslow or occasionally another servant daily sluiced cold water from a large overhead tank fed from the garden well over his naked body.

Eighteen months after his return from Malvern, Darwin's eldest and favorite daughter Annie became increasingly unwell. Darwin feared her debility might be a result of the marriage of first cousins, a recurring concern whenever his children were ill. In March 1851, she became so ill that Darwin took her to Dr. Gully's spa at Malvern, on this occasion leaving Emma, heavily pregnant with Horace, at home. Darwin left Annie at Malvern with her sister Henrietta for company, under the care of Annie's nurse, Brodie, and the family's governess. However, the ten-year-old's condition deteriorated quickly, and in mid-April Darwin returned, again alone, to Malvern. He found Annie in great distress and vomiting, and he worried she might have an "exaggerated" form of his own illness.

Annie died peacefully shortly afterward. Darwin was distraught and confined to bed with stomach problems. Unwell and feeling the need to return home to console Emma and perhaps unable to face the emotion of the burial, as he may have avoided his father's burial, Darwin returned to Down quickly, accepting the offer of his brother-in-law Hensleigh

Wedgwood and his wife Fanny, who happened to be staying nearby, to arrange and attend the funeral. Annie's death certificate gave the cause of death as "Bilious fever with typhoid character." Darwin fretted that there might be a hereditary stomach weakness in the family. The death of "angelic" Annie further diminished whatever faith he retained in a just, moral, loving, and interventionist God. Years later he would write in his autobiography that it killed his Christian faith.

Annie's death understandably also diminished his confidence in Dr. Gully and his methods as a remedy for his own illness, which seemed, perhaps predictably, to worsen again. He tried other therapies, including allowing himself to be draped with brass and zinc "electric chains," which were then doused with vinegar to create a small electric charge. This had no obvious effect beyond producing some tingling and some marks on his skin.

Whole books have been written about the nature and causes of Darwin's illnesses—there was probably more than one. In 1865 he reported, "For 25 years extreme spasmodic daily and nightly flatulence, occasional vomiting . . . tongue crimson in morning ulcerated . . . eczema—(now constant) lumbago." He also reported that after 1838, he suffered "shivering, hysterical crying, dying sensations or half-faint . . . singing of ears, rocking, treading on air, focus and black dots—All fatigues, specially reading, brings on these Head symptoms? Nervousness when E. [Emma] leaves me." Bowel and stomach pain and gas often woke him at night. He regularly recorded his bodily functions, including the color and volume of his urine and the frequency of his bowel movements. He was sometimes constipated but rarely suffered diarrhea. From the time he joined the Beagle he suffered occasional heart palpitations. He also sometimes experienced numbness, tingling, or burning in his extremities.

Among the diagnoses proposed over the years have been anxiety, rumbling appendicitis, poisoning from remedies such as arsenic, calomel (which contains mercury), and bismuth, allergies to the preserving spirits he used, most often alcoholic spirit but sometimes formaldehyde or phenol, brucellosis, Chagas disease resulting from a tropical bug bite (perhaps when he was ill in Chile during the Beagle voyage), Crohn's disease, irritable bowel syndrome, cyclical vomiting syndrome, depression, gastritis, hepatitis, lupus, lactose intolerance, mitochondrial disease, neurasthenia (nervous disorder), obsessive-compulsive disorder, panic attacks, paroxysmal tachycardia (rapid heartbeat), pyroluria (blood disorder), middle ear infection, gastroesophageal reflux disorder, celiac disease (allergy to wheat gluten), and hypoglycemia.

No definitive diagnosis can ever be made at this distance in time. However, some element of anxiety or stress seems highly likely to have been a factor. His wife Emma was certain that "his health was always affected by his mind." He himself recognized a link between what was going on in his head and his stomach. His illnesses often erupted under stress, whether the illness of family or friends or corrosive worries about the reception or reaction to any of his publications, but in particular the prospect of that on the publication of his transmutation theory. He once confessed to his cousin William Darwin Fox, "My abstract [On the Origin of Species] is the cause, I believe of the main part of the ills to which my flesh is heir." As he got older, attendance at social functions was another dread. In particular any requirement to make public appearances or speeches directly related to his work brought on illness and prostration. He wrote that speaking for a few minutes at a scientific society meeting produced "24 hours vomiting." However, anxiety and panic attacks are unlikely to fully explain the seriousness of his illnesses without some underlying physical weakness(es) on which to play, of which perhaps the most likely are bowel or stomach complaints.

On September 9, 1854, with his barnacle work finally completed, published and in his archives, Darwin wrote in his diary, "Began sorting notes for species theory." He was soon thinking how to amplify his theory, fleshing out the arguments, filling lacunae where necessary and seeking further evidence. Among the most important addition to his thinking was the notion that in a crowded environment, variants that differed significantly from the norm had a wider range of opportunities to prosper than those that differed only a little from the original stock. This would help explain the lack of evidence of a steadily evolving evolutionary process. To substantiate his belief that seeds could travel great distances across oceans and still germinate on reaching new land, where their properties might give them

Charles Darwin in middle age

an advantage over endemic varieties—a question he had debated during the *Beagle* voyage at the Keeling Islands—he began a series of experiments soaking seeds of various plants for extended periods in brine. Most sank to the bottom of the brine. Somewhat to his surprise, however, many germinated after a month's immersion, and a few after four months.

On the basis of his results, Darwin hypothesized that seeds did not often make the journey alone but attached to floating leaves, twigs, or branches perhaps blown off in storms. He tried similar experiments with nuts—hazelnuts floated for three months and still germinated. He fed the carcasses of sparrows whose crops were stuffed with wheat to birds of prey at London Zoo. He then extracted from the latter's feces and coughed-up pellets any seeds he found, checking that they would germinate to demonstrate that seeds could be carried long distances over the water inside ocean-traveling birds.

Darwin also resumed his interest in selective domestic breeding and, in 1855, began keeping pigeons and chickens at Down House, where he built a pigeon loft, which at its height housed ninety birds. He also corresponded with pigeon fanciers and even occasionally visited some of them or their meetings—one he attended was in a London gin palace. He collected the corpses of birds and small animals of different ages, steeped the carcasses in potash and silver oxide, then boiled them up to remove any remaining "small fragments of putrid flesh" from the bones—a process that "made my servant and myself . . . retch so violently that we were compelled to desist" and appalled his wife and family. He eventually began dispatching the carcasses to be skeletonized professionally. His purpose in skeletonizing them was to ascertain differences between the same varieties at different ages which might later lead to variations.

All the time he was corresponding with fellow scientists in Britain and abroad. In the UK, in addition to his old friends Lyell and Henslow,

among his closest confidants over the next
years were Joseph Hooker and a new acquaintance, Thomas Huxley. Hooker had returned
to Britain after four adventurous years plant
hunting in the Himalayas and married one of
Henslow's daughters, who, like FitzRoy's and
Sulivan's future wives, had waited quietly for
his return. Hooker was now employed at Kew
Gardens, where his father remained director.
Huxley, sixteen years Darwin's junior, came
from a modest background. His father was a

Thomas Huxley

mathematics teacher at Ealing Grammar School. After qualifying as
a doctor at Charing Cross Hospital, where he gained the gold medal,
Huxley served for four years on HMS *Rattlesnake* in Australia as surgeon
and naturalist, sending home papers on hydrozoa. Returning to Britain
he took a series of poorly paid scientific posts, waiting five years before
he could afford to bring over to Britain and marry the young woman
he had met in Sydney. Darwin congratulated him on the marriage but
warned, "Happiness, I fear is not good for work." Darwin and Huxley
had initially gotten to know each other through discussion of Darwin's
work on barnacles. Huxley had praised Darwin's work as "most beautiful
and complete" and "the more remarkable" being from a geologist "not an
anatomist, *ex professo.*"

Perhaps Darwin's key contact overseas was the celebrated American
botanist Asa Gray, professor of botany at Harvard University and an expert
on the distribution of plant species around the world. Darwin pumped him
for information in flattering terms: "It is extremely kind of you to say that
my letters have not bored you very much, for I am quite conscious that my
speculations run beyond the bounds of true science." In response, Gray
willingly provided a mass of information on American plants. As their
correspondence continued, Darwin found Gray both a "cautious reasoner"
and "a very loveable man" and gradually began to expose elements of his
evolutionary theory. "As an honest man I must tell you that I have come to
the heterodox conclusion that there are no such things as independently

Asa Gray

created species—that species are only strongly defined varieties. I know that this will make you despise me." Gray responded that he had long believed "that there is some law, some power inherent in plants" that caused the creation of variants. On September 5, 1857, Darwin wrote to Gray revealing all but swearing Gray to secrecy: "As you seem interested in [the] subject, and as it is an *immense* advantage to me to write to you and to hear *ever so briefly* what you think. I . . . enclose the briefest abstract of my notions on the *means* by which nature makes her species." Darwin attached a one-thousand-word synopsis of his theory to his letter.

Another frequent overseas contact was Edward Blyth at the Asiatic Museum in Calcutta, who supplied bird carcasses and much other useful information and comment. Two decades previously Blyth had postulated but rejected a theory of transmutation occasioned by environmental change.

Darwin also corresponded with scientists in Europe, particularly in Germany—where he had more than one hundred different contacts—and in France. He always wrote in formal English and apologized politely if chauvinistically for his linguistic deficiencies in much the same way that many Anglophones do today.

In May 1856, Darwin had already begun setting down his theories for publication. He had done so on the urging of his friends and of Lyell in particular during a visit to Down House in mid-April. Although not fully convinced by what Darwin told him about his theory, Lyell had seen and heard enough in the academic press and in academic conversation to know that if Darwin did not publish soon, others might intrude on his field. On May 1, he wrote to Darwin, "I wish you would publish some small fragment of your data . . . and so out with the theory and let it take date—and be cited—and understood." Darwin replied, "To give a fair sketch would be absolutely impossible, for every proposition requires such an array of facts . . . But I do not know what to think: I rather hate the idea of writing for priority, yet I certainly should be vexed if anyone were to publish my doctrines before me." On May 14, Darwin wrote in his diary, "Began by Lyell's advice writing species sketch." He told his cousin Fox in 1857, "I am like Croesus overwhelmed with my riches in facts and I mean to make my book as perfect as ever I can. I shall not go to press at soonest for a couple of years." As a result of his plethora of information, the "sketch" quickly morphed into what he told Fox would be "quite a big book." By spring 1858, Darwin had written a quarter of a million words and was still only roughly two-thirds of the way through.

CHAPTER TWENTY-FOUR

"I Shall Be Forestalled"

On June 18, 1858, the postman delivered to Down House a letter addressed to Darwin from the Moluccan island of Tenarate in the Dutch East Indies. Its contents would turn Darwin's world upside down. The sender was another naturalist, thirty-five-year-old Alfred Russel Wallace, of whom Darwin had known for some three years and with whom he had occasionally corresponded without realizing the similarities in their thinking. Wallace had been born and raised with none of the advantages of privilege, opportunity, and security that Darwin's affluent family brought him. The eighth of nine children, he was the son of a country solicitor who had fallen on hard times and died when his son was only eleven, leaving the family to subsist precariously on his mother's small inheritance.

Wallace had started work at fourteen—unlike Darwin he would have no university education—helping an older brother surveying for the booming railway industry in which the Darwins so successfully invested. Tramping the countryside in his work gave Wallace a love of nature, and when he could, he attended lectures at the increasing number of libraries and mechanics' institutes dedicated to the education of the less privileged. There, in addition to becoming imbued with a lifelong socialism and religious skepticism, Wallace began to acquire an academic knowledge of nature to supplement that gained practically in his day-to-day field work. In the Mechanics' Institute in Leicester, he met Henry Walter Bates, who became a close friend and with whom he went on local plant-, beetle-, and butterfly-collecting expeditions.

In 1848, both being without ties and with few prospects, they traveled to the Amazon to collect botanical and zoological specimens to ship home to sell to the growing number of more wealthy but less adventurous collectors in Britain. There, although often ill and often hungry, they amassed a considerable number of specimens. After four years, including a quarrel with and separation from Bates, Wallace decided to return to Britain with his hoard only to lose nearly all of his specimens and journals when his ship caught fire and sank in August 1852. Rescued after some days in an open boat, Wallace eventually reached home, where he wrote two unsuccessful books, one on his travels, and one on palm trees. Darwin read the first and thought it lacked data. Suffering what he later described as a "want of self-confidence," Wallace found it difficult to penetrate the world of affluent science and was given little encouragement to do so by its inhabitants.

Still without personal ties or regular income, he set out in 1853 alone to the East Indies, again to collect specimens. He spent some time first in Sarawak in Borneo, the personal fief of the British "white raja" James Brook, whom he "delighted and instructed by his clever and inexhaustible flow of talk—really good talk." There he made a close study of orangutans in the wild, just as Darwin had of Jenny in the London Zoo, noting the similarities between them and humans. He also made a successful start to his collecting mission, which again drew him to the attention of Darwin, who asked him to supply him with specimens and raw data, which he did, even if Darwin complained that his shipping charges were "costing me a fortune!"

Wallace wrote a paper for the *Annals and Magazine of Natural History*, published in 1855, containing some thoughts on the diversification and development of species. It ended with the coded words, "Every species has come into existence coincident both in space and time with pre-existing closely allied species." Both Lyell and Darwin read it. Darwin, perhaps with the stereotyped preconception that Wallace was a mere collector or perhaps overconfident in the uniqueness and daring of his own thesis, saw nothing in it to disturb the quiet

A. R. WALLACE SOON AFTER HIS RETURN FROM THE EAST.

Alfred Russel Wallace

pace of his own work, scribbling on his copy of the article, "nothing very new. Uses my simile of a tree, it seems all creation with him." However, Lyell seems to have seen enough similarities in Wallace's thinking to that of Darwin for it to be a major reason why, in spring 1856, he urged Darwin to publish.

Over the next years as Darwin occasionally corresponded with Wallace, somewhat condescendingly praising him and encouraging his collecting activities, Wallace continued to amass specimens across the Malay archipelago and New Guinea while developing his own ideas on variation of species. In February 1858, shaking and shivering with malaria on Tenarate, in what he described as "a sudden flash of light" he "thought of [Malthus's] clear exposition of 'the positive checks to increase'—disease, accidents, war and famine—which keep down the population of savage races to so much lower an average than that of more civilized peoples . . . It then suddenly flashed upon me that this self-acting process would necessarily improve the race, because in every generation the inferior would inevitably be killed off and the superior would remain—that is, the fittest would survive." He immediately wrote out his theory and dispatched the manuscript on its four-month-long journey to Down House.

On opening and reading it, Darwin quickly realized the shattering importance of Wallace's essay with its clearly stated aim to show "that there is a general principle in nature which will cause many varieties to survive the parent species, and to give rise to successive variations departing further and further from the original type." His immediate and very human reaction was that Wallace's views, although not identical, were so similar to his that all his years of pioneering research had been wasted and that he would not receive the recognition he deserved. Full of "trumpery feelings," he admitted to Hooker, perhaps in a plaintive plea for reassurance that it wasn't the case, "It is miserable in me to care at all about priority."

In 1912, Robert Falcon Scott would make a similar confession on being frustrated by Roald Amundsen in his bid to be the first at the South Pole, writing, on reaching it, "Great God! This is an awful place and terrible enough for us to have laboured to it without the reward of priority." However, unlike Scott in his physical challenge, Darwin in his intellectual one still had the possibility of establishing at least equal priority with Wallace since Wallace had written to him knowing his interest in the field, rather than having the self-confidence to submit his paper directly to a journal. Darwin immediately consulted Hooker and Lyell— to whom, in any case, Wallace had asked Darwin to forward the essay. Resigned to his fate, in his accompanying letter to Lyell he wrote, "Your words have come true with a vengeance that I should be forestalled. You said this when I explained to you . . . very briefly my views of 'natural selection' depending on the struggle for existence . . . Please return me the MS which he [Wallace] does not say he wishes to publish; but I shall of course at once write and offer to send to any Journal. So all my originality, whatever it may amount to, will be smashed."

Together, Hooker and Lyell devised a solution that, while publishing Wallace's essay, would recognize Darwin's long-standing research. They would themselves jointly present a paper to the Linnaean Society, which would include dated extracts from Darwin's 1844 unpublished work on evolution and from his letter of September 5, 1857, to Asa Gray at Harvard developing his arguments further, as well as Wallace's essay. Hooker and Lyell would emphasize in the paper, which in total ran to eighteen printed pages, ten of which were Wallace's essay, that Darwin and Wallace "having, independently and unknown to one another, conceived the same very ingenious theory to account for the appearance and perpetuation of varieties and of specific forms on our planet, may both fairly claim the merit of being original thinkers in this important line of enquiry; but neither of them having published his views, though Mr Darwin has for many years past been repeatedly urged by us to do so . . ." and that Darwin and Wallace had placed the work in their hands for publication.

With Darwin preoccupied with the illness of his nineteen-month-old youngest son Charles Waring, who had caught the scarlet fever running through Downe village—he died on June 28, 1858—Hooker and Lyell got their paper inserted at a day's notice onto the agenda of the Linnaean Society meeting on July 1. According to Hooker, when the society's secretary read the late submission out loud it prompted little discussion.

Indeed, when in May 1859 the society president, Thomas Bell, a misogynistic dentist, reviewed the preceding twelve months, he stated the year
had not "been marked by any of those striking discoveries which at once
revolutionize, so to speak, [our] department of science."

Like Wallace, who remained in southeast Asia, Darwin did not attend
the Linnaean Society meeting. Instead, while grieving for his son, he
was busy making hasty preparations to remove himself and his family to
the Isle of Wight, away from the scarlet fever epidemic. Starting on the
isle and continuing after his return six weeks later to Down House, he
had no need of the urging of Hooker, Lyell, and other friends to begin
setting down his theories in a short, definitive, and assimilable form for
wider publication. What had originally been conceived as an essay quickly
grew into the draft of a book. Despite frequent recurrences of "severe
vomiting" and several recourses to water cures, by late spring 1859 he
had completed a draft of some 155,000 words, much of it distilled from
the far larger uncompleted work on which he had been laboring when
Wallace's bombshell package arrived.

He sent his new work draft chapter by draft chapter to Hooker and
Lyell and found their responses encouraging. However, the contribution
of Hooker's children was less helpful. Hooker confessed they had "made
away with upwards of a quarter of the MS. By some screaming accident
the whole bundle, which weighed over 1lb when it came . . . got transferred to a drawer where my wife keeps paper to draw upon and they have
of course had a drawing fit ever since——I feel horrified if not brutalised
for poor D. is so bad that he could hardly get up steam to finish what he
did.——How I wish he could stamp and fume at me——instead of taking it
so good naturedly as he will." Darwin was able to be as relaxed as Hooker
predicted since he had preserved a previous draft.

Darwin called the book "one long argument," but in fact it contained
a wealth of data, drawn in considerable part from the *Beagle* voyage to
support the theories he expounded. This abundance of carefully recorded
information and sorted supporting information distinguished his work
from the unsubstantiated and wide-ranging hypotheses of Chambers and
also from Wallace's short paper. Although Darwin had dreaded telling

Wallace of his intention to write a book, Wallace, when informed, had been nothing but helpful and encouraging, "admirably free from envy or jealousy." Darwin decided "he must be an amiable man."

Darwin's central thesis was that variations occurred naturally and frequently without any particular pattern within all species and that, when they did, those that best fitted their possessors' environments would, in his words, allow them "the best chance of being preserved in the struggle for life; and from the strong principle of inheritance they will tend to produce offspring similarly characterised." He expanded on this mechanism of change, which was his, and indeed Wallace's, greatest innovation: "This principle of preservation, I have called, for the sake of brevity, Natural Selection . . . Amongst many animals sexual selection will give its aid to ordinary selection by assuring to the most vigorous and best adapted males the greatest number of offspring. Sexual selection will also give characters useful to the males alone, in their struggle with other males . . . Natural Selection leads to divergence of character; for more living beings can be supported on the same area the more they diverge in structure, habits and constitution, . . . Therefore during the modifications of the descendants of any one species, and during the incessant struggle of all species to increase in numbers, the more diversified these descendants become, the better will be their chance of succeeding in the battle of life. Thus the small differences distinguishing varieties of the same species, will steadily tend to increase till they come to equal the greater differences between species of the same genus, or even of distinct genera . . . On these principles, I believe, the nature of the affinities of all organic beings may be explained."

Darwin argued that "descent with modification"—the phrase he used at this point for evolution, the latter word only becoming current later—was always a gradual process, just as Lyell had proposed for geological change, and never preceded by sudden leaps. Natural selection was "daily and hourly scrutinising throughout the world, every variation . . . rejecting that which is bad, preserving and adding up all that is good; silently and insensibly working whenever and wherever opportunity offers, at the improvement of each organic being . . ."; "Natural

Selection leads to divergence of character and to much extinction of the less improved and intermediate forms of life." He included only one diagram in his draft. It illustrated what he called "the great tree of life which fills with its dead and broken branches the crust of the earth [i.e., fossils] and covers the surface with its ever branching and beautiful ramifications . . . At each period of growth all the growing twigs have tried to branch out on all sides, and to overtop and kill the surrounding twigs and branches, in the same manner as species and groups of species have tried to overmaster other species in the great battle for life."

In his introduction Darwin acknowledged the primacy of the *Beagle* voyage in his thinking. "When on board H.M.S. 'Beagle,' as naturalist, I was much struck with certain facts in the distribution of the inhabitants of South America, and in the geological relations of the present to the past inhabitants of that continent. These facts seemed to me to throw some light on the origin of species—that mystery of mysteries . . ." Later in the text he emphasized the importance in particular of three discoveries he made during the voyage—the fossils he discovered near Baia Blanca in Patagonia, the distribution patterns of the two species of the South American rhea, and the animals and birds of the Galápagos. He did not, however, specifically mention the famous finches in the book.

Darwin's "long argument" stretched over fourteen clearly written chapters (fifteen in later editions). He included one chapter discussing difficulties with his theories, such as the lack of evidence of transitional forms due to the (then) very incomplete fossil record, admitting, "We see nothing of these slow changes in progress, until the hand of time has marked the long lapse of ages, and then so imperfect is our view into long past geological ages, that we only see that the forms of life are now different from what they formerly were." He avoided any explicit discussion of both human origins and the origin of life itself, knowing full well the controversy such topics would cause, telling Wallace, "I think I shall avoid whole subject, as so surrounded with prejudices."

Darwin did, however, tentatively suggest, "Analogy would lead me one step further, namely, to the belief that all animals and plants have descended from some one prototype . . . probably all the organic beings which have ever lived on this earth have descended from some one primordial form into which life was first breathed." "All the organic beings" implicitly included humans, but Darwin pushed any discussion of that possibility into the future, commenting, "In the distant future I see open fields for far more important researches. Psychology will be based on a new foundation, that of the necessary acquirement of each mental power and capacity by gradation. Light will be thrown on the origin of man and his history." He always used the words *natural selection* to describe the all-important mechanism for change and never the notorious and emotive phrase *survival of the fittest*, which one of his supporters, Herbert Spencer, did not coin until 1864.

Just as in the case of Einstein's paper on his theory of $E = MC^2$, *On the Origin of Species* contains no bibliography and no footnoted references to other works. In Einstein's case this was solely because of the novelty of his theory. This is less so in Darwin's. It partly reflects that *On the Origin of Species* was originally only intended as an abstract of his proposed longer work. It perhaps also reflects a degree of egotism about "my theory" and perhaps also about what FitzRoy identified as a disinclination to give due credit to others. In his introduction, Darwin did, however, refer to Wallace's "excellent memoir," which reached "almost exactly the same general conclusions as I have on the origin of species," and later in the text to what he called Wallace's "ingenious paper."

With the book completed in draft, Darwin needed a publisher and again called on his friends for advice. When Lyell suggested John Murray, Darwin persuaded him to approach Murray on his behalf. Murray's father of the same name had founded the publishing house with among his authors Byron and Jane Austen. Murray the son had himself published all Lyell's works and the second and subsequent editions of Darwin's *Journal of Researches*. During a visit by Lyell, Murray agreed to look at Darwin's draft, and the Darwins' butler Joseph Parslow hurried up to London to Murray's Albemarle Street offices with the manuscript wrapped in brown paper. Despite serious reservations expressed by one of the "readers," a clergyman, to whom he passed it for comment who thought it "a wild and foolish piece of imagination," Murray agreed to publish the work with Darwin to take half the net profit.

Soon Darwin was feeling increasingly stressed and ill as he scrutinized successive proofs, making additions and correcting them with Emma Darwin's patient help. At Murray's urging he agreed to change the book's title from *An Abstract of an Essay on the Origin of Species and Varieties through Natural Selection* to *On the Origin of Species by Means of Natural Selection, or the Preservation of Favoured Races in the Struggle for Life.*

Immediately on completing the proof reading on October 1, Darwin, erupting in boils and eczema, headed north for a spa on the edge of the Yorkshire moors at Ilkley. He wrote to his cousin William Darwin Fox, "My abominable volume . . . has cost me so much labour that I almost hate it," and told Lyell, "I am in a very poor way . . . and useless for everything . . . Hydropathy and rest . . . perhaps that will make a man of me." While at Ilkley undergoing the cure, Darwin orchestrated what, in modern terms, would be called a networking campaign to secure as sympathetic a reception as possible for his pioneering volume. He arranged for Murray to send copies, most at Darwin's own expense, to anyone he thought might have academic or public influence or be inclined to review. He himself sent accompanying letters—some eighty in total—carefully crafted both to charm and to respond to what he thought would be each individual recipient's likely reaction. Most of the world's best-known naturalists received copies. However, Darwin excluded from his munificence those who had supplied useful information or specimens but lacked the requisite influence or place in society. Among those excluded was his former Edinburgh mentor Robert Grant, still teaching at University College London, who nevertheless praised Darwin as "with one fell swing of the wand of truth scattering the pestilential vapours" of the Creationists.

On the Origin of Species was published on November 24, 1859, priced at fourteen shillings. The 1,250 copies of this first edition, which ran to more than five hundred pages, sold out within a day, in large part due to a prepublication sales push by John Murray, which secured a preorder of five hundred copies from Mudie's Circulating Library to be read by its twenty-five thousand subscribers. Also on November 24, Murray published Samuel Smiles's *Self-Help* and Leopold McLintock's *Narrative of the Discovery of the Fate of Sir John Franklin*, describing his adventures during his search for the remains of Franklin's ill-fated expedition to find the Northwest Passage. Both greatly outsold Darwin on publication.

On his return to London, Darwin negotiated with Murray the contract for future editions of *On the Origin of Species*, whose second edition of three thousand copies was already underway. The terms featured what Murray then called an unusual arrangement but one that is now common—an advance on estimated future sales. According to Murray, to ensure his fair reward, Darwin constantly inquired into details of the calculation of the advance rather than blindly accepting them.

Even before publication and while he was still in Yorkshire, reaction to his book had already begun to reach Darwin from his fellow natural historians, friends, family, and acquaintances to whom he had sent copies. His brother Erasmus wrote, "For myself I really think it is the most interesting book I ever read, and can only compare it to the first knowledge of chemistry, getting into a new world . . ." Historian Thomas Carlyle, Darwin's longtime acquaintance, commented, "A humiliating discovery and the least said about it the better." Not unexpectedly, the now fundamentalist Robert FitzRoy was violently opposed. He began, "My dear old friend, I, at least, cannot find anything 'ennobling' in the thought of being a descendant of even the most ancient ape." This comment of course immediately evoked the question of human descent that Darwin had scrupulously avoided. He quickly identified by its content a critical letter to *The Times* under the pseudonym "Senex" as being by FitzRoy and commented, "It is a pity he did not add his theory of the extinction of Mastodon etc from the door of ark being made too small. What a mixture of conceit and folly, and the greatest newspaper in the world, inserts it!"

Adam Sedgwick, now seventy-four, was also critical, as was Sir John Herschel, whose book on the philosophy of science Darwin had admired when he was at Cambridge and whom he had visited in Cape Town. Sedgwick wrote to Darwin, "If I did not think you a good tempered man and truth loving man I should not tell you that . . . I have read your book with more pain than pleasure. Parts of it I admired greatly; parts I laughed at till my sides were almost sore; other parts I read with absolute sorrow; because I think them utterly false and grievously mischievous. You have . . . started up a machinery as wild I think as Bishop Wilkin's locomotive that was to sail with us to the moon." He would later discuss *On the Origin of Species* in an unsigned article in *The Spectator*, in which he dismissed Darwin's thinking by stating, "Each series of facts, is laced together by a series of assumptions, and repetitions of the one false principle. You cannot make a good rope out of a string of air bubbles."

Henslow was also unconvinced: "The Book is a marvellous assemblage of facts and observations—and no doubt contains much legitimate inference—but it pushes hypothesis (which is not real theory) too far . . ." He later told a friend, "Darwin attempts more than is granted to Man, just as people used to account for the origin of Evil—a question past finding out." Lyell too, whose gradualist geological theories so influenced Darwin and who had urged him to publish, also retained reservations, in particular about the implicit proposition that humans were descended from apes. It would take him until 1863 to go "the whole orang" with Darwin, and even then he only fully did so privately. His later published work dismayed Darwin by remaining noncommittal.

However, the Reverend Charles Kingsley, author of *The Water Babies* and *Westward Ho!*, a liberal-thinking, socially reforming clergyman and keen natural historian, wrote to Darwin, "All I have seen of [your book] *awes* me; both with the heap of facts, and the prestige of your name, and also with the clear intuition, that if you be right, I must give up much that I have believed and written."

Darwin, who remained an incorrigible and inveterate lister and was compiling lists of supporters and opponents—"our side" and "outsiders"— was also heartened by a favorable anonymous review in *The Times* on Boxing Day, December 26, 1859. He rightly believed that Huxley was the author and wrote to him that his piece meant more than "a dozen reviews in common periodicals." Even before the formal publication of *On the Origin of Species*, Huxley had written to Darwin praising his work:

"As for your doctrines, I am prepared to go to the stake if requisite in support . . . And as to the curs which will bark and yelp—you must recollect that some of your friends . . . are endowed with an amount of combatitiveness [sic] . . . I am sharpening up my claws and beak in readiness." In his *Times* review Huxley included a slighting reference to the work of one of those he considered a "cur"—Richard Owen, the man who had examined and identified Darwin's fossil specimens from the *Beagle* and was now the superintendent of the natural history collection at the British Museum and Huxley's bitter rival for preeminence in comparative anatomy.

Darwin had already worried about Owen's likely reaction. He was so concerned that exceptionally he had paid a visit to the anatomist, during which Owen had been politely noncommittal about Darwin's work. Owen's later critique in the *Edinburgh Review* was corrosively and personally hostile. Over forty-five pages he attacked the "vagueness and incompleteness" of Darwin's theories, his intellect, methods, and data and criticized his "short-sighted" supporters Hooker and Huxley for good measure. Darwin found Owen's views "extremely malignant, clever . . . and very damaging." The review marked a permanent estrangement between them, stemming—Darwin believed—from Owen's "jealousy" of his success, which had made him into a "bitter enemy."

Darwin took all criticisms—not just Owen's—personally, often blaming himself for being a "very bad explainer." Elsewhere, even though many made the connection between man and ape, with the *British Quarterly Review* going furthest by raising the specter of an ape proposing marriage to a demure Victorian damsel, in general the reviews, although often critical, were a little more restrained on religious grounds than Darwin might have expected. Perhaps one reason was that in the second edition published in January 1860, only six weeks after the first, among other amendments, Darwin introduced into the concluding sentence of the whole book a reference to the potential involvement of "the Creator," so that it now read, "There is grandeur in this view of life with its several powers, having been originally breathed by the Creator into a few forms or into one; and that, whilst this planet has gone cycling on according to the fixed law of gravity, from so simple a beginning endless forms most beautiful and most wonderful have been, and are being, evolved."

On the Origin of Species was debated not only across Britain but also worldwide, as it would be translated into eleven languages in Darwin's lifetime.[1] In Britain, Queen Victoria obtained a copy. She told her daughter (Vicky) she did not expect to understand it fully. However, Lyell recorded after meeting her, "She asked me a good deal about the Darwinian theory . . . She has a clear understanding, and thinks quite fearlessly for herself." In the United States, Asa Gray enthusiastically praised Darwin's work and attacked his opponents. Darwin wrote appreciatively that every one of Gray's attacks had the effect of "a 32-pound shot" and that "no one other person understands me so thoroughly as Asa Gray. If I ever doubt what I mean myself, I think I shall ask him!"

With no international copyright as yet in operation, Gray helped Darwin subdue two pirated American editions of *On the Origin of Species* and secured on his behalf a legitimate publishing contract from Appleton of New York. Darwin was delighted. "I never dreamed of my book being so successful . . . [once] I should have laughed at the idea of sending the sheets to America." He offered Gray a share of his twenty-two-pound royalties, which Gray declined.

From his explorations in Africa, David Livingston shortsightedly informed Darwin he could see no struggle for existence on African plains. Karl Marx described *On the Origin of Species* "as a basis in natural science for the class struggle in history" and was amused that "Darwin recognises among beasts and plants his English Society."

By June 1860, another critical if much delayed review was being readied for publication in the July issue of *The Quarterly Review*, owned by John Murray himself. The author, paid sixty pounds for his trouble, was the worldly, ambitious Bishop of Oxford, Samuel Wilberforce, the son of William Wilberforce, the anti-slavery campaigner, and himself the reputed prototype for Archdeacon Grantly in Anthony Trollope's Barchester novels. Nicknamed "Soapy Sam" for his slippery, unctuous

---

1. Today *On the Origin of Species* is available in at least thirty-six languages.

manner, he was a flamboyant speaker and writer with a talent for exaggerated comparisons and humor. In his review he would write, "Now we must say at once and openly . . . that such a notion [as evolution] is absolutely incompatible not only with single expressions in the word of God but with the whole representation of that moral and spiritual condition of man which is its proper subject matter." The very idea of "a brute origin of him who was created in the image of God" was degrading and entirely opposed to man's God-appointed position at the head of the hierarchy of creation. He included some scientific arguments, probably supplied by his friend Richard Owen, and launched one of his signature sallies when he suggested that "if transmutations were happening . . . the favourable varieties of turnip" were "tending to become men."

Vanity Fair cartoon of Samuel Wilberforce, Bishop of Oxford

However, Wilberforce's review would be overtaken both at the time and in history by one of the most legendary events in science, perhaps almost akin to Archimedes's exit from his bath, Newton's falling apple, Heisenberg's meeting with Bohr in Copenhagen, and Robert Oppenheimer's quote from the *Bhagavad Gita* at the Trinity atom bomb test, "I am become Death, the shatterer of worlds"—the Oxford debate of Saturday, June 30, 1860, in which Wilberforce and Huxley were the chief but by no means only protagonists. The context was the annual weeklong conference of the British Association for the Advancement of Science, where leading thinkers presented science to a wider public. The precise venue was the wood-paneled lecture hall in the new Oxford University Museum of Natural History. The titular headliner was John William Draper, the Liverpool-born head of the medical school of New York City University. His lecture was titled "On the Intellectual Development of Europe, Considered with Reference to the Views of Mr. Darwin."

But the subsequent discussion was what everyone was looking forward to. Among the highly anticipated participants were expected to be Wilberforce, Huxley, and Hooker. Henslow would chair the session. Darwin himself had been debating whether to attend and was not at all relishing the personal contacts and eyeball-to-eyeball confrontations that would almost inevitably be involved. His stomach problems flared up as the date of the conference approached, probably due to nervous stress, and instead of traveling to Oxford, Darwin headed to Richmond in Surrey and a water cure spa. Huxley, although in Oxford, nearly decided not to attend but was persuaded to do so by Robert Chambers, who told him if he didn't he would be deserting the cause. The British Association for the Advancement of Science took no formal record of the session, and the several accounts that exist differ in the details of the proceedings and oratory. Even the estimates of the number of people present, sweating in the summer heat in the crowded hall, vary considerably, ranging from four hundred to seven hundred to more than one thousand.

Henslow, as chairman, introduced Draper, who talked for fully an hour and a half using copious references from the classics to illustrate his controversial view that human progress depended on science overcoming theology. Hooker thought his performance "flatulent stuff." When Draper finally finished, Henslow looked inquiringly across to Huxley, sitting nearby on the platform. Huxley shook his head, preferring to keep his powder dry for the present, and so Henslow turned to Wilberforce, who—needing no second invitation—stood, and after glancing round the room, began. He had dined with Richard Owen—who was not in the hall—the previous evening and had undoubtedly been briefed by him on the science. Drawing on that briefing and his own forthcoming piece in the *Quarterly Review*, he inveighed against *On the Origin of Species* and such theories in a speech full of humorous quips, including his review reference to turnips turning into men. The audience laughed appreciatively—this was what they had come to hear. As he drew to his conclusion after about thirty minutes, he looked across to Huxley and delivered what he thought might prove a witty, theatrical coup de grace: "I should like to ask Professor Huxley . . . Is it on his grandfather's or his grandmother's side that the ape ancestry comes in?"

Already well on his way to earning his soubriquet of "Darwin's Bulldog," Huxley instantly saw his opportunity, reputedly telling the man sitting next to him, "The Lord hath delivered him into mine hands," and quickly rose to his feet. First he refuted Wilberforce's scientific criticism before, saving his best for last, he answered Wilberforce's own final question about his ape ancestry: "If the question is put to me would I rather have a miserable ape for a grandfather or a man highly endowed by nature and possessed of great means of influence [an obvious reference to Wilberforce] and yet who employs these faculties for the mere purpose of introducing ridicule into a grave scientific discussion—I unhesitatingly affirm my preference for the ape." Hooker then added his views to the onslaught, later telling Darwin, "My blood boiled . . . I hit [Wilberforce] in the wind at the first shot in 10 words taken from his own ugly mouth—and then proceeded to demonstrate in as few more 1. that he could never have read your book and 2. that he was absolutely ignorant of the rudiments of Botanical Science . . . Sam was shut up—had not one word to say in reply."

With the intellectual and physical heat rising, a pregnant woman, Lady (Jane) Brewster, the wife of the inventor of the kaleidoscope, fainted and was carried out. Robert FitzRoy tried to make himself heard over the hubbub. He had been in Oxford the previous day to speak about work he was undertaking on meteorology and had decided to stay on for this debate rather than return immediately home. According to several accounts, waving a heavy Bible above his head, he shouted that he "regretted the publication of Mr Darwin's book, and denied Professor Huxley's statement that it was a logical arrangement of facts." He was sorry that he had given Darwin "the opportunities of collecting facts for such a shocking theory" and had "often expostulated with his old comrade of the *Beagle* for entertaining views which were contradictory to the first chapter of Genesis." Before sitting down distressed, he implored the meeting to believe in the Bible, not Darwin. Others then rose to offer their contrasting views until, after what Hooker thought was about four hours in all, Henslow brought the meeting to an end with "an impartial benediction."

When he heard about "poor FitzRoy's" intervention at the meeting, Darwin told Henslow, "I think his mind is often on the verge of insanity." Darwin congratulated Huxley on his performance, asking ironically, "How durst you attack a live Bishop in that fashion? I am quite ashamed of you! Have you no reverence for fine lawn sleeves? By Jove, you seem to have done it well." Then he confessed, "I would as soon have died as tried to answer the Bishop in such an assembly."

The story of the debate resonated throughout Britain and beyond. Perhaps its greatest consequence was to bring the controversy to public attention—any thinking person had to have a view on it. "Darwin's book is in everybody's hands," one commentator wrote.

Natural Selection

In the years that followed, Darwin worked on further British editions of *On the Origin of Species*—he produced six in total during his lifetime, the last in 1872—selling eighteen thousand copies in all. Each involved considerable revision. In undertaking them, Darwin, continuing to prefer conciliation to controversy and confrontation, attempted to accommodate some of his critics' views. In doing so, he slightly modified his theory of natural selection, rendering it somewhat less accurate. A case in point was his response to the views on the age of the Earth of the physicist William Thomson (later Lord Kelvin), the father of thermodynamics, who suggested, incorrectly as was later proved, that the Earth might be only one hundred million years old, not long enough to support a process of entirely gradual natural selection. Although he considered Thomson's views "an odious spectre," Darwin amended his sixth edition to allow the presence of a degree of Lamarckian modification by environment, which would increase the speed of change.[1]

While he was preparing these amended editions, Darwin was also continuing networking by correspondence and through intermediaries to promote his views on natural selection. In *On the Origin of Species* he had lamented, "I must here treat the subject [natural selection] with extreme brevity though I have the materials prepared for an ample discussion." In 1862 he embarked on readying for publication this mass of evidence he had accumulated on natural selection, much of which he had included in the drafts of his uncompleted three-volume magnum opus from which he had drawn the material for *On the Origin of Species*. He sought out even

---

1. The now commonly accepted age of the Earth is some 4.5 billion years.

more information to augment the material. For example, he wrote to
Thomas Bridges, a missionary working with the Fuegians, as part of his
research about the relationship between developed and less-developed
people, to ask, "Do the Fuegians or Patagonians . . . nod their heads
vertically to express assent, and shake their heads horizontally to express
dissent? Do they blush? And at what sort of things? Is it chiefly or most
commonly in relation to personal appearance, or in relation to women?
Do they express astonishment by widely open eyes, uplifted eyebrows
and open mouth?"

He tried to clarify how varieties among domestic rabbits compared
to wild ones by boiling rabbit carcasses, then deboning them and laying
out their skeletons on his billiard table to identify variations. He again
studied and dissected pigeons from his own pigeon coops and sought
information from a wide range of sources. These included professional
animal and bird breeders, paying them five pounds to have particular
breeds crossed to check their offspring's fertility. He consulted his fellow
natural scientists at home and abroad, quizzing Gray, for example, about
how and when invasive British weeds spread in the United States at the
expense of endemic plants, as he had observed happening on St. Helena
during the *Beagle* voyage. He cluttered the house as well as his garden with
growing plants and good-naturedly corralled his children, nephews, and
nieces to collect and sort plant specimens. He built a hothouse at Down
House and pestered Hooker at Kew for specimens to fill it, offering to
send over his own cart complete with protective wrapping material on
a warm day to collect them.

Darwin became fascinated by orchids and the processes by which the
fertilization of these diverse and beautiful plants occurred. He cultivated
orchids himself and wandered around the countryside surrounding the
village of Downe searching for wild specimens. In May 1862 he published
a monograph, *On the Various Contrivances by Which British and Foreign Orchids
Are Fertilised by Insects*, showing how those orchid species best adapted to
attract insects to cross-fertilize them thrived at the expense of others.
His work was generally praised for its depth of detail and observation,
even if by no means every botanist was prepared to go along with his

view of "adaptation." He had corresponded extensively with both Gray and Hooker in writing this work to draw out their knowledge and experience of plants and orchids in particular.

As the American Civil War broke out, Darwin—a strong supporter of Abraham Lincoln, his exact contemporary, and the northern states in their opposition to slavery in the South—added in a letter to Asa Gray at Harvard about plant development, "I have not seen or heard of a soul who is not with the North. Some few, and I am one, even wish to God, though at the loss of millions of lives, that the North would proclaim a crusade against slavery. In the long run, a million horrid deaths would be amply repaid in the cause of humanity."

In early summer 1862, Alfred Wallace, recently returned after eight years in the East Indies with more than 125,000 plant and animal specimens, and Darwin met for the first time, though neither left a record of their meeting. No close friendship developed, but Darwin frequently reassured Wallace that the theory of natural selection was "just as much yours as mine." Wallace's work on geographical distribution defined what soon became known as "The Wallace Line" through southeast Asia, marking the boundary between Australian and Asian fauna. Darwin helped Wallace enter scientific society and publish papers on his studies in journals of societies of which Darwin, unlike Wallace—still an outsider in his own view and probably in that of others—was a member. Wallace was not elected a Fellow of the Royal Society until 1893. He found socializing an even greater trial than Darwin, writing that to him, "talking without having anything to say and merely for politeness or to pass the time was most difficult and disagreeable." Lyell's always snobbish wife in turn thought Wallace "shy, awkward and quite unused to good society."

In early 1863, Darwin heard of the finding of a fossil of a half-reptile, half-bird creature at Solenhofen in Bavaria in southern Germany. It had feathers but also teeth, a long bony tail, and clawlike fingers on its wingtips. The British Museum purchased the fossil for the large sum of £450. Darwin thought it money well spent for an example of a rare transitional form to bolster his adaptation theory—"a grand case for me, as no group was so isolated as Birds; and it shows how little we know what lived during former times," he wrote.

Throughout this period Darwin suffered further frequent bouts of illness. In 1861, then aged fifty-two, he had felt too ill to respond to pleas to visit Henslow, his mentor and the man who had secured the *Beagle* voyage for him, who was dying. He excused himself to Hooker, who was with Henslow: "I shd. be certain to have serious vomiting afterwards, but that would not much signify, but I doubt whether I could stand the agitation at the time. I never felt my weakness a greater evil. I have just had specimen for I spoke a few minutes at Linn. Soc. on Thursday and though extra well, it brought on 24 hours vomiting." He tormented himself with guilt after Henslow's death that May. One of his worst health crises came in September 1863, when he visited the water cure at Malvern with Emma. They searched for their daughter Annie's grave, which neither had seen, and could not find it at first. After some days Emma eventually discovered it, overgrown with vegetation. Darwin became so debilitated and weak he could scarcely walk and was too ill to partake of any water cure. Emma feared he might have an epileptic fit. He left Malvern feeling worse than when he arrived.

His ill health persisted after his return home. He became depressed and unable to work. His hair began to turn grey. He was confined to bed, vomiting regularly into a chamber pot for weeks at a time. He suffered flatulence, headaches, dizziness, and skin problems so severe he could not shave, leaving his beard to grow. He could not work, however febrilely anxious he was to do so, never mind see visitors, becoming of necessity for some time a partial recluse. Emma nursed him uncomplainingly, discouraging callers. With her daughter Henrietta she took care of all Darwin's correspondence, becoming adept at forging his signature. Darwin recalled his nervousness "when Emma leaves me." When his unmarried sister Susan and his sister Catherine both died in 1866, Darwin was grateful that his brother Erasmus took responsibility for the arrangements and the disposal of the family home, The Mount, and its contents.

His health recovered for short periods in between frequent relapses. Not until around 1867 did a sustained improvement begin. On doctor's orders Darwin took up riding again, only for a bad fall from his favorite horse to deter the man who had happily ridden the pampas with the gauchos from any such further adventures. When he again visited the Royal Society and his friends there, Emma recalled, "He was obliged to name himself to almost all of them as his beard altered him so much."

While he had been ill, Darwin's supporters, with Huxley prominent among them, had continued to press the case for natural selection. They established several short-lived magazines to promote their cause and founded an informal club—the X Club—in which they discussed how best to frustrate their opposition. In 1863, Huxley confronted the question of the links between humans and apes, publishing a series of lectures showing their similarities and suggesting their ancestry was linked. From his sickbed Darwin welcomed Huxley's "monkey book." In 1864, Wallace put a paper to the Anthropological Society exploring shared instincts and social behavior in humans and primates. Darwin urged him to publish more on the relationship, offering him his own notes which he would have liked to take further, "but I have not strength." Wallace did not follow up his offer.

Nevertheless, on his recovery, Darwin did not pursue the topic but continued his studies of variations between domestic and wild plants and animals and on the nature of heredity, though it was of course long before the understanding of genetics. He finally published his work under the title *Variations of Animals and Plants under Domestication*. It achieved steady if unspectacular sales and was widely praised for its mass of data.

Then in 1869 Wallace suggested in *The Quarterly Review* that natural selection was in itself insufficient to account for human evolution. Natural selection was only able to bring development to the brink of humanity, whereupon other factors such as an ill-defined intellectual or spiritual power had become dominant. Darwin was appalled, telling Wallace, "I hope you have not murdered too completely your own and my child." Wallace's views were influenced by his growing belief in spiritualism, a fad that attracted many eminent Victorians such as Arthur Conan Doyle. Darwin himself was immune to its attractions, only ever attending one séance. His son George and his brother Erasmus, both of whom were sympathizers, organized it. Emma, Huxley, Mary Ann Evans (the writer George Eliot), and her partner George Lewis, among others, attended. The latter two quit in the middle, as did Darwin—"I found it so hot and tiring that I went away before all those astounding miracles or jugglery took place . . . The Lord have mercy on us all, if we have to believe in such rubbish." He considered it imposture and indeed the medium involved, Charles Williams, was later exposed as a fraud.

Now, feeling disappointed with Wallace and that he had been elsewhere "taunted with concealing my opinions," Darwin felt compelled to turn to the origins of man.

The Fuegians were often in Darwin's mind as he struggled in the draft of *The Descent of Man* to explain how humans descended from the apes and how, while being a single species, "civilized and savage" races diverged. He wrote, "It is notorious that man is constructed on the same general type as other mammals. All the bones in his skeleton can be compared with corresponding bones in a monkey, bat or seal." Every fissure and fold in the human brain had its analogy in the orangutan. He used illustrations to emphasize the similarities between human and animal embryos—for example, that human embryos had a vestigial tail. He suggested with examples from Jenny the orangutan that primates shared nearly all human instincts—"the same senses, intuitions and sensations, similar passions, affections and emotions, even the more complex ones such as jealousy, suspicion, emulation, gratitude and magnanimity . . . though in very different degrees."

He described how human ancestors of both sexes would have been covered with hair and have tails and prehensile big toes. They would have lived in trees and only gradually climbed down to the ground, graduating from being quadrupeds to walking on two feet, leaving their hands free to wield a tool or weapon. He speculated that Africa had been the cradle of this early evolution and that "all the higher mammals are probably derived from an ancient marsupial animal and this through a long line of diversified forms either from some reptile-like or some amphibian-like creature and this again from some fish-like animal. In the dim obscurity of the past we can see that the early progenitor of all the Vertebrata must have been an aquatic animal provided with branchiae with the two sexes united in the same individual, and with the most important organs of the body (such as the brain and heart) imperfectly developed."

The divergence between races in humans was, in Darwin's view, due to sexual selection leading to differing choices of sexual partners according to local perceptions of beauty. As background to the topic that took up nearly three-quarters of his completed two-volume, thousand-page work, he instanced how the peacock's tail, although a useless impediment to flight, was not an example of the watchmaker design theory of William Paley—whose work he had so admired when he was at Cambridge—that such beauty must have a designer and that designer must be God. Instead it was a naturally selected adaptation to attract peahens. Those males with the greatest adornment bred more often, and over successive generations their young by sexual selection developed even greater tails. Among primates he noted the bright blue and red faces of mature mandrills, a West African primate. "No other member of the whole class of mammals is coloured in so extraordinary a manner as the adult male mandrill . . . The face at this age becomes of a fine blue, with the ridge and tip of the nose of the most brilliant red." Male lions' manes were another such example.

Darwin thought that in humans some local female populations had preferred certain skin tones, and over time, due to more successful and frequent breeding among those possessing them, certain tones had become dominant. He suggested that another form of sexual selection—fighting between males to mate with available females—had led to increases in size among males compared to females, as in the case of elephant seals, and the development of fighting adaptations such as enlarged canine teeth in male baboons, spurs on cocks, and antlers on male deer.

Based on his observation of social feelings in animals, Darwin considered that such feelings were likely to be natural and instinctive in humans, not simply imposed externally by convention and law: "Ultimately a highly complex sentiment, having its first origin in the social instincts, largely guided by the approbation of our fellow-men, ruled by reason, self-interest, and in later times by deep religious feelings, confirmed by instruction and habit, all combined, constitute our moral sense or conscience."

He suggested that as society developed, qualities other than appearance—such as intellectual ability, bravery, and foresight—had become involved and that the male had become dominant both in selection of his reproductive partner and in society, stating that "man has ultimately become superior to woman." Despite his friendship with prominent members of the feminist movement such as Harriet Martineau, he would write to one woman, "I certainly think that women though generally superior to men in moral qualities are inferior intellectually; and there seems to me to be a great difficulty from the laws of inheritance . . . in their becoming the intellectual equals of man." He told his publisher John Murray about one of his books, "the subject . . . is treated in my paper so that any woman could read it."

In *The Descent of Man*, he would write, "The chief distinction in the intellectual powers of the two sexes is shewn by man attaining a higher eminence in whatever he takes up, than woman can attain whether requiring deep thought, reason, or imagination, or merely the use of the senses and hands." His ability to stereotype also led him into unflattering comparisons between nationalities in the same way as when, aboard the *Beagle*, he compared the Portuguese Brazilians unfavorably with those of Spanish stock elsewhere in South America. The Irish suffered when he cited with approval the following comparison: "The careless, squalid, unaspiring Irishman multiplies like rabbits; the frugal, foreseeing, self-respecting ambitious Scot stern in his morality . . . sagacious and disciplined in his intelligence, passes his best years in struggle and in celibacy, marries late, and leaves few behind him."[2]

In contrast to his Victorian views on the relative attributes of males and females or of different nationalities, Darwin never diverged—as others would, often in an attempt to justify slavery or imperialism—from his fundamental view that all humans, whatever their state of development and however superficially diverse and apparently "bestial," such as the

2. American feminist Antoinette Brown Blackwell wrote a series of essays titled *The Sexes in Nature*, challenging Darwin's comments about the female intellect. She had earlier sent him a copy of her book *Studies in General Science*, taking issue with views expressed in *On the Origin of Species*. Darwin sent her a courteous letter of acknowledgment but addressed it "Dear Sir."

Fuegians, had come from the same root and were one species. Given the right opportunities, they were equally capable of development. After all, Jemmy Button had shown his ability to integrate into "civilization" even if he had ultimately rejected it. He wrote, "The Fuegians rank amongst the lowest barbarians; but I was continually struck with surprise how closely the three natives on board H.M.S. 'Beagle,' who had lived some years in England, and could talk a little English, resembled us in disposition and in most of our mental faculties."

As Darwin continued to labor on *The Descent of Man*, one of Emma's nieces wrote about her, "I don't know any wife quite so absorbed in a husband—of course there is not often so much to be absorbed in—and her time is quite taken up by ministration to him, and yet there is something peculiar in her clear sightedness to his narrowness of view." Emma herself wrote to her daughter Henrietta, "I think it [his new book] will be very interesting, but that I shall dislike it very much as again putting God further off."

In 1870, the Franco-Prussian War broke out. Ever worried, despite his liberal views, about the threat from France and the socialism and anarchy for which the country was renowned to Victorians, Darwin made clear where his sympathies lay when he wrote to a friend in Germany, "I have not yet met a soul in England, who does not rejoice in the splendid triumph of Germany over France: it is a most just retribution against that vain-glorious, war-loving nation."

Just as he had with *On the Origin of Species*, Darwin worried what reception his book, the writing of which "half killed me," as he wrote to his old *Beagle* colleague Bartholomew Sulivan, would receive. He again did his best to influence opinions by sending out advance copies, together with tailor-made letters, and by encouraging his supporters to exert themselves on his behalf.

John Murray published the two-volume *Descent of Man and Selection in Relation to Sex* on February 24, 1871. With its confident assertion that human beings were descended from "the lower animals" and his theory about the role of sexual competition and selection in promoting the spread of variations best suited to their surroundings, it marked Darwin as one of the fathers of anthropology. Unsurprisingly, the book drew heavily on his encounters with Fuegians during the *Beagle* voyage—for example, describing how differences in food availability and in climate between east and west Tierra del Fuego influenced the stature of the inhabitants and drawing examples to support his theories from how Fuegians chose their sexual partners. In his conclusion he wrote, "The astonishment which I felt on first seeing a party of Fuegians on a wild and broken shore will never be forgotten by me for the reflection at once rushed into my mind—such were our ancestors."

In general, although cartoons of Darwin as an ape-man proliferated, and few endorsed his view fully, reviews and responses were more muted and less violent and virulent than Darwin expected. He put this down to "the increasing liberality of England" and was pleased that "evolution [is] talked of as an accepted fact." Even if people disagreed with Darwin's theses—and many still did—they respected his scientific standing, integrity, and ability to assemble, manipulate, and display coherently a mass of indisputably valuable data to support his argument. The twenty-five hundred copies of the first edition of *The Descent of Man* quickly sold out, and five thousand more were ordered.

Critics included a writer in *The (London) Times* who noted that in France "loose philosophy" had damaged moral principles and that analogously in Britain, "a man [such as Darwin] incurs a grave responsibility when with the authority of a well-earned reputation, he advances at such a time the disintegrating speculations of this book." Another reviewer suggested, "Society must fall to pieces if Darwinism be true," while the *Edinburgh Review* believed "the constitution of society would be destroyed . . . Never perhaps in the history of philosophy have such wide generalisations been derived from such a small base of fact." Wallace believed "there are plenty of points open to criticism but it is a marvellous contribution to the history of the development of the forms of life."

Darwin had accumulated so much material on human and animal behavior and in particular on the expression of emotion that he could not include it all in *The Descent of Man*, so he immediately set to work on a further book to be published by John Murray in 1872, *The Expression of*

Cartoon of Charles Darwin as "A Venerable Orang-outang," published in *The Hornet*, March 1871

the Emotions in Man and Animals. It drew on his observations of apes, his careful listing of the childhood development of his sons and daughters, and many details acquired from friends and experts at home and abroad. These included photographs of inmates from what were then known as lunatic asylums. His thesis was that how humans expressed their feelings had its origins in animals. The curling of the lips into a sneer might derive from the snarling action used to bare teeth to an opponent by animals who used teeth as their weapon. The book was one of the first commercial volumes to incorporate photographs. Emma Darwin was correct when she wrote, "I don't think it is a book to affront anybody . . . . I think it will be generally interesting." The book sold well—nine thousand copies were bought in the first four months.

Darwin had now become accepted as a leading scientific authority. In 1858 Huxley had correctly predicted, "You will have the rare happiness to see your ideas triumphant during your lifetime."

CHAPTER TWENTY-SIX

"The Clerk of the Weather"

In 1865, Darwin had heard of the tragic death of Robert FitzRoy, without whom he would never have made the voyage on the *Beagle* that enabled him to collect the information to frame his theories.

Few records—other than dry official ones—survive of FitzRoy in the years immediately following his return to Britain after his dismissal as Governor of New Zealand. In 1848, the Admiralty appointed him captain of HMS *Arrogant*, an experimental vessel, the navy's first to use a steam-driven screw propeller rather than a stream-driven paddle wheel to augment her sail power. The *Arrogant* had many teething troubles that FitzRoy, methodical and almost obsessively hardworking as usual, did his best to solve. While on a proving voyage to Lisbon, pressure of work, together with mounting financial difficulties—in part accentuated by his continuing willingness to spend his own money on naval business—seem to have produced a reoccurrence of the nervous exhaustion and depression of the type he suffered in Valparaiso on the *Beagle*. This time, in early 1850, he did indeed resign his commission. He slowly recovered, in part bolstered by his election to the Royal Society—where among his sponsors were Darwin and Beaufort—on the grounds of his achievements as a "scientific navigator" skilled in "nautical astronomy" and "hydrography."

Then, in April 1852, his wife Mary died aged only thirty-nine. "I do not grieve," he wrote to a friend, "as one without hope. My beloved wife was so sincere and consistent a Christian that I know she is safe—and permanently happy—while I am just as certain that if I do my best to follow her blessed example and earnestly work out my own salvation I shall rejoin her sooner or later."

Darwin met FitzRoy a little later and wrote, "He looked very well and was very cordial to me. Poor fellow, I fear besides his other misfortunes, he is rather poor; at least he has given up House-keeping." Perhaps partly pragmatically to secure a mother for his children and to remedy his housekeeping difficulties, less than two years after Mary's death, FitzRoy, then forty-eight, married a much younger cousin, Maria Smyth, who bore him a fifth child. His appointment in 1854 as "meteorological statist" or statistician, on the foundation of what was to become Britain's Meteorological Office, provided FitzRoy the opportunity to make a lasting name in the history of science. Just as the time was ripe through the advances of the Age of Enlightenment for Darwin's work on geology, biology, and thence evolution, so it was for FitzRoy's on meteorology. More reliable and robust instrumentation was becoming available, together with much more data on weather patterns worldwide through pioneering and well-documented voyages such as that of the *Beagle* itself. Among the instruments, as well as the mercury barometer and the mercury thermometer introduced by Gabriel Fahrenheit in 1714, were newer inventions such as the rain gauge and the hygrometer (used to measure humidity and the amount of water vapor in the atmosphere). Another major—if at first thought seemingly unrelated—development was that of the telegraph in the 1830s and of Morse code a few years later. These innovations allowed weather data to be transferred quickly to be collated and analyzed centrally.

FitzRoy's interest in meteorology was known both to the Admiralty and to the wider scientific community from his work on the *Beagle*, including his use of Beaufort's innovative scales and his institution of meticulous twelve-hourly—and sometimes even more frequent—barometer and thermometer readings. In 1843, just before departing for New Zealand, FitzRoy proposed to a parliamentary committee on shipwrecks the establishment of weather stations around Britain to use barometers to provide storm warnings.

However, it took an American proposal a decade later to hold an international conference to discuss collaboration on recording and sharing weather data to give the impetus required to create the UK statist post. The conference, held in Brussels in 1853, was the initiative of a US naval officer Lieutenant Matthew Maury, who had already collected substantial amounts of historical meteorological data. While praising Maury's work,

FitzRoy, perhaps typically, was also critical, querying how much was original: "He has a large reputation among men of my cloth who have not heard quite as much of old Dampier—Cook—Flinders . . . as educated men in England have generally. Maury's adoption of these men's ideas—and non-recognition of their origin—is sad."

The British government was somewhat reluctantly represented at the conference but in 1854 established the new Meteorological Department of the Board of Trade with FitzRoy as its first head—a post for which he had already been lined up with his enthusiastic concurrence. FitzRoy would have three assistants. His remit was to provide data to allow the UK shipping industry to make more efficient and economic use of their vessels. The emphasis following Maury's precedent was data recording, although FitzRoy would soon on his own initiative expand it into weather forecasting, of which he became the world's pioneer.

FitzRoy began by establishing a network of agents in British ports who persuaded ships' officers to take on their voyages carefully standardized barometers to record readings and other data on wind, temperature, and humidity on pro forma logs that were to be sent to FitzRoy. Building on work by Maury, FitzRoy invented charts showing what he called "wind stars" for each ten-degree square of ocean that provided information on prevailing wind directions for each quarter of the year. By spring 1855, eighty vessels—both merchantmen and naval ships—were carrying FitzRoy's barometers. He also issued them to fishing villages around the country, together with a fifty-page guide to their use entitled *A Barometer and Weather Guide*. He suggested "difficult as it is to foretell weather accurately, much usual foresight may be acquired by combining the indications of instruments with atmospheric appearance." The manual contained a deal of weather lore, sometimes in rhyme, such as:

When glass falls low
Prepare for a blow.
When it rises high,
Let all your kites fly

and at others in more didactic prose:

> Small inky-looking clouds foretell rain; a light scud driving across
> heavy clouds presages heavy wind and rain

In April 1856, family tragedy again struck FitzRoy. His eldest
daughter, eighteen-year-old Emily, died. Darwin, who knew the pain
of losing children, wrote a letter of consolation. In early 1857, Darwin
and FitzRoy—now promoted to rear admiral on the reserve list—met
for, as far as is known, the last time when FitzRoy and his wife visited
the Darwins at Down House. The two men apparently enjoyed chatting and reminiscing. Emma Darwin told her son William, "Papa was
much awestruck with the honour and the admiral was very gracious and
friendly." However, she added, perhaps a little jealously, "Mrs FitzRoy is
a cold, dry stick and I could not find a word to say to her. Papa says she is
a *remarkably* nice woman (because she laughed at his jokes I say)." FitzRoy
began to work even harder in spite of further career disappointment. Also
in 1857 his immediate superior at the Board of Trade, the Chief Naval
Officer, resigned. FitzRoy applied for the post. So too did Sulivan, by
now knighted. Sulivan got the job.

Out of respect for his old captain's feelings, and perhaps fearing out-
pourings of "hot coffee," Sulivan immediately secured acceptance that the
Meteorological Office should no longer report to the Chief Naval Officer
but directly to the Secretary of the Board of Trade. As well as avoiding
embarrassment to both men, this solution allowed FitzRoy greater freedom of action. The need for his efforts was highlighted when in autumn
1859 the modern iron-hulled 2,719-ton merchantman *Royal Charter*, carrying 480 passengers and £322,000 (around £40 million today) of gold bul-
lion from the Western Australian gold fields to Liverpool, was wrecked
on the Welsh coast with the loss of all but thirty-nine of those aboard.

Conscious and conscience-stricken that on the basis of the information available a storm warning could have been issued that might
well have saved the ship, FitzRoy campaigned successfully to establish

a system to issue such warnings. This was the mission that had brought him to Oxford in summer 1860 when he was caught up in the historic debate between Huxley, Hooker, and Wilberforce. By early autumn the following year, he had put in place eighteen stations around the British and Irish coasts and six more on the continent. He supplied each with improved standardized barometers designed to his specification. They were placed in the charge of local telegraph operators who took readings at nine o'clock in the morning and telegraphed them immediately to FitzRoy's London offices. Based on the data, in February 1861 he instituted the world's first large-scale storm warning system, issuing alerts to ports that in turn signaled shipping by the hoisting of combinations of cones and cylinders—each three feet six inches high and three feet in diameter so they could be seen from a distance. At night a pattern of lanterns was displayed.[1]

FitzRoy soon used his initiative to issue on August 1, 1861, the world's first "weather forecast"—a term he coined. It read:

> *General weather probably for the next two days in the*—
> North—Moderate westerly wind; fine.
> West—Moderate south-westerly; fine.
> South—Fresh westerly; fine.

He went on to provide daily forecasts, soon covering the east point of the compass as well as the other three, to *The London Times* before passing them on to five other newspapers. Those who benefited from and praised his forecasts included both naval and merchant captains, harbor masters, and individual travelers. A dry dock owner ascribed a fall in his profits to fewer ships being damaged in storms as a result of FitzRoy's warnings, causing a drop in demand for repairs. Queen Victoria consulted him on the weather for her crossings to the Isle of Wight, and when two of her daughters were planning to cross the Channel from Folkestone to

---

1. Although FitzRoy's was the first large-scale storm warning system, the Dutch had instituted a smaller system based on four weather stations a little earlier.

Boulogne in March 1863 he provided the following forecast: "Weather on Friday favourable for crossing—moderate—mild—cloudy, fine, perhaps showery at times." The humorous magazine *Punch* labeled FitzRoy "The Clerk of the Weather."

FitzRoy's forecasts were not always correct. Some owners of fishing boats or coasting vessels castigated him and his storm warnings for causing their captains to remain unnecessarily in harbor with consequent loss of profits. FitzRoy, as forthright as ever, condemned them in his official report as "those pecuniary interested individuals and bodies who would leave the coasters and the fishermen to pursue their precarious occupation heedlessly without regard to risk—lest occasionally a day's demurrage should be caused unnecessarily or a catch of fish missed for the London market." Adding to FitzRoy's worries was that Maury (now Captain Maury), self-exiled in Europe as a Confederate Virginian from the American Civil War, criticized some of FitzRoy's forecasting methods. The two, however, continued to meet and correspond, with FitzRoy sympathizing with Maury's personal predicament.

In 1862 FitzRoy published his masterwork, a 440-page volume titled *The Weather Book*, subtitled *A Manual of Practical Meteorology*. The opening paragraph stated, "This small work is intended for many rather than for few, with an earnest hope of its utility in daily life." He reiterated "that the state of the air foretells coming weather rather than indicates weather that is present" and detailed his views on the weather and how to forecast it, such as his belief that storms occur along the unstable fault lines between hot and cold air masses. Unlike Darwin, who kept the copyright of his books and thus profited from their popularity, FitzRoy sold the copyright of *The Weather Book* outright to his publisher, Longmans, for 200 pounds, thus failing to capitalize on its success.

Robert FitzRoy

By now *The Times*, to whom FitzRoy had given priority in publishing his forecasts, was publicly criticizing their accuracy. The criticism, the demands of his job, and continuing financial worries began to weigh on the increasingly deaf and febrile FitzRoy. One symptom was that his handwriting, so neat and precise on the *Beagle*, degenerated into a large, hasty scrawl with frequent extravagant crossings out. In early 1865, he moved with his family from South Kensington to the outer London suburb of Upper Norwood, perhaps partly for financial reasons, perhaps partly for the calm. His wife Maria became increasingly worried about him, writing, "The Doctors unite in prescribing total rest, and entire absence from his office . . . Leave has been given him, but his active mind and over-sensitive conscience prevent him from profiting by this leave, as he does not like to be putting the work he is paid to do upon others, and it keeps him in a continual fidget to be at his post, and the moment he feels at all better he hastens back, only to find himself unequal to work satisfactorily when he gets there."

FitzRoy studied his Bible and prayer book and unsuccessfully attempted to work at home and made short trips to London and his office, which exacerbated his nervous exhaustion. He became unable to sleep properly. On Saturday, April 29, he briefly visited Maury, who was about to leave for the Caribbean. No record exists of what they talked about, but he returned home somewhat agitated and remained restless throughout the remainder of the day and evening before going to bed at midnight. FitzRoy woke early the next morning. Then, at around 8:00 A.M., after kissing his youngest daughter Laura, he retreated into his bathroom and after a brief pause bolted the door and cut his throat with his razor.

Darwin's initial reaction was realistic if a little harsh: "I was astounded at news about FitzRoy; but I ought not to have been, for I remember once thinking it likely; poor fellow his mind was quite out of balance once during our voyage. I never knew in my life so mixed a character. Always much to love and I once loved him sincerely; but so bad a temper and so given to take offence, that I gradually quite lost my love and wished only to keep out of contact with him. Twice he quarrelled bitterly with me, without any just provocation on my part. But certainly there was much noble and exalted in his character." In the autobiography Darwin wrote for his family, his judgment was a little kinder—a mixture of frustration, exasperation, and affection. "FitzRoy's temper was a most unfortunate one . . . He was also somewhat suspicious and occasionally in very low spirits . . . He seemed to me often to fail in sound judgement or common sense. He was extremely kind to me but was a man very difficult to live with on an intimate basis." He had "many very noble features; he was devoted to his duty, generous to a fault, bold, determined, indomitably energetic and an ardent friend to all under his sway."

Highly ironically, unlike the agnostic Darwin who would be buried in Westminster Abbey, because of the then ban on suicides being buried in hallowed ground, the Creationist Christian FitzRoy had to be buried just outside the cemetery wall of his local church. Darwin contributed 100 pounds to a fund to clear FitzRoy's debts and support his widow.

CHAPTER TWENTY-SEVEN

"All Soon to Go"

Just before the publication of his book on the expression of the emotions, Darwin had written to Wallace, "I shall now try whether I can occupy myself, without writing anything more on so difficult a subject, as evolution." He was as good as his word. With evolution behind him and his health improving, perhaps as a consequence, he turned his researches first toward plants. Following his previous work on orchids, in 1875 he produced a volume titled *The Movement and Behaviour of Climbing Plants*, suggesting their ability to climb gave such plants an advantage, particularly in thick vegetation in the struggle for life. His conclusion was, "It has often been vaguely asserted that plants are distinguished from animals by not having the power of movement. It should rather be said that plants acquire and display this power only when it is of some advantage to them; this being of comparatively rare occurrence, as they are affixed to the ground, and food is brought to them by the air and sun."

In 1877 he published *The Different Forms of Flowers of Plants of the Same Species*, a book dedicated to his Harvard friend Asa Gray, in which he examined the fertilization, cross-fertilization, and hybridization of plants. His last botanical work, published in 1880, was *The Power of Movement in Plants*, extending and amplifying his previous work. The comments of one of his gardeners about Darwin show his abstraction during his botanical studies: "I often wish he had something to do. He moons about in the garden, and I have seen him stand doing nothing before a flower for ten minutes at a time."

In much of his later botanical work Darwin had the help of his son Francis, who also assisted on what proved to be his last book, which was about earthworms: how they lived and how their burrowing changed the soil. He would go out at night with Francis into the garden at Down

House and collect worms, testing their sensitivity to light with a lantern and subsequently experimenting as to what they were prepared to eat and what not. He even had his wife play the piano loudly to them to see if they might be able to hear—they couldn't. In the book, he showed the part worms played in reshaping the landscape and in the slow build-up of soil leading, for example, to the burial of ancient ruins. When published in 1881, it sold well and Darwin received many a postbag of letters from interested gardeners and others.

After the publication of *On the Origin of Species* and *The Descent of Man*, Darwin had become somewhat of a celebrity, even if a reluctant one. He almost perforce entertained more visitors at Down House than previously. Among those he was perhaps most pleased to see, as a lifelong liberal Whig, was the once and future Prime Minister William Gladstone, the leader of the Liberal Party. Gladstone visited in 1877 with a large group. Darwin told a friend, "I never saw him before and was much pleased with him: I expected a stern, overwhelming sort of man, but found him as soft and smooth as butter and very pleasant. He asked me whether I thought that the United States would hereafter play a much greater part in the history of the world than Europe. I said that I thought it would." Another of those present related that in fact Gladstone's main contribution had been to read out loud and at length his latest pamphlet on Turkish atrocities in the Balkans. Beyond noting Darwin's "pleasing and remarkable appearance," Gladstone himself made no known comment on the visit.

Four years later, when Gladstone was again Prime Minister, Darwin successfully lobbied him to award a state pension to Wallace, who had fallen on hard times through poor investments. When Gladstone granted Wallace an annual government pension of two hundred pounds, Darwin was delighted. Two months later Gladstone offered Darwin a position as a trustee of the British Museum, which Darwin declined, citing his poor health and lack of strength.

Among others who contacted Darwin was Karl Marx, who, in 1873, sent him a copy of a German edition of *Das Kapital,* inscribed, "Mr Charles Darwin on the part of his sincere admirer Karl Marx." Darwin responded, "I heartily wish that I was more worthy to receive it, by understanding

Charles Darwin in old age

more of the deep and important subject of political economy. Though
our studies have been so different, I believe that we both earnestly desire
the extension of knowledge, and that this in the long run is sure to add
to the happiness of mankind." Darwin never read *Das Kapital*—the book
remained in his library with its pages uncut.

Darwin rejected an approach by the radical socialists Charles Brad-
laugh and Annie Besant to speak in their defense at their trial for obscen-
ity for publishing a book advocating contraception. Darwin told them
his judgment was strongly opposed to theirs, being against any artificial checks on the natural rate of human increase. Such practices would
"spread to unmarried women and would destroy chastity on which the
family bond depends; and the weakening of this bond would be the greatest of all possible evils to mankind."

Darwin lamented, "Half the fools throughout Europe write to ask me the stupidest questions." Some correspondents felt emboldened to inquire what Darwin's religious beliefs were and indeed whether he had any. In his later years, his opinions seem to have swayed between what might be regarded as rationalism and hopeful optimism. According to his son Francis, "he felt strongly that a man's religion is an essentially private matter and one concerning himself alone." He refused an invitation to attend a conference proposed by the Archbishop of Canterbury to reconcile science and religion because he could not see "any benefit" arising from it.

He did respond to some inquiries, and among the views he expressed during the latter part of his life were: "I know well the feeling of life being objectless and all being vanity of vanities"; "I do not believe in the Bible as a divine revelation, and therefore not in Jesus Christ as the Son of God"; "I can hardly see how anyone ought to wish Christianity to be true; for if so, the plain language of the text seems to show that the men who do not believe, and this would include my father, brother and all my best friends, would be everlastingly punished. And this is a damnable doctrine." He had long ago, of course, expressed his disbelief in the Old Testament. On the relationship between science and Christianity, he wrote, "Science has nothing to do with Christ, except in so far as the habit of scientific research makes a man cautious in admitting evidence. For myself, I do not believe that there has ever been any Revelation. As for a future life, every man must judge for himself between conflicting vague probabilities." Whether there was a God was "beyond the scope of man's intellect." He considered "the mystery of the beginning of all things is insoluble by us."

Perhaps his most accurate expression of his views was, "I may state that my judgement often fluctuates . . . In my most extreme fluctuations I have never been an atheist in the sense of denying the existence of a God.——I think that generally (and more and more so as I grow older) but not always, that an agnostic would be the most correct description of my state of mind."

Cambridge University offered Darwin an honorary Doctorate of Letters, which he accepted. To attend the degree ceremony, he hired a private railway carriage so that he and his family could be carried from Kent to Cambridge without changing trains at the crowded London station and without any contact with others. It was an extravagance that he now sometimes employed and that he could easily afford, since at this stage of his life his income exceeded his expenditure by some four thousand pounds per annum (nearly half a million pounds in today's values). He successfully endured a raucous two-hour degree ceremony, which began with some undergraduates who packed the galleries on either side of the hall throwing a rope across and propelling along it a monkey figure above the worthies crowded below. A proctor had to climb up to remove it before Darwin, clad in red doctorial robes, received his honor fifty years after his matriculation at Christ's College. The awed welcome he received from dons underlined his status as a great man of science.

Another symptom of Darwin's fame was that he was asked several times to sit for portraits or photographs. On one occasion the painter Walter Ouless produced twin portraits of Darwin and his wife. Emma Darwin disliked them both, and hers so much that it has never been seen since. Presumably she destroyed it, just as Winston Churchill or his wife destroyed the portrait of him painted in 1954 by Graham Sutherland, which both detested. Other portraits of Darwin or oil copies by the original artists hang in his club the Athenaeum, Christ's College Cambridge, the National Portrait Gallery in the United Kingdom, and elsewhere.

In these later years Darwin's life was punctuated by the marriage of his children, the birth of grandchildren, and deaths among his own generation of friends and family. The Darwins' eldest daughter Henrietta married in 1871. Darwin worried about her health—she had never been strong and herself suffered hypochondriacal tendencies. Darwin as usual was concerned that her weakened state resulted from his own cousin-marriage. His son Francis married in 1874 and moved to live in the village of Downe as he became Darwin's assistant.

Charles Lyell died in February 1875, less than two years after his wife. His health and always poor eyesight had been declining for some time. A depressed Darwin felt "as if we were all soon to go." He refused an invitation to act as a pallbearer at Lyell's burial in Westminster Abbey—"I should so likely fail in the midst of the ceremony and have my head whirling off my shoulders"—and did not attend.

Francis's wife Amy died in February 1876, two days after the birth of their only child and the Darwins' first grandchild, a boy named Bernard. On this occasion, Darwin stood up to the tragedy better than either his son or his own wife. Francis moved into Down House with Bernard, and both Charles and Emma Darwin enjoyed having their grandson around them. Their oldest son, William, married in late 1877, and their youngest, Horace, in 1880.

During these years, probably prompted by the birth of his grandson, Darwin found time to work intermittently on his autobiography. He destined it only for his family. As with all such works it contains a mixture of fact and recollections, selected and shaped to give coherence and to reflect how he wished his life to be perceived. In it, he suggested, probably correctly, where his greatest abilities lay: "I am superior to the common run of men in noticing things which easily escape attention, and in observing them carefully. My industry has been nearly as great as it could have been in the observation and collection of facts. What is far more important, my love of natural science has been steady and ardent. This pure love has, however, been much aided by the ambition to be esteemed by my fellow naturalists . . . I have had the strongest desire to understand or explain whatever I observed—that is, to group all facts under some general laws. These causes combined have given me the patience to reflect or ponder for any number of years over any unexplained problem."

The turn of the decade saw the death of more family members—Emma's brother Jos and her sister Elizabeth, Darwin's cousin William Darwin Fox, and then, in August 1881, his brother Erasmus, without whom Darwin had never known life and whose home had provided a refuge on visits to London and a source of friendly intellectual conversation and society. Erasmus was buried in the churchyard at Downe, where Darwin himself intended to be interred.

Darwin suffered acute chest pains in the street on a visit to London in mid-December 1881. He seemed to recover, but the pains returned two months later. On March 7, he had a severe attack while walking slowly around the Sand Walk but managed to stagger back to the house. He again recovered somewhat, only to collapse unconscious a few days later at dinner on Saturday, April 15. Another partial recovery followed, but in his son Francis's words, "During the night of April 18th about a quarter to twelve, he had a severe attack and passed into a faint, from which he was brought back to consciousness with great difficulty. He seemed to recognise the approach of death, and said, 'I am not the least afraid to die . . .' He died at about four o'clock on Wednesday, April 19th, 1882, in the 74th year of his age."

In their obituaries *The New York Times* called him "the greatest scientific discoverer of his age and country"; *The Tribune* hailed "a giant among his fellows"; *The Herald* claimed "his life was like that of Socrates." *The (London) Times* wrote, "One must seek back to Newton or even Copernicus to find a man whose influence on human thought . . . has been as radical.

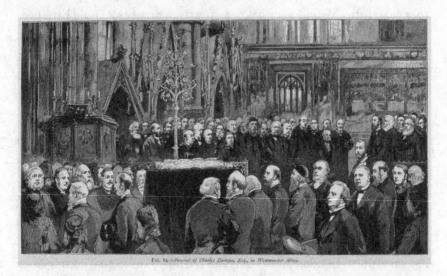

FIG. 62—*Funeral of Charles Darwin, Esq., in Westminster Abbey.*

Charles Darwin's funeral, Westminster Abbey

Mr Darwin will in all the future stand out as one of the giants in scientific thought and scientific investigation." The (London) *Morning Post* thought "he passed that life in elaborating one central idea, and he remained in the world long enough to see the whole course of modern science altered by his speculations." To *The* (London) *Daily Telegraph* he was "the greatest naturalist of our time, and perhaps all time."

Darwin was not buried in Downe churchyard next to Erasmus as he had wanted. Pressure from the scientific community, some of his friends close to government, and others persuaded his family to allow him to be buried in Westminster Abbey near the grave of Sir Isaac Newton. The pallbearers included Joseph Hooker and Alfred Russel Wallace, and the choir sang from the anthem "Let Us Now Praise Famous Men": "His body is buried in peace, but his name liveth for evermore."

DARWIN'S LEGACY

Whatever his views on religion, Darwin attained a kind of immortality. He changed our views on ourselves, our world, and the relationship between the two. The untried, untested young man who embarked on the *Beagle* in 1831 could scarcely have imagined how ideas the voyage would seed in his mind would alter the intellectual landscape in his lifetime and beyond.

Darwinism—a term coined by his supporter Thomas Huxley—came to be applied in every conceivable context—scientific, social, political, economic, and, of course, religious. His ideas seemed to justify confidence in progress—a concept so dear to Victorian hearts. Imperialists found justification for colonialism. Anarchists found evidence in Darwin's ideas that social characteristics—such as a willingness to cooperate—were instinctive and inherent, improving a species's chances of survival, and used that evidence to argue that the state should have little or no role in people's lives. Socialists quoted his view that altruism improved the lives of all, including the altruist. Feminists were among the few to find less encouragement, given his views expressed in *The Descent of Man*, that evolution made man intellectually and artistically superior to woman. Even before his death, Darwin—so prolific, such a fertile thinker, with ideas probing in so many directions—was in many ways already becoming all things to all people.

That carries a risk. Of all the ways in which people have sought to use Darwin's work to justify their own ideas, theories promoting "social engineering" have been the most dangerous to his reputation. Considering he did not even coin the term *survival of the fittest*, this is ironic. However, for some, the subtitle of *On the Origin of Species*—"the preservation of favoured races in the struggle for life"—provided justification for a range of racist views. Not that such thinking was anything new. Well before Darwin published *On the Origin of Species*, a number of scientists were suggesting that race was the dominant influence on human culture and behavior. The Scottish anatomist Robert Knox maintained in his 1850 work *The Races of Men* that "the human character, individual and national, is traceable solely to the nature of that race to which the individual or nation belongs . . . it is simply a fact . . . Race is everything . . ."

Knox seems not to have corresponded with Darwin, but one scientist whose work would have a bearing on how some would later perceive Darwin was Francis Galton, the founder of eugenics. He was also Darwin's cousin—like him, a grandson of Erasmus Darwin, but by a different wife. When he read *On the Origin of Species* Galton was greatly impressed, writing, "It made a marked epoch in my own mental development, as it did in human thought generally." Darwin's work on artificial selection, including his studies of selective breeding in pigeon varieties, particularly interested Galton, who saw a correlation between how to improve specific breeds of animals and how to improve the human race.

In 1865, Galton published a well-received paper titled *Hereditary Talent and Character*, suggesting clever and distinguished fathers would have clever and distinguished sons. (Clever mothers and daughters were not considered.) Galton followed this up four years later with *Hereditary Genius*, a book in which he argued that physical, mental, and even moral attributes were inherited. He based his proposition on an analysis of eminent men from which he again concluded that talented men were likely to have gifted sons. A system of arranged marriages between distinguished men and women of wealth (he did not mention intellect) would, he argued, in time, produce a gifted race.

Darwin wrote courteously to Galton, who had sent him a copy of the book, that although he had only read fifty pages, "I do not think I ever in all my life read anything more interesting and original . . ." but without offering any detailed comment. At the time, he himself was exploring a central question that had long exercised him—the mechanism by which characteristics passed from one generation to the next. He never knew of the massive and recent breeding study of the edible pea conducted by Gregor Mendel, an Augustinian monk at his monastery in Brunn in Moravia (what is today Brno in the Czech Republic). By systematically tracing how specific traits passed from one generation to the next of his plants, Mendel produced a mathematical formula to explain the frequency with which each trait appeared and noted dominant and recessive traits. His definitive study, *Experiments on Plant Hybridization*, written in German, was published in 1865 in the *Proceedings of the Natural History Society of Brunn*. It would not become known to the wider scientific community until the turn of the twentieth century—the first translation into English was by biologist William Bateson in 1901—when Mendel's explanation of how traits are inherited—and its application to humans—would finally be appreciated and would become the cornerstone of modern genetics.

In the second volume of *The Variation of Animals and Plants under Domestication*, published in 1868, Darwin suggested his own ideas of how characteristics were passed on. In his theory of "pangenesis," particles—or "gemmules"—were collected from over all over the body and concentrated into the sexual organs for transmission to the next generation.

Darwin's theory of pangenesis interested Galton, who challenged some of his cousin's ideas about gemmules while incorporating others into his own theories of inherited characteristics. However, some of his resultant scientific papers were so opaque that even the courteous, eager-to-please Darwin wrote to him, "You probably have no idea how excessively difficult it is to understand."

After Darwin's death, Galton continued gathering material that he hoped would suggest how best to improve the human race. For that he needed human data, and he set up an anthropological laboratory in the summer of 1884, where he examined parents and children, recording heights, arm length, and other physical parameters. The previous year Galton had coined the expression *eugenics*, meaning "well born," and he began arguing for a national eugenics program. Under the system he proposed, it would be the state's duty to encourage the most able and successful to have the most children, while the least successful members of society—including the poorest and those with mental disabilities— should be discouraged or even stopped from breeding.

Eugenics found plenty of supporters from writers like George Bernard Shaw, who advocated for selective breeding to create a socialist society, to politicians like Teddy Roosevelt, who suggested sterilizing criminals and people with mental disabilities, to business magnate John D. Rockefeller Jr., who believed that female convicts with mental disabilities should remain in jail, whatever the length of their sentence, until their child-bearing years were over to prevent them "perpetuating their kind." The theory was also used against suffragettes, many of them well-educated, middle-class women, to argue it was their duty to turn from politics to procreation. One of the attractions of eugenics was that it played to growing fears, as the new twentieth century began, that society had lost its vigor and was degenerating.

The First International Congress of Eugenics was held in London in 1912, a year after Galton's death. Darwin's son Leonard, like his brothers George and Horace, an enthusiastic disciple of Galton and eugenics, chaired the meetings during which speaker after speaker warned of the dangers of allowing the unfit to breed. However, many found some of the ideas distasteful. Former Prime Minister Arthur Balfour, invited to speak at the Congress banquet, remarked neatly, "We say that the 'fit' survive. But all that means is that those who survive are fit." Member of Parliament Josiah Wedgwood—Leonard Darwin's cousin and as liberal a thinker as his Wedgwood forebears—also vigorously opposed eugenic policies. The movement never gained traction in Britain but found fervent supporters in Nazi Germany, with unspeakable results that Galton would never have imagined.

Darwin would have been appalled that, in later generations, some would cite his comments about "favoured races" to claim respectability for their ideas of racial inferiority and superiority. Throughout his life, from his years aboard the *Beagle* to his days of fame and acclaim, he never deviated from his view that there was no hierarchy of humanity. Some people might be at a more advanced stage of development than others because of the conditions in which they lived—as he suggested in his *Beagle* diary when comparing the Fuegians with the Tahitians—but all belonged to the same human family, which, in turn, belonged to the great tree of life. His criticism of the slavery he witnessed in Brazil, his regret that a man of color he met at one of Rosas's *postas* would not sit down and eat with him, his admiration for former slave John Edmonstone, who taught him taxidermy in Edinburgh, his opposition to "artificial checks" on the natural rate of human increase, his telling comments that a civilized society had a duty to protect the vulnerable—"the evil which would follow by checking benevolence and sympathy in not fostering the weak and diseased would be greater than by allowing them to survive and then to procreate"—and many other examples underline that the principles of eugenics were not his.

In our time, we have abundant proof of Darwin's theory that some randomly generated varieties prove better adapted to their environment than others and thus survive and prosper. The way the coronavirus has mutated, with the more easily transmitted varieties or strains quickly supplanting others, is perhaps the most striking demonstration. We also have many examples of how creatures are adapting to climate change and other threats. For instance, a scientific report in September 2021 demonstrated how, as temperatures rise, some warm-blooded species— wood mice, masked shrews, and bats—are developing longer tails and legs or larger ears to help them shed excess heat and regulate their body temperature better. Another recent study detailed how the bills of several species of Australian parrot—like the gang-gang cockatoo and red-rumped parrot—have increased in size from 4 to 10 percent since 1871 to help dissipate body heat, the increase correlating with rising summer temperatures over these years. One can imagine Darwin sharpening his pencil and opening one of his field notebooks to analyze such findings.

Darwin was a pioneer of ecology, identifying the symbiotic relationship between creatures and their environment. During the *Beagle* voyage he noted many examples of the interdependence of living organisms for food, shelter, and hence life itself—from the flies, ticks, and moths living as parasites on birds and the crabs stealing fish from birds on St. Paul's

Rocks, to the mass of creatures depending for survival on the kelp of the Falkland Islands to the "little living world" he discovered while poking in the mud of a South American *salina*. He also well understood the impact of humans on their environment, both indirectly by their introduction—as in New Zealand and St. Helena—of new plants pushing out native ones, and more directly, as in the Falkland Islands, where he predicted—correctly—that man would soon hunt the Falkland fox to extinction. In Australia, he predicted less accurately that the kangaroo would soon be extinct, though more than a dozen larger species of Australian animal have vanished since his time. He was also clear-eyed about the impact of humans on one another, from the decline in the numbers of indigenous peoples in Tasmania, pushed out of their traditional hunting grounds by colonists, to the military campaigns against some of the indigenous peoples of South America.

Darwin of course knew nothing about another impact of human activity on the environment—global warming and its consequences—although during the *Beagle* voyage he noted evidence of changes in climate. In *On the Origin of Species* he wrote, "Climate plays an important part in determining the average numbers of a species, and periodical seasons of extreme cold or drought, I believe to be the most effective of all checks." The greenhouse effect, produced by the carbon dioxide and other gases pushed into the atmosphere from the time of the industrial revolution in which Darwin's grandfather Josiah Wedgwood was a leading figure, is having an ever-accelerating impact that both Darwin—and FitzRoy, as an expert in weather patterns—would have found both fascinating and alarming. NASA scientists estimate that since 1880—two years before Darwin died—the average global temperature on Earth has risen by a little more than 1°C (2°F). Two-thirds of that warming has occurred since 1975. Ice caps are melting, and sea levels are rising. Catastrophic events, from wildfires and droughts to floods, cyclones, and hurricanes, are seldom out of the media.

To Darwin, the "mystery of mysteries" was the origin of species, and he provided answers. His revelation that all life on Earth descends from a single common ancestor—now generally, if not universally, accepted—is potent. It goes to the core of how we humans see our world and our place within it and challenges us to consider our responsibility to the living things we share it with. In the twenty-first century, the greatest challenge is how best to protect our planet. Darwin suggested that cooperation plays a part in survival and that among we humans, those groups willing to cooperate are more likely to flourish than communities composed of more selfish individuals focused on their own personal needs. Perhaps the answer to our environmental dilemma lies somewhere in those thoughts.

"Weep for Patagonia"

The Fuegians had been the reason FitzRoy was so eager to have a second voyage of the *Beagle* to South America commissioned and to command it. His encounters with them underlay a great part of Darwin's thinking as he drafted *The Descent of Man*. In subsequent years, up till his death, Darwin donated money, despite his religious skepticism, to missionary efforts to help "civilize" the Fuegians. During the *Beagle* voyage, both he and FitzRoy had praised the "civilizing" effect of missionaries in Tahiti, New Zealand, and South Africa, and he hoped the same would be true in Tierra del Fuego. However, such missionary endeavors had already proved fraught with difficulty, as shocking news Darwin and FitzRoy had received from Tierra del Fuego in 1860 underlined.

For the first few years after the *Beagle* departed in 1834, the only other visitors to go ashore in Tierra del Fuego were occasional seamen from merchant vessels, sealers, and whalers. Then, in the early 1840s, the attention of Allen Gardiner, a devout missionary in search of a mission, alighted on Tierra del Fuego after reading FitzRoy's account of the region and its inhabitants. In 1841 he put out this "Plea for Patagonia":

> Weep! weep for Patagonia!
> In darkness, oh! how deep,
> Her heathen children spend their days;
> Ah, who can choose but weep?

The tidings of a saviour's love
Are all unheeded there,
And precious souls are perishing
In blackness of despair.

Together with the Reverend George Despard, Gardiner founded the Patagonian Missionary Society. Gardiner perished in 1851 with six companions—all as inexperienced in the conditions as himself—in an underfunded and ill-prepared attempt to found a mission station on Tierra del Fuego. Notes affirming their steadfastness in their faith and their unwavering purpose were found with their rotting bodies.

When the story of their missionary martyrdom reached Britain, funds flowed into the Patagonian Missionary Society. Despard soon had enough to fund the building of a two-masted schooner to be named after Allen Gardiner and to follow in his wake to Tierra del Fuego. Among his prominent supporters were Bartholomew Sulivan and Robert FitzRoy. Sulivan, a member of the mission's committee, suggested the "deaths were the appointed means for carrying on their mission."

Despard, while remaining in the UK himself, dispatched the *Allen Gardiner* under its captain William Parker Snow to South America. In January 1855 the vessel reached the Falkland Islands, where the Patagonian Missionary Society intended to establish a base on Keppel Island in the West Falkland Islands to be named Cranmer after the sixteenth-century martyred Protestant archbishop. Buying land, securing permissions from the skeptical Falkland authorities, and beginning construction work—the first building was named after Sulivan—took time, and the *Allen Gardiner* did not enter the Beagle Channel until November.

One of Snow's optimistic aims was to contact Jemmy Button, whom the missionaries hoped might be a useful conduit to his fellow Fuegians. When two canoes crammed with Fuegians put out toward the *Allen Gardiner* as it headed for Woollya Cove, Captain Snow, in his own words, "sang out to the natives, 'Jemmy Button, Jemmy Button?' To my amazement and joy—almost rendering me . . . speechless—an answer came from one of the four men in the first canoe, 'Yes, yes; Jam-mes Button,

Jam-mes Button!' at the same time pointing to the second canoe. A stout, wild and shaggy-looking man stood up in the second canoe. As it approached he shouted, 'Jam-mes Button, me! Where's the ladder? '"

Helped onboard twenty-one and a half years after he had said good-bye to the *Beagle*, Jemmy Button again stood seminaked on the deck of a British ship. When he discovered that a woman—Snow's wife—was on board, he immediately asked for trousers, then said, according to Snow, "'Want braces . . .' he appeared suddenly to call to mind many things. His tongue was loosened and words . . . came to his memory expressive of what he wanted to say . . . no connected talk . . . but broken sentences abrupt and pithy. Short enquiries and sometimes painful efforts to explain himself . . . with, however, an evident pleasure in being again able to converse . . . in the 'Ingliss talk.'" Snow tried to persuade Jemmy Button to return with him to the Falkland Islands but he refused and left the next morning loaded with gifts.

In August 1856, Reverend Despard himself arrived in the Falkland Islands. He soon quarreled with Snow and dismissed him. Despard, an autocratic disciplinarian, brought greater order to the Cranmer settlement, introducing twelve-hour working days punctuated only by breaks for prayer and food. In June 1858, he set out for Tierra del Fuego in the *Allen Gardiner*, determined to lure Jemmy Button to Keppel Island as the catalyst for his evangelical ambitions. He located Jemmy, but it took him and two companions eight days, alternating pressure and promises of gifts, to persuade him to accompany them to the mission station in the Falkland Islands for "five moons." The older of Jemmy's two wives, Lassaweea, a son, daughter, and young baby accompanied him.

The five months they spent at the Cranmer mission proved to be a purgatory rather than a heaven for Jemmy and his family, and perhaps also for the missionaries. Jemmy was compelled to attend church every day and was castigated for his occasional failure to do so and for his laziness in refusing to take part in hard manual labor such as peat cutting. The missionaries suspected him and his family of theft. When his wife was wrongly accused of stealing fence paling for firewood, Jemmy proclaimed, "I not stay Keppel Island," before relenting. He became truly angry when

Mrs. Despard suggested she should adopt his eight-year-old daughter, Passawullacuds, whom the missionaries called Fuegia, and bring her up at the mission.

The missionaries found it difficult to persuade Jemmy Button to teach them his language, Yamana, and became frustrated at their own inability to pick up what little he was prepared to tell them and by some linguistic lacunae in the Yamana tongue, such as an apparent lack of a word for heaven. In all probability there was relief on both sides when, in November, Jemmy, his wife, and all three children boarded the *Allen Gardiner* with Despard to return home. Once in Woollya, Jemmy Button showed Despard where FitzRoy had built the three wigwams and laid out his vegetable garden.

Despite his reservations about his treatment, Jemmy Button helped persuade nine other Fuegians, including two of his own brothers, Macooallan and Macalwense—known to the missionaries as Tommy and Billy, respectively—to go to the Cranmer mission station on the promise they would return before "wild bird egg season," when the Fuegians feasted on birds' eggs collected from the cliffs. This meant a stay of about nine months with the missionaries. Unlike Jemmy Button, none of the nine had previous experience of the Europeans' ways, and the party soon grew restless and homesick.

Their treatment was more heavy-handed than Jemmy Button's. Despard insisted both on daily church attendance and heavy labor on the mission's behalf. He restricted the Fuegians' movements as the "savages" frightened the mission women by appearing unexpectedly in their homes. The missionaries overreacted to small thefts and made matters worse by sometimes making false accusations, which increased resentment. This resentment boiled over when, at the end of September, before the Fuegians boarded the *Allen Gardiner* to return home, the missionaries systematically searched them and their baggage and seized some pilfered items such as hammers and chisels. This indignity so angered the Fuegians that some, women among them, stripped off their clothes and threw them into the sea or onto the jetty.

Everyone calmed down, superficially at least, and the *Allen Gardiner* finally put to sea with the nine Fuegians and a crew of nine led by the ship's new captain, Robert Fell, and the mission's catechist, Garland Phillips.

The journey to Woollya took more than three weeks at a time when the wild bird egg season the Fuegians so much looked forward to had already begun. When the *Allen Gardiner* anchored off Woollya, Robert Fell recorded in his diary, "A canoe came off, and we saw poor Jemmy Button naked, and as wild looking as ever. It was almost too trying to behold him. It seemed to prove that all our labours with him had been thrown away." When he came aboard Jemmy Button made clear his disappointment at the paucity of the gifts awaiting him.

Hearing reports that some of the crew's possessions were missing, Fell ordered another search of the Fuegians' baggage before allowing them ashore. One of them, Schwaiamugunjiz, known to the Britons as Squire Muggins, resisted the search and seized Captain Fell by the throat, pinning him against the bulkhead. Others simply abandoned their bundles, climbed into the waiting canoes, and paddled ashore. Phillips and Fell again managed to patch up relations with the Fuegians, and over the next few days some helped construct a mission shelter on the shore. Jemmy Button, although remaining somewhat querulous, was placated with further gifts. However, all the time more and more Fuegians were arriving. Where there had at first been only a few, now about three hundred sat grouped around their fires or floating in their canoes.

On Sunday, November 6, 1859, Fell and Phillips decided it would be feasible to hold Sunday services ashore in the hope of attracting the interest of the Fuegians. They and six of the other seven members of the *Allen Gardiner*'s crew, smartly dressed and all wearing their woolen jerseys embroidered with the logo "Mission Yacht," lowered themselves into the ship's launch and rowed the more than three hundred yards to the shore. Only Alfred Coles, the twenty-three-year-old cook, remained on board, ready to prepare lunch. He watched his companions beach the launch and walk up to the nearly completed new building where they intended to hold their service.

Soon after they entered they began singing a hymn. Then, to Coles's increasing horror, he saw a group of Fuegians seize the *Allen Gardiner*'s launch and another mob rush toward the mission building. The hymn-singing stopped abruptly as the Fuegians stormed in armed with clubs and stones to attack Coles's eight companions. Seven of them managed to escape from the hut, only to find themselves in the midst of the baying mob. Five fell there under the ferocious onslaught, battered to death. Two—the catechist Garland Phillips and a tall Swedish sailor named August Petersen—ran toward the water's edge beneath a hail of stones. There a stone felled Petersen, whom the Fuegians surrounded and beat to death, leaving only Phillips alive up to his knees in the water and trying desperately to scramble into a canoe. Moments later a stone, thrown, according to Coles, by Jemmy Button's brother Billy, hit him on the temple, and he crumpled facedown into the cold waters of the Beagle Channel. Coles grabbed three loaves of bread and a gun and scrambled into one of the ship's smaller boats and rowed away in the opposite direction as fast as he could.

Back at the Cranmer settlement, Despard waited in increasing anxiety for the return of the *Allen Gardiner*, expected in early December. When the schooner had still not appeared by February, Despard hired Rhode Islander William Smyly to take his brigantine the *Nancy* to search for the mission ship. When the *Nancy* nosed into Woollya Cove on March 1, 1860, the crew saw the *Allen Gardiner* floating at anchor. Several Fuegian canoes quickly approached. A naked white man stood in one of them—Alfred Coles. The *Nancy*'s crew pulled him onboard while Jemmy Button clambered onto the deck from another canoe and went below to the galley in search of food. Smyly quickly gave Coles a blanket and some food and began to question him.

The cook recounted his story, which he would repeat to the Falkland authorities. After rowing away from the *Allen Gardiner*, he had landed on an island and evaded his pursuers by running into a dense thicket and climbing up a tree. He had dropped his three loaves during his headlong flight. After a few days surviving on berries and without matches to make a fire to cook or to warm himself, he became so desperate that despite all that had happened, he hailed a Fuegian passing in his canoe and was taken aboard. He then lived among the Fuegians for three months. The Fuegians he had known before, including Jemmy Button and his brothers, in general treated him well. Strangers, however, stole his clothes, and some plucked out his beard and eyebrows or "shaved" them off with sharp seashells. Coles was allowed to board the *Allen Gardiner*, finding it a virtual wreck, stripped of everything useful. Some of the Fuegians had indicated that Jemmy Button had slept in the captain's cabin after the massacre. Although he had not been able to identify him among the attackers, Coles suggested Jemmy Button was the ringleader of the killings. "The cause . . . was Jemmy Button being jealous that he did not get as much as he thought he had a right to."

Believing Coles and convinced his small crew might be in danger, Smyly immediately upped anchor and set sail for the Falkland Islands, imprisoning Jemmy Button below. Back in the Falkland Islands, Jemmy Button narrowly escaped lynching as news of the massacre spread. Brought before the Falkland colonial secretary in the recently established settlement of Port Stanley, he told his story. As he had so often done on the *Beagle* for other misdeeds, he blamed the Oens men for the massacre: "Oens country boys say we no kill you you go away we kill them . . . I no sleep in schooner run about on main land—no more sleep run about." He went on to suggest that some of the killers came from York Minster's land and were cannibals—"York man eat man."

The Falkland governor, Thomas Moore, wisely decided that the evidence against Jemmy Button's brother Billy was weak. Coles had been more than three hundred yards from the killing and at that distance could scarcely have picked him out with any certainty. He concluded that the only evidence against Jemmy Button himself was secondhand hearsay from Coles, who spoke little Fuegian, and that any retribution, as some demanded, against the Fuegians as a whole would be both immoral and pointless. He therefore took no action and arranged the return of Jemmy Button to Tierra del Fuego.

There, for the last time, a tragic figure caught between two societies and finally choosing the one into which he was born, Jemmy Button climbed down from a foreign ship into a Fuegian canoe and paddled away. He died of disease in early 1864.

Of the other Fuegians whom the *Beagle* had returned home—Fuegia Basket and York Minster—there were a few sightings of Fuegia Basket over the years. In 1841, an English ship met a Fuegian woman who said to the crew, "How do? I have been to Plymouth and London." In the second edition of his *Journal of Researches*, Darwin wrote, "Captain Sulivan heard from a whaler in 1842 that when in the western part of the Strait of Magellan, he was astonished by a native woman coming on board, who could talk some English. Without doubt this was Fuegia Basket. She lived (I fear the term probably bears a double interpretation) some days on board."

In 1873, the missionary Thomas Bridges met her on Tierra del Fuego. She still remembered some English, calling Bridges's children who were with him "little boy and little girl." She told Bridges that York Minster had been murdered in revenge for his killing of another Fuegian. Bridges met Fuegia once more in 1883 when she was in her sixties and in poor health, being cared for by her daughter.

Bridges was the adopted son of the Reverend Despard, one of the founders of the Patagonian Missionary Society, who had left South America in 1861. Bridges continued the Society's missionary activities with a little more success. He produced a Yamana-English dictionary and supplied both Darwin and Sulivan with more information about the Fuegians. In 1884, he witnessed the arrival of the Argentinian navy in Tierra del Fuego on a mission to claim the territory for their country. They made their base at Ushuaia on the Beagle Channel. The sailors brought with them the measles virus, to which the Fuegians had not yet been exposed, and which killed more than half the population.

In the next two decades, mineral prospectors and settlers eager for land deliberately shot down many Fuegians, just as Rosas's men had killed the indigenous people in Patagonia, as witnessed by Darwin during the *Beagle* voyage. Estimates suggest that in the 1850s there were between 7,000 and 9,000 Fuegians. When a count was made in 1947, only around 150 of pure Fuegian heritage remained and perhaps the same number of people of mixed race.

Since then, numbers have fallen further, and it is doubtful whether any people of pure Fuegian heritage remain. Their tragic demise in many ways epitomizes Darwin's theory of how changing conditions and the arrival of better-adapted competition can cause the disappearance of those who do not adapt quickly enough.

ACKNOWLEDGMENTS

As in all my work, my husband, Michael, took an equal part with me in every aspect of the writing and researching of this book.

I've long been interested in Darwin's story—Emma Darwin is my first cousin many generations removed—and for many years I've wanted to know more about the voyage of the *Beagle*. Michael and I also became curious about Darwin while writing *A Pirate of Exquisite Mind*, a biography of the seventeenth-century buccaneer-turned-naturalist William Dampier, who was a pioneer of descriptive botany, the first to identify the concept *subspecies*, was cited by Darwin, and also admired by FitzRoy.

Much of this book was written during the ever-evolving COVID-19 pandemic. At its start, before borders began closing, we were fortunate enough to visit Chile. On the island of Chiloe it rained on us with all the force Darwin complained about. In Valdivia we walked in forests where Darwin experienced his first earthquake. We replicated Darwin's journey from Santiago to Mendoza over the high Andes when he found fossilized trees that had once flourished at sea level. However, as we arrived in the Atacama desert to retrace some of Darwin's journey there, our luck ran out. We were advised to leave Chile at once. Twenty-four hours later we were in Rio, and forty-eight hours after that back in a gray London.

Two years passed before we could return to the Atacama's pale, lunar landscapes, smoking volcanos, and brilliantly colored rock strata. It was worth the wait. Up in the high Andes we found spurting geysers. In nearby salt flats flamingos picked for food—the sight that prompted

Darwin to reflect on how efficiently creatures could adapt to harsh conditions. Guanacos grazed on almost vertical green slopes, and we glimpsed a rhea—the northern relation of the smaller southern rhea nearly consumed by the hungry *Beagle* crew at Port Desire before Darwin realized its significance to his thoughts on the distribution of species.

We revisited the Galápagos. Crunching across the cindery islands explored by Darwin, we saw for ourselves the tame mockingbirds and finches, the marine iguana—those "imps of darkness"—almost indistinguishable from the black lava on which they basked, huge land iguanas munching cacti, and giant tortoises, like the one on which Darwin rode. One night we watched the Wolf Volcano on Isabela (Albemarle) Island—the Galápagos's highest peak—erupt. The shimmering gleam on the sea from the streams of red-gold lava recalled the "long bright shadow" Darwin saw cast by the erupting Osorno Volcano in southern Chile.

These travels helped us, at least a little, to see things through Darwin's eyes. So did our visits to many other of the *Beagle*'s destinations, from Brazilian jungles, Patagonian plains, and remote Tierra del Fuego to the Falklands, Tahiti, Australia, New Zealand, Mauritius, St. Helena, the Azores, Cape Town, and Cape Verde.

Equally evocative were visits to Down House, the Darwins' home for forty years, where he found the tranquility to think and write. The study where he worked—retreating behind a curtain to be ill—is filled, cluttered even, with his possessions. Everything from the Sandwalk where he took his daily exercise to the greenhouse where he cultivated orchids conjures his presence.

Along the way, many people generously gave us their time and expertise. In Chile, Dr. Daniel Melnick d'Etigny (Instituto de Ciencias de la Terra, Universidad Austral de Chile in Valdivia) shared research on seismic activity in Chile, and Cristina Eftimie (Universidad Austral de Chile in Valdivia) helped us plan our visit. On Chiloe, Ben Castro and Pablo Dutilh of the Senda Darwin—"Darwin's Path"—environmental project at Chacao showed us that though the nearly impenetrable forests Darwin described are gone—victims of logging—the project is using Darwin's descriptions to propagate native shrubs and trees like the alerce, olivillo, and meli. Jeffery Hayden showed us vertiginous Valparaiso and the Campana mountain that Darwin climbed. We're grateful to our expert and enthusiastic guides Francisco Morales in the Atacama desert and Hanzel Martinetti in the Galápagos.

In Australia, our thanks go to the State Library of New South Wales. In Britain, we're grateful to the London Library, to Cambridge University for the collection of Darwin correspondence made available online through the Darwin Correspondence Project, and for the manuscripts and books made available through the complementary and wonderfully comprehensive Darwin Online Project founded and directed by Dr. John van Wyhe. We're grateful to Pamela Hunter, Hoares Bank's archivist, for FitzRoy's correspondence with the bank, and to Dr. Oliver Strimpel for very helpful geological advice and insights.

As always, we would like to thank our publisher, George Gibson of Grove Atlantic, and his colleague Emily Burns, and our agents, Bill Hamilton of A. M. Heath in London and Michael Carlisle of Inkwell Management in New York, for all their help and encouragement.

NOTES AND SOURCES

ABBREVIATIONS

ADM   Admiralty Papers, UK National Archives, Kew, London
Auto   Darwin's Autobiography
CO   Colonial Office Papers, UK National Archives, Kew, London
DCP   Darwin Correspondence Project:

- letters from Darwin are referred to as D/recipient/date, DCP ref; and letters to him as sender/D/date, DCP ref.
- SD is Susan Darwin, CATD is Catherine Darwin, CARD is Caroline Darwin, RWD is Robert Waring Darwin, JH is John Henslow, and WDF is William Darwin Fox.

In some places, spelling and capitalization have been modernized. Ampersand has always been replaced by "and."

After the two voyages of the *Beagle*—the first under the overall command of Philip Parker King and the second under Robert FitzRoy—three volumes and an appendix reporting on the voyages were published under the overall title of *Narrative of the Surveying Voyages of HM Ships Adventure and Beagle, 1826–1836.*

- The first volume was entitled *Proceedings of the First Expedition, 1826–1830*, under the command of Captain P. Parker King, R. N., F. R. S. and is referred to in these notes as *Proceedings of the First Expedition, 1826–1830*;
- The second volume was entitled *Proceedings of the Second Expedition, 1831–1836, under the Command of Captain Robert FitzRoy, R. N.* and is referred to in these notes as *Proceedings of the Second Expedition, 1831–1836*;
- The third volume was entitled *Journal and Remarks, 1832–1836, by Charles Darwin, Esq.* and is referred to in these notes as *Journal of Researches.* Since there is more than one edition of this Journal, quotes are from the first edition unless otherwise stated.

NB: In chapters 1 to 20, any unsourced quotes from Darwin are from his *Beagle* diary for the period. Quotes from FitzRoy in these chapters, if unsourced, are from *Proceedings of the Second Expedition, 1831–1836.*

INTRODUCTION

"great . . . myself": Darwin, *Beagle* diary, 13.

"a man . . . curiosity": J.Wedgwood/RWD/31/8/31, DCP-LETT-109.

"There . . . man": Extract from Sedgwick's letter quoted in SD/D/22/11/1835, DCP-LETT-288.

"fiddler . . . cabin": FitzRoy, Log of the *Beagle*, ADM51.

"by far . . . career": Darwin, *Auto*, 76.

CHAPTER ONE   The Selection of Darwin

"All . . . self": Darwin, "Life," autobiographical note August 1838, Darwin Online Project, CUL-DAR91.56-63.

"partly . . . state": *Auto*, 22.

"stolen . . . for . . . excitement": *Auto*, 23.

"at . . . quarrelsome": Darwin, "Life," August 1838, Darwin Online Project, CUL-DAR91.56-63.

"first . . . *Beagle* . . . 'Gas . . .' showed . . . science": *Auto*, 44–46.

"unjust": Francis Darwin, *The Life and Letters of Charles Darwin*, 3.

"the largest . . . saw": *Auto*, 29.

"intolerably dull": *Auto*, 47.

"is so . . . sense": D/CARD/6/1/26, DCP-LETT-20.

"dry and formal": *Auto*, 49.

"that the . . . larvae": *Auto*, 50.

"the jealousy of scientific men": Charles Darwin's comment in 1871 as recollected by his daughter Henrietta Lichfield, Darwin Online Project, CUL-DAR262.23.3.

"a negro . . . man": *Auto*, 51.

"intimate . . . many . . . character . . . how . . . ours": Darwin, *The Descent of Man*, vol. 1, 232.

"revolting . . . well . . . Edinburgh": Quoted in Desmond and Moore, *Darwin's Sacred Cause*, 19.

"old . . . stick": D/Hooker/29/5/54, DCP-LETT-1575.

"incredibly dull . . . never . . . science": *Auto*, 52.

"my father . . . sisters": *Auto*, 56.

"for nothing . . . family": *Auto*, 28.

"I did . . . Bible": *Auto*, 57.

After . . . straight": Quotes in this para are *Auto*, 44 and 60.

"unknown . . . palate": universityarms.com.

"Upon . . . life": *Auto*, 68.

"dawdling . . . examinations": Quoted in Thomson, *Young Charles Darwin*, 97.

"rare species": *Auto*, 63.

The drawing of Darwin's beetle appeared in Stephens's *Illustrations of British Insects*, vol. 3, 266.

"the magic . . . Esq.'": *Auto*, 63.

Also . . . adopt": All quotes in these two paragraphs are *Auto*, 64.

The Reverend Adam Sedgwick: Sedgwick, a Yorkshireman, was one of the group of Cambridge academics known as the "Northern Lights."

"more as . . . *gentleman*": JH/D/24/8/31, DCP-LETT-105.

Henslow . . . search of": All quotes in this paragraph, JH/D/24/8/31, DCP-LETT-105.

"If . . . consent": *Auto*, 71.

"My father . . . FitzRoy": D/JH/30/8/31, DCP-LETT-107.

"one of . . . world": *Auto*, 72.

"Charles . . . advice": RWD/J.Wedgwood/30/8/31, DCP-LETT-108.

Where . . . few": All quotes in this paragraph are from J.Wedgwood/RWD/31/8/31, DCP-LETT-109.

"I am . . . uncomfortable . . . that . . . uncomfortable": D/RWD/31/8/31, DCP-LETT-110.

"I should . . . *Beagle* . . . But . . . clever": *Auto*, 72.

"I am . . . well": Quoted J. Gribbin and M. Gribbin, *FitzRoy*, 24.

Admiral Sir Robert Otway: Otway had served at the Battle of Copenhagen during the Napoleonic Wars, where he tried to dissuade Admiral Sir Hyde Parker from sending the signal to withdraw, which Nelson famously ignored, pleading his "blind eye."

"We are . . . employment?" Quoted in Browne, *Charles Darwin*, vol. 1, 147.

"it opens . . . exertions": Quoted in J. Gribbin and M. Gribbin, *FitzRoy*, 42.

"the dreary sea . . . inhospitable shores . . . the soul . . . in him": Stokes's handwritten journal, State Library NSW, quoted in *Sydney Morning Herald*, June 27, 1909.

"of drawings . . . dirty": *Proceedings of the First Expedition, 1826–30*, 216.

"that so . . . estimation": *Proceedings of the First Expedition, 1826–30*, 405.

"I incurred . . . undertaking": *Proceedings of the First Expedition, 1826–30*, 459.

"I could not . . . existence": *Proceedings of the Second Expedition, 1831–1836*, 10.

"If I ever . . . hydrography": Quoted in *Beagle Record*, ed. R. D. Keynes, 5.

"hereditary predisposition": Darwin referred to FitzRoy's fears about his mental health in D/CATD/8/11/34, DCP-LETT-262.

"a most zealous . . . history": Quoted in *A Voyage Round the World*, ed. A. M. Pearn, 110.

"a Mr. Darwin . . . idea": Beaufort/FitzRoy/1/9/31, DCP-LETT-113.

"a sudden . . . vessel . . . to throw . . . scheme": D/SD/9/9/31, DCP-LETT-122.

"more . . . kind": D/SD/5/9/31, DCP-LETT-117.

"bumpology": Darwin described FitzRoy's doubts about his physiognomy in *Auto*, 72.

"bear . . . Devil": D/SD/5/9/31, DCP-LETT-117.

"You cannot . . . me": D/JH/5/9/31, DCP-LETT-118.

"I like . . . workshop": FitzRoy/Beaufort/5/9/31, quoted in F. Darwin, "FitzRoy and Darwin, 1831–1836," *Nature* 88 (February 1912).

CHAPTER TWO   "A Birthday for the Rest of My Life"

Darwin . . . opposition: All quotes in these two paragraphs are from D/SD/6/9/31, DCP-LETT-119 and D/SD/9/9/31, DCP-LETT-122.

Capstan: A capstan was a drum-shaped device rotated by sailors pushing wooden bars fitted into slots and used to hoist weights or raise the anchors.

Beagle crew: FitzRoy recorded the total number of crew at seventy-four, but it is possible he double-counted one crewmember, Syms Covington, in which case the total was seventy-three.

"a capital . . . home": D/SD/17/9/31, DCP-LETT-127.

"no . . . dinners": D/SD/5/9/31, DCP-LETT-117.

"I will . . . very bad": D/SD/9/9/31, DCP-LETT-122.

"my beau . . . as one": D/SD/14/9/31, DCP-LETT-126.

"Captain Wentworth": CARD/CATD/SD/D/20–31/12/31, DCP-LETT-153.

"a slight . . . handsome": D/SD/6/9/31, DCP-LETT-119.

"Mr. Thunder . . . Harris": D/JH/15/11/31, DCP-LETT-147.

"What a . . . life": D/FitzRoy/10/10/31, DCP-LETT-142.

"to look . . . can . . . I have . . . child": D/JH/15/11/31, DCP-LETT-147.

What . . . advantages": D/JH/30/10/31, DCP-LETT-144.

"bridle . . . rebuke . . . a mistake . . . into": JH/D/20/11/31, DCP-LETT-150.

"like . . . freshmen": D/JH/15/11/31, DCP-LETT-147.

"Pick . . . find": Sulivan, *Life and Letters of the Late Admiral Sir Bartholomew James Sulivan*, 38.

"I'm Arthur . . . anyone!" Mellersh, *Fitzroy of the Beagle*, 72.

"My friend . . . him": D/JH/30/10/31, DCP-LETT-144.

"a large . . . button": *Proceedings of the First Expedition, 1826–30*, 44.

The orders . . . merchant ships: All quotes and information in these five paragraphs about FitzRoy's orders are from *Proceedings of the Second Expedition, 1831–1836*, 22–39. As the *Beagle* prepared to sail, James Cook's wife Elizabeth—a widow for more than fifty years—was still alive and hosting weekly dinner parties at which the toast was always to "Mr Cook." She died in May 1835 while the *Beagle* was homeward bound.

"most . . . Mahogany . . . I grieve . . . years": D/JH/15/11/31, DCP-LETT-147.

"I look . . . anxiety": D/JH/3/12/31, DCP-LETT-152.

"the most . . . time . . . troubled . . . heart-disease . . . resolved . . . hazards": *Auto*, 79–80.

"I should . . . disobliging . . . You . . . blackguard": *Auto*, 74–75.

"the prettiest . . . possesses": D/WDF/24/12/28, DCP-LETT-54.

Fanny Owen . . . long": All quotes in this paragraph are from Fanny Owen/D/6/10/31, DCP-LETT-141 and Fanny Owen/D/2/12/31, DCP-LETT-151.

"in his . . . gun": Sulivan, *Life and Letters of Admiral Sir Bartholomew James Sulivan*, 37.

"date . . . sublime": D/WFD/19/9/31, DCP-LETT-132.

CHAPTER THREE "Like Giving a Blind Man Eyes"

Darwin . . . pay: All quotes in this paragraph from FitzRoy's voyage log, December 18, 1831, *Proceedings of the Second Expedition 1831–1836*, 40.

"you came . . . hands": D/FitzRoy/28/10/46, DCP-LETT-1014.

"a death . . . Jib'": D/FitzRoy/28/10/46, DCP-LETT-1014.

"The invariable . . . eating": D/SD/14/7-7/8/32, DCP-LETT-177.

"preeminently . . . new . . . the right . . . others": D/JH/18/5-16/6/32, DCP-LETT-171.

"the size . . . soap": Quoted in Herbert, *Charles Darwin, Geologist*, 103.

"very . . . simple . . . it . . . to me": *Auto*, 81.

"the . . . Tree": McCormick's diary reproduced in Steel, *He Is No Loss*, 54.

"a dark . . . ink" . . . a very . . . green": Darwin, *Journal of Researches*, 6.

"final . . . creatures": Quoted in Gregor, *Charles Darwin*, 43.

"living . . . cylinder": Erasmus Darwin, *Zoonomia*.

"The final . . . improved": Erasmus Darwin, *Zoonomia*.

"high . . . evolution . . . in . . . me": *Auto*, 49.

"to free . . . Moses": Quoted in Lyell, *Life, Letters and Journals of Sir Charles Lyell*, vol. 1, 26.

"from . . . hereafter": Hutton, *Theory of the Earth*, vol. 1, section 1.

"smooth . . . Lithodomus": Lyell, *Principles of Geology*, vol. 1, 453.

"The very . . . geology": *Auto*, 77.

"a perfectly . . . band": Darwin, *Journal of Researches*, 5.

"the investigations . . . world": Quoted in Thomson, *Young Charles Darwin*, 111–112.

"on no . . . advocated": *Auto*, 101.

"I do . . . heart": D/John Lubbock/15/11/59, DCP-LETT-2532.

"your money . . . correct": RWD/D/7/3/33, DCP-LETT-201.

"continues . . . work": D/RWD/8/2-1/3/32, DCP-LETT-158.

Darwin . . . History": All quotes in this paragraph are from D/RWD/10/2/32, DCP-LETT-159.

"after . . . voyage": Quoted in Nicholas and Nicholas, *Charles Darwin in Australia*, 8.

"English . . . gardening": Quoted in Hazlewood, *Savage*, 81.

"Her Majesty . . . country": Hazlewood, *Savage*, 94–95.

While being dressed . . . required": All quotes from the phrenologist's report are from Hazlewood, *Savage*, 87–89.

"Is it . . . Volcanic?": D/JH/18/5-16/6/32, DCP-LETT-171.

"numerous . . . sharks." Darwin, *Journal of Researches*, 10.

Midshipman King . . . here!' ": Midshipman King's account quoted in *A Narrative of the Voyage of HMS Beagle*, ed. Stanbury, 51.

"high . . . captain . . . readily . . . fun": Quoted in *A Narrative of the Voyage of HMS Beagle*, ed. Stanbury, 52.

"By the . . . expectations": All quotes from Darwin, *Beagle diary*, 39.

CHAPTER FOUR "Red-Hot with Spiders"

"many . . . ground . . . every . . . leaf . . . the efforts . . . wonderful": Darwin, *Journal of Researches*, 39.

While . . . idea: All quotes in paragraph are from Dampier, *Voyages*, vol. 2, ed. Masefield, 389, 394–395.

"the remarkable . . . buccaneer . . . the subsequent . . . little": Quoted in Preston, *A Pirate of Exquisite Mind*, 18.

"so odious . . . it": Lord Mansfield's historic judgment of June 22, 1772, prevented the forced return to the West Indies of James Somerset, brought to England in 1769 by his "master," Bostonian Charles Stewart of Boston. Somerset escaped in 1771 but was recaptured and put on a ship bound for Jamaica.

Darwin's . . . together": All quotes in this paragraph from *Auto*, 73–74.

Having . . . temper?": *Auto*, 73–74.

"After . . . him": *Auto*, 74.

"a very . . . likes him'": FitzRoy to Beaufort March 5, 1832, quoted in *Beagle Record*, ed. R. D. Keynes, 42.

"cheery . . . companionship . . . by . . . Harry . . . beyond belief": John Lort Stokes's letter reminiscing about Darwin, September 16, 1882, CUL-DAR112.A97-A98, Darwin Online Project.

"gneiss . . . strata": *Proceedings of the Second Expedition, 1831–1836* quoted in Darwin, *Beagle* diary, 48.

"nests . . . ground': *Proceedings of the Second Expedition, 1831–1836* quoted in Darwin, *Beagle* diary, 48.

"appears . . . God": David Gelber, *History Today* vol. 66, no. 8 (August 2016).

"gaudy . . . Cathedrals . . . studded . . . nation": D/CARD/2-6/4/32, DCP-LETT-164.

"beautiful . . . discipline": D/CARD/2-6/4/32, DCP-LETT-164.

"Mr Darwin . . . him'": Philip Gidley King's Reminiscences of Darwin during the Voyage of the *Beagle*, 16.

"becoming . . . etc.": D/CARD/2-6/4/32, DCP-LETT-164.

"Send . . . below . . . every . . . duty": D/CARD/2-6/4/32, DCP-LETT-164.

"awful . . . event": Fanny Owen/D, DCP-LETT-103.

"I hope . . . Charley": CATD/D/8/1/32, DCP-LETT-154.

"if . . . night": D/CARD/2-6/4/32, DCP-LETT-164.

"overwhelmed . . . flowers": D/CARD/2-6/4/32, DCP-LETT-164.

"enormous . . . high": Darwin, *Journal of Researches*, 24.

"In a . . . obstinacy": Darwin, *Journal of Researches*, 19.

"Instantly . . . animal": Darwin, *Journal of Researches*, 24.

"the extreme . . . mimosae": Darwin, *Journal of Researches*, 24.

"I have . . . it . . . returning . . . Wickham . . . He . . . loss": D/CARD/25-26/4/32, DCP-LETT-166.

"an ass": D/JH/30/10/31, DCP-LETT-144.

"philosopher . . . last": D/JH/18/5-16/6/32, DCP-LETT-171.

"Having . . . home": Extract from McCormick's memoirs, *Voyages of Discovery*, reproduced in footnote, Darwin, *Beagle* diary, 61.

"so much . . . wasted": McCormick, *Voyages of Discovery*, vol. 1, 218.

If . . . with": All quotes in this paragraph are from D/CARD/25-26/4/32, DCP-LETT-166.

Darwin . . . possible": FitzRoy's orders described in this paragraph are from *A Narrative of the Voyage of HMS Beagle*, ed. Stanbury, 69–70.

"I like . . . travelling . . . when . . . sand . . . although . . . Palms": D/CARD/ 25-26/4/32, DCP-LETT-166.

"Whilst . . . morality": *Auto*, 85.

"We . . . different . . . that . . . running": Darwin, *Journal of Researches*, 37–38.

"I am . . . Cambridge . . . The Captain . . . Slavery . . . does . . . consideration:" D/JH/18–16/6/32, DCP-LETT-171.

"My mind . . . primitive": D/WDF/May/32, DCP-LETT-168.

"most . . . news": D/CATD/May–June 1832, DCP-LETT-169.

Sir Ronald Ross: Ross also founded the celebrated London School of Tropical Medicine.

"You . . . them ": D/CATD/5/7/32, DCP-LETT-176.

"we . . . reform . . . I long . . . before . . ." D/CATD/5/7/32, DCP-LETT-176.

CHAPTER FIVE "Gigantic Land Animals"

"my face . . . to it": D/SD/14/7-7/8/32, DCP-LETT-177.

"clear . . . action . . . prepare . . . boats": Darwin, *Beagle* diary, 85.

"a military usurpation": Darwin, *Beagle* diary, 85.

"to convince . . . property": D/SD/14/7-7/8/32, DCP-LETT-177.

Montevideo: Some claim the name "Montevideo" comes from *monte vide eu*, Portuguese for "I see a mountain." Others believe that the Spanish recorded the location of the mountain as Monte VI de Este a Oeste, or 'the sixth mountain from east to west."

Darwin . . . merrier": D/SD/14/7-7/8/32, DCP-LETT-177.

"We Philosophers . . . more": D/F. Watkins/18/8/32, DCP-LETT-181.

"begged . . . troops": Darwin, *Beagle* diary, 89.

"to afford . . . protection": *Proceedings of the Second Expedition, 1831–1836*, 95.

"to hoist . . . boats . . . heavily . . . Pistols": Darwin, *Beagle* diary, 90.

"like . . . ice": FitzRoy to Beaufort, 15/8/32, *Beagle Record*, ed. Keynes, 77.

"It was . . . town": D/SD/14/7-7/8/32, DCP-LETT-177.

"animals . . . sorts: D/JH/23/7-15/8/32, DCP-LETT-178.

"a very . . . Flycatcher": Recollection of Darwin's daughter, Henrietta Litchfield, Darwin Online, DAR 262.23.7.

"a good . . . rock": D/JH/23/7-15/8/32, DCP-LETT-178.

After . . . soot": All quotes in this paragraph are from JH/D/15-21/1/33, DCP-LETT-196.

"our . . . work": D/JH/23/7-15/8/32, DCP-LETT-178.

"gigantic . . . animals": Darwin, *Journal of Researches*, 2nd ed., 76.

"osseous": D/CARD/24/10-24/11/32, DCP-LETT-188.

"The cabin . . . seats": Sulivan, *Life and Letters of Admiral Sir Bartholomew James Sulivan*, 39–40.

In Montevideo . . . proved": All quotes in this paragraph are from D/CARD/24/ 10-24/11/32, DCP-LETT-188.

"For . . . hung": WDF/D/30/6/32, DCP-LETT-175.

"vote . . . it . . . I . . . travel": Erasmus Darwin/D/18/8/32, DCP-LETT-182.

"angels . . . how . . . Aires": All quotes in this paragraph are from D/CARD/24/ 10-24/11/32, DCP-LETT-188.

"for I . . . him": D/CARD/24/10-24/11/32, DCP-LETT-188.

The letters . . . confirmed: All quotes in this paragraph are from D/CARD/24/ 10-24/11/32, DCP-LETT-188.

He also . . . return": All quotes in this paragraph are from D/JH/26/10–24/11/32, DCP-LETT-192.

"Anything . . . Cam": D/CARD/24/10-24/11/32, DCP-LETT-188.

"I am . . . south": Letter December 4, 1832, quoted in Hazlewood, *Savage*, 117.

CHAPTER SIX Land of Fire

"a fine beetle": Darwin, *Journal of Researches*, 150.

"as savage . . . desire": D/WDF/23/5/33, DCP-LETT-207.

"in the . . . creature": D/CARD/30/3/33, DCP-LETT-203.

"certainly . . . sounds": Darwin, *Journal of Researches*, 229.

"was . . . classic": D/CARD/30/3/33, DCP-LETT-203.

"in . . . Durham!": Darwin, *Journal of Researches*, 232.

"The Captain . . . happy": Syms Covington's journal.

"a gale . . . seaboat . . . to wait . . . sea": D/CARD/30/3/33, DCP-LETT-203.

"always . . . eventualities . . . found . . . ship": Sulivan, *Life and Letters of Admiral Sir Bartholomew James Sulivan*, 38.

CHAPTER SEVEN "Truly Savage Inhabitants"

"in spite . . . safety": Quoted in Hazlewood, *Savage*, 122.

"miserably . . . bodies": Hamond diary, quoted in Hazlewood, *Savage*, 124.

"were . . . numbers": Darwin, *Journal of Researches*, 239.

"savages . . . fire-arms . . . appears . . . sling . . . by . . . you": Darwin, *Journal of Researches*, 239.

"We . . . way": Hamond diary, quoted in Hazlewood, *Savage*, 125.

"The dusky . . . forest . . . truly horizontal . . . the . . . beach": Darwin, *Journal of Researches*, 240.

"whilst . . . child": Darwin, *Journal of Researches*, 235.

"snow-capped . . . Fuego": Quoted in Hazlewood, *Savage*, 130.

"strange . . . relatives": Quoted in Hazlewood, *Savage*, 131.

"no . . . affection": Darwin, *Journal of Researches*, 2nd ed., 211.

"telescopes . . . eyes": Quoted in Browne, *Charles Darwin*, vol. 1, 238.

"distinct . . . life": Darwin, *Journal of Researches*, 204.

"a . . . favourite . . . poor . . . fellow!": Darwin, *Journal of Researches*, 2nd ed., 197.

"the . . . equality . . . must . . . civilization . . . till . . . authority?"; Darwin, *Journal of Researches*, 242.

"Things . . . better": Quoted in Hazlewood, *Savage*, 141.

CHAPTER EIGHT   Res Nullius

"closer . . . migration": Darwin, Falklands field notebook, March 2, 1833.

"very . . . eating . . . wild oxen": Syms Covington's journal.

"the enormous . . . eggs . . . wonderful fecundity": Darwin, zoology notes, March 7, 1833.

"As . . . interesting": D/JH/11/4/33, DCP-LETT-204.

"indicate . . . warmer": Quoted in Armstrong, *Darwin's Other Islands*, 92.

"She . . . materials . . . the wise . . . me . . . I hope . . . me": FitzRoy's letter to Hoare's bank, November 30, 1833.

"I believe . . . sum": FitzRoy's letter to the Admiralty, ADM/1/1819.

"Now *pray* . . . old plan": FitzRoy's letter to Beaufort, May 10, 1833, reproduced in *The Beagle Record*, ed. Keynes, 131–133.

"send . . . mended": Quoted in Armstrong, *Darwin's Desolate Islands*, 35.

Waiting . . . Montevideo: All quotes in this paragraph are from CARD/D/1/9/33, DCP-LETT-214.

"It does . . . Slavery": D/Herbert/2/6/33, DCP-LETT-209.

"to . . . house": Darwin, *Journal of Researches*, 47.

"it was . . . teeth": Darwin, *Journal of Researches*, 47.

"I . . . was . . . England . . . London!": Darwin, *Journal of Researches*, 47.

"Back in . . . allow me this": All quotes in these three paragraphs are from D/CATD/22-14/7/33, DCP-LETT-206, other than *"exceedingly . . . sooner . . . Pray . . . sure,"* which are from SD/D/15/10/33, DCP-LETT-219.

"Here . . . money": Syms Covington's journal.

"stammering . . . Service": D/CARD/30/3-12/4/33, DCP-LETT-203.

"the largest . . . world": Darwin, *Journal of Researches*, 57.

"my . . . anticipated . . . The work . . . *for* . . . another . . . result": FitzRoy's letter to Beaufort, 16/7/33, reproduced in *The Beagle Record*, 143–145.

"I have . . . specimens": D/JH/18/7/33, DCP-LETT-210.

"remarkable . . . composed . . . Andes": Darwin, *Journal of Researches*, 73.

"still . . . man": Darwin, *Journal of Researches*, 73.

"these . . . growth": Darwin, *Journal of Researches*, 75.

"Thus . . . habitable!": Darwin, *Journal of Researches*, 2nd ed., 62–63.

"the wandering . . . country": Darwin, *Journal of Researches*, 2nd ed., 63.

CHAPTER NINE   El Naturalista Don Carlos

"hammer . . . shirt": D/JH/12/11/33, DCP-LETT-229.

"A more . . . animals: Darwin, *Journal of Researches*, 83.

"perfect . . . races": Darwin, *Journal of Researches*, 93.

"Finally . . . *careful*": All quotes in this paragraph are from FitzRoy/D/24/8/33, DCP-LETT-212.

"the teeth . . . rock": Sulivan, *Life and Letters of Admiral Sir Bartholomew James Sulivan*, 42.

"in that . . . Gauchos": D/CARD/20/9/33, DCP-LETT-215.

"The father . . . away": Darwin, *Journal of Researches*, 2nd ed., 100.

"gave . . . dogs": Darwin, *Journal of Researches*, 129.

"the most . . . line . . . brilliant conflagrations . . . any . . . Indians . . . In grassy . . . growth": Darwin, *Journal of Researches*, 132.

"I am . . . air": D/CARD/20/9/33, DCP-LETT-215.

"the Geology . . . pampas": D/CARD/20/9/33, DCP-LETT-215.

"a lady . . . tea": D/CARD/20/9/33, DCP-LETT-215.

FitzRoy . . . recommendation": All quotes in this para are from FitzRoy/D/4/10/33, DCP-LETT-218.

"sea-like pampas": D/CARD/23/10/33, DCP-LETT-222.

"two . . . skeletons": Darwin, *Journal of Researches*, 2nd ed., 120.

Still . . . children: All quotes in this paragraph are from Darwin, *Journal of Researches*, 121.

"teeth . . . Mastodon": Darwin, *Journal of Researches*, 123.

"I . . . America": Darwin, *Journal of Researches*, 149.

"One of . . . shores!": Darwin, *Journal of Researches*, 164.

"a furious . . . set": D/CARD/23/10/33, DCP-LETT-222.

"in . . . pickle": D/CARD/23/10/33, DCP-LETT-222.

"nearly . . . completely": D/CARD/23/10/33, DCP-LETT-222.

"the confounded . . . gentlemen . . . like . . . left": D/CARD/23/10/33, DCP-LETT-222.

CHAPTER TEN   "Great Monsters"

"un grande galopeador . . . a fine . . . S. America . . . an immense . . . you": D/JH/12/11/33, DCP-LETT-229.

"a mortification . . . road": D/WDF/25/10/33, DCP-LETT-223.

"turned out . . . to do . . . Send . . . fossils": JH/D/31/8/33, DCP-LETT-213.

"He stripped . . . seated": Darwin, *Journal of Researches*, 168.

"Like . . . monsters": All quotes in these two paragraphs from Darwin, *Journal of Researches*, 2nd ed., 144–145, 147–148, 165.

"I can . . . there . . . The Captain . . . capable of": D/SD/3/12/33, DCP-LETT-233.

"everything . . . sea-shells": Darwin, *Journal of Researches*, 2nd ed., 163.

"forgetting Petises . . . a . . . sort": Darwin, *Journal of Researches*, 87.

"dived . . . copper . . . Wishing . . . sure": Letter from Sulivan, quoted in *A Narrative of the Beagle*, ed. Stanbury, 177.

"a heavy . . . gun . . . Too . . . move . . . More . . . shore . . . stoop . . . it": *Proceedings of the Second Expedition, 1831–1836*, 319–320.

"a snow-white . . . cubes!": Darwin, *Journal of Researches*, 200.

"The thirsty . . . affair": *Proceedings of the Second Expedition, 1831–1836*, 320.

"some . . . animal": D/JH/March 34, DCP-LETT-238.

"a very . . . reception": Darwin, *Beagle* diary, 217.

"a very . . . operation . . . behaved . . . gentlemen": Darwin, *Beagle* diary, 217.

"the ship . . . down": Syms Covington's journal.

"What . . . globe?": Darwin, *Journal of Researches*, 237.

"How . . . country": Darwin, *Journal of Researches*, 2nd ed., 205–206.

"Dear . . . country": Darwin, *Beagle* diary, 224.

"vainly . . . course": Darwin, *Journal of Researches*, 2nd ed., 215.

"naked . . . devoured": D/SD/3/12/33, DCP-LETT-233.

"a last . . . farewell": Darwin, *Journal of Researches*, 2nd ed., 218.

CHAPTER ELEVEN   The Furies

"such . . . atmosphere": Darwin, *Beagle* diary, 228.

"a desperate character": *Proceedings of the Second Expedition, 1831–1836*, 331.

"streams . . . stones": Darwin, *Journal of Researches*, 254.

"like . . . cathedral": Darwin, *Journal of Researches*, 255.

"A large . . . London": Darwin, *Journal of Researches*, 247.

With . . . perish": All quotes in this paragraph are from Darwin, zoology notes, April 1834.

The jackass . . . determined": All quotes in this paragraph are from Darwin, *Journal of Researches*, 256, 257.

"wolf-like": Darwin, *Journal of Researches*, 249.

"the three . . . colour": Darwin's zoology notes, April 1834.

"confined . . . archipelago": Darwin, *Journal of Researches*, 250.

"of . . . creation": Darwin's zoology notes, April 1834.

"many . . . itself": Darwin, *Journal of Researches*, 250.

"within . . . earth": Darwin, *Journal of Researches*, 250.

During . . . gentlemen": D/JH/March 1834, DCP-LETT-238.

"how . . . coquettishly . . . Has . . . me?": CARD/D/1/9/33, DCP-LETT-214.

"He wrote . . . numbering": D/CATD/6/4/34, DCP-LETT-242.

"a glorious . . . interesting": D/CATD/6/4/34, DCP-LETT-242.

"a host of Indians": Darwin, *Beagle* diary, 232.

"beef . . . biscuit": Martens, *Journal of a Voyage from England to Australia, 1833–35*, April 18, 1834.

"While . . . Geo-graphy": FitzRoy's letter to sister Fanny, April 1834, quoted in *Voyage Round the World*, ed. Pearn, 76.

"it would . . . eaten": Martens, *Journal of a Voyage*, 18/4/34.

"the inside . . . within": Martens, *Journal of a Voyage*, April 21, 1834.

"which . . . ground . . . Indians . . . night": Darwin, *Beagle* diary, 233.

"the stragglers . . . close . . . we . . . scouts": Martens, *Journal of a Voyage*, April 21, 1834.

"difficult and dangerous": Darwin, *Beagle* diary, 235.

"retreating . . . regions . . . war . . . extermination": Martens, *Journal of a Voyage*, April 21, 1834.

"a true harbinger": Darwin, *Journal of Researches*, 2nd ed., 171.

"suddenly . . . abundant . . . the angular . . . platform": Darwin, *Journal of Researches*, 2nd ed., 171.

"encumbered . . . masses": Darwin, *Journal of Researches*, 2nd ed., 171.

"the whole . . . elevated": Martens, *Journal of a Voyage*, April 25, 1834.

"on . . . scale": Darwin, *Journal of Researches*, 217.

"fine bold crags": Martens, *Journal of a Voyage*, April 26, 1834.

"great blocks of lava": Darwin, *Beagle* diary, 236.

"five . . . deep": Darwin, Banda Oriental Santa Cruz notebook, 81.

"he was . . . left . . . were . . . consumption": Martens, *Journal of a Voyage*, May 2, 1834.

"Erasmus . . . room": CARD/D/28/10/33, DCP-LETT-224.

"*little wife . . . little* parsonage . . . the good . . . *postillion*": F. Biddulph (nee Owen)/D/21/10/33, DCP-LETT-221.

"an enormous . . . structures": D/CATD/20-29/7/34, DCP-LETT-248.

"elegant . . . coralline": Darwin, *Journal of Researches*, 2nd ed., 192.

"perfect . . . will": Darwin, *Journal of Researches*, 2nd ed., 192.

CHAPTER TWELVE   "The Very Highest Pleasures"

"a . . . hole": D/CATD, 20/7-29/7/34, DCP-LETT, 248.

"reduced . . . simplicity": Martens, *Journal of a Voyage*, July 13, 1834.

Waiting . . . home": All quotes in these two paragraphs are from SD/D/12/2/34, DCP-LETT-237.

Adding . . . purposes": All quotes in this paragraph are from D/CATD/20-29/7/34, DCP-LETT-248.

"nursing . . . affection . . . ones": D/C. Whitley/23/7/34, DCP-LETT-250.

"cannot . . . end . . . father . . . father": D/JH/24/7-7/11/34, DCP-LETT-251.

"a sort . . . Paris": D/CATD/20-29/7/34, DCP-LETT- 248.

"innumerable . . . paths": Martens, *Journal of a Voyage*, August 8, 1834.

"most . . . countries": D/CARD/9-12/8/34, DCP-LETT-253.

"to admire . . . signoritas": D/CARD/12/8/34, DCP-LETT-253.

"some . . . wine . . . half-poisoned": D/CARD/13/10/34, DCP-LETT-259.

"very miserable": D/CARD/13/10/34, DCP-LETT-259.

"one little . . . vibrate": D/CATD/8/11/34, DCP-LETT-262.

"some . . . Andes": D/CARD/13/10/34, DCP-LETT-259.

"very . . . predisposition": D/CATD/8/11/34, DCP-LETT-262.

"We shall . . . America": D/CARD/13/10/34, DCP-LETT-259.

"At . . . deeply": *Proceedings of the Second Expedition, 1831–1836*, 361.

"if her . . . country": Letter November 30, 1833, to Hugh Hoare, Hoares Bank.

"all . . . fail": *Proceedings of the Second Expedition, 1831–1836*, 361.

"I have . . . wind": Letter August 14, 1834, quoted in Gribbin and Gribbin, *FitzRoy*, 158.

"I am . . . unhappy": Letters, September 26, 1834, and September 28, 1834, reproduced in *The Beagle Record*, ed. Keynes, 238.

Darwin . . . on board": *Auto*, 75.

"Very disinterestedly . . . promotion . . . to do . . . withdrawn . . . his . . . lost": D/CATD/8/11/34, DCP-LETT-262.

"I am . . . London": Letter 6/11/34, quoted in Gribbin and Gribbin, *FitzRoy*, 161.

"our . . . painter . . . to . . . world": D/CARD/13/10/34.

"picturesque and venerable": Darwin, *Journal of Researches*, 338.

"chosen . . . prophet": Darwin, *Journal of Researches*, 2nd ed., 276.

"although . . . upwards": Darwin, *Journal of Researches*, 275.

"a great . . . body": Darwin's Port Desire notebook, EH 1.8, Darwin Online.

On . . . Day": All quotes in this paragraph are from Sulivan's letter, January 9, 1835, reproduced in Sulivan, *Life and Letters of Admiral Sir Bartholomew James Sulivan*, 44.

"These . . . head": Quoted in Armstrong, *Darwin's Other Islands*, 119.

"cut off . . . miles": Quoted in *Voyage Round the World*, ed. Pearn, 77.

CHAPTER THIRTEEN   **"Skating on Very Thin Ice"**

"something . . . seaway . . . wave . . . fro": Syms Covington's journal.

"Aboard the *Beagle* . . . insecurity": Darwin, *Beagle* diary, 292–293.

"one . . . parts": Darwin, *Beagle* diary, 297.

"the three . . . Concepcion": D/CARD/10/3/35, DCP-LETT-271.

"always . . . adventures . . . do . . . come . . . blundering": CARD/D/30/9/34, DCP-LETT-257.

"some . . . me": D/CARD/10/3/35, DCP-LETT-271.

"horse . . . spurs": D/CARD/10/3/35, DCP-LETT-271.

"old steady mare": Darwin, *Journal of Researches*, 384.

"that a . . . nature": Darwin, *Journal of Researches*, 385.

The shingle . . . Pacific": All quotes in this paragraph are from Darwin, *Journal of Researches*, 2nd ed., 303.

"and . . . arm": Darwin, Santa Fe field notebook, EH 1.13, Darwin Online.

"the famous . . . Countries": D/JH/18/4/35, DCP-LETT-274.

"eggs . . . animals": Darwin, Santa Fe field notebook, EH 1.13, Darwin Online.

"that nothing . . . earth": Darwin, *Journal of Researches*, 2nd ed., 307–308.

"have existed . . . sea": Darwin, *Journal of Researches*, 399–400.

"a distant . . . ocean": Darwin, *Journal of Researches*, 400.

"heavy smoke": Darwin, Santa Fe field notebook, EH 1.13, Darwin Online.

"horribly . . . blood": Darwin, Santa Fe field notebook, EH 1.13, Darwin Online.

"good . . . once": Darwin, Santa Fe field notebook, EH 1.13, Darwin Online.

"for which . . . officers . . . it . . . suck": Darwin, *Journal of Researches*, 404.

"very fine grapes": Darwin, Santa Fe field notebook, EH 1.13, Darwin Online.

"sad . . . raggamuffins": Darwin, Santa Fe field notebook, EH 1.13, Darwin Online.

"armed . . . spines": Darwin, *Journal of Researches*, 405.

"I . . . succeeded . . . snow white . . . Lot's wife": D/JH/18/4/35, DCP-LETT-274.

"It required . . . heads": Darwin, *Journal of Researches*, 406–407.

"carry . . . stockings": Darwin, Santa Fe field notebook, EH 1.13, Darwin Online.

"such a famous . . . S. America . . . the strata . . . pie . . . half . . . load":
D/SD/23/4/35, DCP-LETT-275.

"Some . . . incredible": D/JH/18/4/35, DCP-LETT-274.

"since . . . journey . . . sure . . . it": D/SD/23/4/35, DCP-LETT-275.

"Plenty . . . many . . . having . . . politics": Quoted in Gribbin and Gribbin, *FitzRoy*,
164–165.

"hard working shipmates": Quoted in Gribbin and Gribbin, *FitzRoy*, 164–165.

In turn . . . wonderful": All quotes in these three paragraphs are from D/
SD/23/4/35, DCP-LETT-275.

CHAPTER FOURTEEN **"Eternal Rambling"**

"bed . . . basin": D/CATD/31/5/35, DCP-LETT-276.

Lord Chesterfield: Chesterfield (1694–1773) was a statesman, diplomat, and fashionable wit about town best remembered for letters to his son advising on etiquette
and how to succeed in society.

"been formed . . . land": Darwin, *Journal of Researches*, 423.

Before . . . England": D/CATD/31/5/35, DCP-LETT-276.

"a mere . . . desert": Darwin, *Journal of Researches*, 430.

"great . . . trees": Darwin, *Journal of Researches*, 2nd ed., 338.

"downright martyrdom": D/CARD/19/7-12/8/35, DCP-LETT-281.

Subsequently . . . rescue them: D/CARD/19/7-12/8/35, DCP-LETT-281.

Ordering . . . Seymour: FitzRoy describes the rescue of the *Challenger* crew in his
Proceedings of the Second Expedition, 1831–1836, 428–475.

Saltpeter: Later in the nineteenth century the export of saltpeter to Europe and North
America for use in manufacturing explosives and fertilizers sparked an economic
boom in northern Chile. Humberstone, east of Iquique, established in 1862 as
a center for saltpeter mining but now a ghost town in the desert, is a UNESCO
World Heritage site.

"money . . . islands": D/CARD/19/7-1/8/35, DCP-LETT-276. Darwin mentioned he
in fact drew fifty pounds in DCP-LETT-286.

"Growl . . . usual": D/SD/3/9/35, DCP-LETT-286.

"a . . . boat . . . Inform . . . subject": All quotes in these two paragraphs are from
ADM/1/3848.

FitzRoy had spent some seven thousand pounds: An article in *Life-Boat Magazine*,
2/10/1865, puts FitzRoy's expenditure on the *Beagle* voyage at £6,100, including
£3,000 for instruments, but not such items as the two brass cannons in Rio and
the salaries of Earle, Martens, and Stebbing.

"has . . . *me* . . . some . . . importance": D/SD/3/9/35, DCP-LETT-286.

"The investigation . . . intelligible": *Auto*, 77.

CHAPTER FIFTEEN **The Enchanted Islands**

"as . . . stones": Quoted in Weiner, *The Beak of the Finch*, 12.

William . . . turtle": Dampier, *Voyages*, vol. 1, ed. Masefield, 128, 132.

"shamefully negligent . . . I . . . strata": D/WDF/August 1835, DCP-LETT-282.
"the thenca . . . ornithology": Darwin, Galápagos field notebook EH1.17, 30b.
"on seeing . . . brought": Darwin's zoology notes, 291.
"in . . . vegetation": Quoted in *Voyage Round the World*, ed. Pearn, 101.
"I have . . . island": Darwin's zoology notes, 298.
In his . . . bitter": All quotes in these two paragraphs are from Darwin's zoology notes, 291–293.
On the . . . bone": All quotes in this paragraph are from Darwin's zoology notes, 294–296.
On . . . fluid.)": All quotes in this paragraph are from Darwin's zoology notes, 293–294.

CHAPTER SIXTEEN   Aphrodite's Island

"a floating island . . . this land . . . outrigger": Quoted in Salmond, *The Trial of the Cannibal Dog*, 39–40.
"the true Utopia": Quoted in Salmond, *Aphrodite's Island*, 109.
"I cannot . . . them": Cook, *Cook's Voyage to the Pacific Ocean*, critical review (July 1784), 14.
"Tahiti . . . in 1769": Quoted in Moorehead, *Fatal Impact*, 89.
"The characteristic . . . strata": Darwin, Tahiti geological diary, DAR/37.798-801, 4.
"high merit": D/CARD/27/12/35, DCP-LETT-289.
"Tahiti . . . harmony": D/JH/January 1836, DCP-LETT-295.

CHAPTER SEVENTEEN   "Not a Pleasant Place"

"in improving . . . character . . . the arts . . . England": D/CARD/27/12/35, DCP-LETT-289.
"I know . . . expect": D/CARD/27/12/35, DCP-LETT-289.
"pretty . . . place": Syms Covington's journal.
"man-slave": Darwin, *Journal of Researches*, 514.
"the soil . . . hills": Darwin, *Journal of Researches*, 2nd ed., 407.

CHAPTER EIGHTEEN   "A Rising Infant"

"I shall . . . cry": D/SD/28/1/36, DCP-LETT-294.
"of a . . . tint . . . one . . . world": Darwin, *Journal of Researches*, 517–518.
"as the . . . offspring": Darwin, *Journal of Researches*, 520.
"the most . . . seen": Darwin, Australia field notebook, EH1.3, p. 10a.
"I consider . . . animal": D/P.P.King/21/1/36, DCP-LETT-293.
"half . . . heat": D/P.P.King/21/1/36. DCP-LETT-293.
"I . . . novel": Nicholas, *Charles Darwin in Australia*, 18.
"My dear . . . amiable": Quoted in Nicholas and Nicholas, *Charles Darwin in Australia*, 20.
"He has . . . well": Letter from Philip Parker King to Beaufort, February 2, 1836, reproduced in *Proceedings of the Second Expedition, 1831–1836*, 337.

"some . . . charts": Quoted in Nicholas and Nicholas, *Charles Darwin in Australia*, 69.

"Could . . . offences?": Quoted in Nicholas and Nicholas, *Charles Darwin in Australia*, 70.

"The . . . up": Quoted in Nicholas and Nicholas, *Charles Darwin in Australia*, 71.

"I regret . . . service": Letter from Philip Parker King to Beaufort, February 2, 1836, reproduced in *Proceedings of the Second Expedition, 1831–1836*, 337.

Returned . . . unhappy one": All quotes in these two paragraphs are from D/SD/28/1/36, DCP-LETT-294.

"This . . . offspring": D/SD/28/1/36, DCP-LETT-294.

"as a . . . history": Darwin, *Journal of Researches*, 532.

With this . . . people: D/CATD/14/2/1836, DCP-LETT-298.

"flying visits": D/WDF/15/2/1836, DCP-LETT-299.

"so full . . . heroes . . . 1800 . . . stomach": D/CATD/14/2/36, DCP-LETT-298.

CHAPTER NINETEEN "Myriads of Tiny Architects"

"complete . . . moonlight": Syms Covington's journal.

"one of . . . animal in . . . nine feet long": Syms Covington's journal.

"the gigantic . . . which . . . if . . . withdraw it": Darwin, *Journal of Researches*, 2nd ed., 442.

Darwin . . . oil": All quotes in this paragraph are from Darwin's zoology notes, 311.

"I shall . . . water": D/JH/28/3/37, DCP-LETT-353.

Exploring . . . pustules": Darwin, *Journal of Researches*, 2nd ed., 445.

"the submarine . . . ocean": Quoted in Herbert, *Charles Darwin Geologist*, 237.

From Mauritius . . . likes it": All quotes in these three paragraphs are from D/WDF/15/2/36, DCP-LETT-299 and D/CARD/29/4/36, DCP-LETT-301.

"exceedingly . . . natured . . . rather awful": D/JH/9/7/36, DCP-LETT-304.

"never . . . dirty": *Auto*, 107.

CHAPTER TWENTY "A Great Name among the Naturalists of Europe"

"blowing . . . spirits": D/JH/9/7/36, DCP-LETT-304.

"simple . . . man . . . a large . . . stone": Syms Covington's journal.

"The many . . . predominant": Darwin, *Journal of Researches*, 2nd ed., 467.

"There . . . insects": Darwin, *Journal of Researches*, 2nd ed., 469.

Darwin . . . eruptions": All quotes in this paragraph are from Darwin, *Journal of Researches*, 2nd ed., 468. While on St. Helena, Darwin related, "The only inconvenience I suffered during my walks was from the impetuous winds," adding that some of these were air currents produced by cliff edge effects. More recently, these winds and currents have proved a great inconvenience to the airport completed on the island by the British authorities in 2014 at a cost of £285 million. Wind sheer problems prevented its use entirely for some time. Even now, only infrequent flights by smaller aircraft than originally planned can operate.

"volcanic . . . fluid . . . At . . . subjected?": Darwin, *Journal of Researches*, 2nd ed., 473–474.

He killed . . . coast": Darwin, *Journal of Researches*, 2nd ed., 472.

"I have . . . Species": Darwin's ornithological notes, 262.

"[Darwin] . . . Europe": SD/D/22/11/35, DCP-LETT-288.

"did not . . . agreeable": CARD/D/29/11/35, DCP-LETT-291.

"after . . . hammer!": *Auto*, 82.

CHAPTER TWENTY-ONE   "A Peacock Admiring His Tail"

"the most . . . spent": *Auto*, 82.

"the stupid . . . England": D/FitzRoy/6/10/36, DCP-LETT-310.

"the shape . . . altered": Quoted in Browne, *Charles Darwin*, vol. 1, 340.

Darwin . . . Darwin": D/FitzRoy/6/10/36, DCP-LETT-310.

"the Zoologists . . . nuisance": D/CARD/24/10/36, DCP-LETT-313.

"snarling . . . gentlemen": D/JH/30-31/10/36, DCP-LETT-317.

"*most* good-natured": D/JH/30-31/10/36, DCP-LETT-317.

"delight . . . theist": *Auto*, 100.

"glorious": Quoted in Browne, *Charles Darwin*, vol. 1, 353.

"The account . . . think": FitzRoy/D/19-20/10/36, DCP-LETT-312.

"a most . . . marry": D/CARD/24/10/36, DCP-LETT-313.

"how . . . begin . . . All . . . do": D/C. Whitley/24/10/36, DCP-LETT-314.

"He . . . himself . . . We . . . on": E. Wedgwood/H. Wedgwood/16/11/36,
    DCP-LETT-322.

"I felt . . . Moon": D/JH/1/11/36, DCP-LETT-318.

"some . . . days!": D/CARD/9/11/36, DCP-LETT-321.

"A curious fact": Quoted in Browne, *Charles Darwin*, vol. 1, 365.

"a peacock . . . tail": D/WDF/7/7/37, DCP-LETT-364.

"a series . . . species": Zoological Society minutes, quoted in Desmond and Moore,
    Darwin, 219.

"so very . . . lady": D/CARD/27/10/39, DCP-LETT-542.

"would . . . species": Darwin's ornithological notes, 262.

"a hodge-podge": D/WDF/12/3/37. DCP-LETT-348.

"dirty . . . London": D/JH/1/10/36, DCP-LETT-317.

"I was . . . is . . . our . . . practice": D/CARD/9/11/36, DCP-LETT-321.

"noon . . . night . . . been . . . rhinoceros": D/SD/1/4/38, DCP-LETT-407.

"simple . . . effective": Quoted in Desmond and Moore, *Darwin*, 206.

"I could . . . continents": Lyell/D/13/2/37, DCP-LETT-343.

"I thought . . . come": D/Jenyns/10/4/37, DCP-LETT-354.

"opened . . . views": Darwin's daily journal, correspondence, vol. 2, 431.

"Heaven . . . Nature": Darwin's B notebook, 44.

"mental rioting": D/Hooker/6/5/47, DCP-LETT-1086.

"It is . . . [highest]": Darwin's B notebook, 74.

"If . . . angels": Darwin's B notebook, 169.

"Animals . . . kind": Darwin's B notebook, 231.

"Propagation . . . proved": Darwin's B notebook, 219.

"animals . . . Fox": Darwin's B notebook, 7.

"Animals . . . change": Darwin's B notebook, 4–6.

"Tree . . . seen": Darwin's B notebook, 25.

"to adapt . . . world": Darwin's B notebook, 4.

"the gleanings . . . work": D/Jenyns/10/4/37, DCP-LETT-354.

"creditable . . . myself": D/Beaufort/16/6/37, DCP-LETT-360A.

"In the . . . character . . . But . . . subject": Darwin, *Journal of Researches*, 472.

"Seeing . . . ends": Darwin, *Journal of Researches*, 2nd ed., 364.

"writing . . . work": D/WDF/28/8/37, DCP-LETT-374.

"sat . . . own volume": D/JH/4/11/37, DCP-LETT-384.

"I will . . . Robt. FitzRoy: All quotes in these three paragraphs from FitzRoy/D/16/11/37, DCP-LETT-387.

"The Captain . . . manner": D/SD/1/4/31, DCP-LETT-407.

"I have . . . heart": D/JH/20/9/37, DCP-LETT-378.

"I find . . . society": D/CARD/May/38, DCP-LETT-411.

"like . . . Lord . . . I enjoy . . . to see": D/Lyell/9/8/38, DCP-LETT-424.

"She . . . imaginable": D/SD/1/4/38, DCP-LETT-407.

"Compare . . . great": Darwin's M notebook, 153.

"Man . . . animals": Darwin's C notebook, 1838, 196–197.

"enjoyed . . . well": D/Lyell/9/8/38, DCP-LETT-424,

"one . . . blunder": D/Lyell/6/9/61, DCP-LETT-3246.

"I never . . . things": D/Lyell/9/8/38, DCP-LETT-424.

"In October . . . work": *Auto*, 120.

CHAPTER TWENTY-TWO   "It Is Like Confessing a Murder"

Darwin's list of the pros and cons of marriage is transcribed in Darwin, correspondence, vol. 2, appendix iv, 444–445, and also in *Auto*, 231–234. He also wrote a brief memorandum on marriage.

"He is . . . humane to animals": Quoted in Healey, *Emma Darwin*, 148, and Browne, *Charles Darwin*, vol. 1, 393.

"on quietness . . . solitude": D/Emma/20/1/39, DCP-LETT-489.

"Since . . . burned": Emma/D/21/10-2/11/38, DCP-LETT-441.

"November 27th . . . erection": Darwin's N 41 notebook, 51–52.

"the awful day": D/Emma/6-7/1/39, DCP-LETT-484.

"I . . . me": Emma/D/30/11/38, DCP-LETT-447.

"the Old . . . Hindoos": Quoted in F. Darwin, *The Life of Charles Darwin*, 58.

"You will . . . him . . . rather . . . patronizing": D/CARD/27/10/39, DCP-LETT-542.

"You . . . achieve": Humboldt/D/18/9/39 (in French), DCP-LETT-534.

"On this . . . depth": Browne, *Charles Darwin*, vol. 1, 418.

"considerable . . . coast": *Edinburgh Review*, no. 69 (1839).

"fund . . . instruction": *United Service Journal*, pt. II (1841).

"a first-rate . . . pen . . . ample . . . thinking": *Quarterly Review*, no. 65 (1839).

"It is . . . egg": Owen/D/11/6/39, DCP-LETT-519.

"this . . . memory": Stokes, *Discoveries in Australia*, vol. 2, 5–6.

"animated . . . shipmate": Quoted in Desmond and Moore, *Darwin*, 314.

"What . . . herself": D/WDF/7/6/40, DCP-LETT-572.

Throughout . . . year": All quotes in this paragraph are from D/Fitzroy/20/2/40, DCP-LETT-555.

"long . . . illness": D/Redfield/22/12/1840, DCP-LETT-585.

"pure . . . dirt . . . misery": D/WDF/28/9/41, DCP-LETT-609.

"more . . . ear": Darwin, *The Structure and Distribution of Coral Reefs*, 103.

In May . . . of this: All quotes in these two paragraphs are from Darwin's *Sketch of Species Theory 1842*, Darwin Online, ref. DAR 6 and F1556.

Down House and the village of Downe: Somewhat confusingly the house and village are spelled differently.

"the publicity . . . intolerable": D/SD/27-28/4/43, DCP-LETT-673.

Over . . . 10.30: All quotes and information in these three paragraphs are from Francis Darwin's reminiscences of his father in his *Life of Charles Darwin*, 69–78, with the exception of the quote about backgammon, which is from Darwin's letter of 28/1/76 to Asa Gray, DCP-LETT-10370.

Cigarette: Cigarettes were first invented in sixteenth-century Seville when beggars picked up discarded cigar butts, shredded the tobacco inside them, and wrapped it in paper to smoke. Late in the eighteenth century the habit became respectable. It was then brought back to their own countries by British and French soldiers fighting in the Napoleonic Peninsula Wars. The French gave these little cigars the name *cigarettes*, and by 1830 cigarette smoking was widespread in Britain.

"She was . . . details": Darwin's daughter Henrietta, quoted in Healey, *Emma Darwin*, 164.

"disgraceful . . . history . . . a liar . . . knave": Quoted in Mellersh, *Fitzroy of the Beagle*, 186.

On . . . muster: All quotes from Mellersh, *Fitzroy of the Beagle*, 190.

"I have . . . immutable": D/Hooker/11/1/44, DCP-LETT-729.

"I shall . . . subject": Hooker/D/29/1/44, DCP-LETT-734.

Darwin's Abstract is available on Darwin Online, reproduced in F1556.

"My dear . . . it is": D/Emma/5/7/44, DCP-LETT-761.

"ordinary reader . . . dogs . . . clergy": Quoted in Desmond and Moore, *Darwin*, 320.

"bad . . . worse": D/Hooker/7/1/45, DCP-LETT-814.

"species . . . immutable": Quoted in Browne, *Charles Darwin*, vol. 1, 461.

"All Things Bright and Beauteous": The comparison between *Vestiges* and the hymn was first made by Patrick H. Armstrong in his book *Darwin's Luck*.

"I . . . unflattered": D/WDF/24/4/45, DCP-LETT-859.

"pointed . . . reasons'": Quoted in Desmond and Moore, *Darwin*, 323.

"I . . . ones": *Auto*, 123.

"a most . . . form": *Auto*, 117.

"of . . . use": *Auto*, 118.

"did his barnacles": Quoted in Keynes, *Creation*, 124.

"whether . . . time": *Auto*, 118.

CHAPTER TWENTY-THREE   "Most Hasty and Extraordinary Things"

de Thierry: Subsequently de Thierry tried his hand unsuccessfully in the California gold rush. He then did persuade the French authorities to appoint him to the staff of their consulate in Honolulu, where he spent nearly two years before returning to New Zealand. There he fell into straitened circumstances teaching music and tuning pianos but befriended the governor George Grey, who funded the writing of his autobiography. He died in New Zealand in 1864.

"the missionaries . . . step in": Quoted in Gribbin and Gribbin, *FitzRoy*, 203.

"He is . . . delicacy": Quoted in Gribbin and Gribbin, *FitzRoy*, 211–212.

"tomahawked": Thomson, *The Story of New Zealand*, vol. 2, 75.

"crushed . . . civilisation": Gribbin and Gribbin, *FitzRoy*, 215.

"a good . . . natives . . . all . . . wrong": Quoted in Mellersh, *Fitzroy of the Beagle*, 212.

"In the . . . Paheka": Mellersh, *Fitzroy of the Beagle*, 216.

"The task . . . follow": Quoted in Gribbin and Gribbin, *FitzRoy*, 226.

"deeply . . . injured": Quoted in Nichols, *Evolution's Captain*, 260.

"*Beagle* . . . things": Sulivan/D/ 13/1/45–12/2/45, DCP-LETT-730.

"I fear . . . Governorship": D/FitzRoy/1/10/46, DCP-LETT-1002.

"Oh Mammy . . . safe": D/Emma/2-/28/5/48, DCP-LETT-1180.

"obediently . . . faith": D/SD/19/3/49, DCP-LETT-1234.

"I have . . . some years": D/Covington/30/3/49, DCP-LETT-1237.

"I am heated . . . quackery": D/Hooker/28/3/49, DCP-LETT-1236.

"exaggerated": D/Emma/18/4/51, DCP-LETT-1400.

Whole books . . . complaints: Quotes in these three paragraphs are from K. Thomson, "Darwin's Enigmatic Health," *American Scientist* 97, no. 3.

"his . . . mind": Letter from Fanny Allen to Elizabeth Wedgwood, April 18, 1851, see note to DCP-LETT-1399.

"My abstract . . . heir": D/WDF/12/2/59, DCP-LETT-2412.

"24 . . . vomiting": D/Hooker/23/4/61, DCP-LETT-3125.

"Began . . . theory": Darwin's pocket diary, September 1854.

"small . . . flesh . . . made . . . desist": Darwin, *Expression of the Emotions*, 259.

"Happiness . . . work": D/Huxley/29/9/55, DCP-LETT-1757.

"most . . . complete . . . the . . . remarkable . . . not . . . *professo*": Huxley praised Darwin's work on barnacles in a published lecture to the Royal Institution in 1857. (See note 9 to letter D/Huxley/11/4/53, DCP-LETT-1514.)

"It is . . . science": D/Gray/18/6/57, DCP-LETT-2109.

"cautious reasoner": D/Gray/18/6/57, DCP-LETT-2109.

"a . . . man": D/Hooker/after 20 January 57, DCP-LETT- 2033.

"As an . . . despise me": D/Gray/20/7/57, DCP-LETT-2125.

"that there . . . plants": Gray/D/August 57, DCP-LETT-2129.

"As you . . . species": D/Gray/5/9/57, DCP-LETT-2136.

"I wish . . . understood": Lyell/D/1/5/56, DCP-LETT-1862.

"To give . . . me": D/Lyell/3/5/56, DCP-LETT-1866.

"I am . . . years": D/WFD/8/2/57, DCP-LETT-2049.
"quite . . . book": D/WDF/3/10/56, DCP-LETT-1967.

CHAPTER TWENTY-FOUR "I Shall Be Forestalled"

"a want . . . confidence": Wallace, *My Life*, vol. 1, 260.
"delighted . . . talk": Quoted in Browne, *Charles Darwin*, vol. 2, 24.
"costing . . . fortune!": D/Tegetmeier/29/11/56, DCP-LETT-2004.
Wallace's paper was "On the Law Which Has Regulated the Introduction of New Species,"
    Annals and Magazine of Natural History, 2nd series, 16 (February 1855): 184–196.
"nothing . . . with him": Quoted in Desmond and Moore, *Darwin*, 438.
"a sudden . . . light . . . thought . . . survive": Wallace, *My Life*, vol. 1, 361–362.
"that . . . type": Wallace's essay is in Hooker and Lyell's paper to the Linnaean Society,
    July 1, 1858, available at Darwin Online.
"trumpery feelings": D/Lyell/25/6/58, DCP-LETT-2294.
"It . . . priority": D/Hooker/29/6/58, DCP-LETT-2298.
"Great . . . priority": Captain Robert Falcon Scott's Polar diary, January 17, 1912,
    http://www.spri.cam.ac.uk/museum/diaries/scottslastexpedition, 7.
"Your . . . smashed": D/Lyell/18/6/58, DCP-LETT-2285.
"having . . . so": Hooker and Lyell's paper to the Linnaean Society, July 1, 1858.
"been . . . science": Quoted in Desmond and Moore, *Darwin*, 470.
"made . . . will": Hooker's letter to Huxley/March 1859, quoted in Desmond and
    Moore, *Darwin*, 473.
"one . . . argument": D/Murray/2/4/59, DCP-LETT-2445.
"admirably . . . jealousy": D/Lyell/18/5/60, DCP-LETT-2806.
"He . . . man": D/Hooker/23/1/59, DCP-LETT-2403.
Darwin . . . explained.": Quotes in this paragraph are from the first edition of *On the
    Origin of Species* (1859), 127–128.
"descent with modification": Darwin, *On the Origin of Species*, 1st ed., 123 et seq.
"daily . . . being": Darwin, *On the Origin of Species*, 1st ed., 84.
"Natural . . . life . . . the great . . . life": Darwin, *On the Origin of Species*, 1st ed., 128–130.
"When . . . mysteries": Darwin, *On the Origin of Species*, 1st ed., 1.
"We . . . were": Darwin, *On the Origin of Species*, 1st ed., 84.
"I think . . . prejudices": D/Wallace/22/12/57, DCP-LETT-2192.
"Analogy . . . breathed": Darwin, *On the Origin of Species*, 1st ed., 484.
"In . . . history": Darwin, *On the Origin of Species*, 1st ed., 488.
"excellent memoir . . . almost . . . species . . . ingenious paper": Darwin, *On the
    Origin of Species*, 1st ed., 1, 355.
"a wild . . . imagination": Quoted in Browne, *Charles Darwin*, vol. 2, 75.
"My . . . hate it": D/WDF/23/9/59, DCP-LETT-2493.
"I am . . . me": D/Lyell/2/9/59, DCP-LETT-2486.
"with . . . vapours": Quoted in Desmond and Moore, *Darwin*, 481.
"For myself . . . world": Erasmus Darwin/D/23/11/59, DCP-LETT-2545.
"A humiliating . . . better": Quoted in Gregor, *Charles Darwin*, 126.

"My dear . . . ape": Quoted in Browne, *Charles Darwin*, vol. 2, 94.

"It is . . . inserts it!": D/Lyell/3/12/59, DCP-LETT-2567.

"If I . . . moon": Sedgwick/D/24/11/59, DCP-LETT-2548.

"Each . . . bubbles": Unsigned article, *Spectator*, quoted in Browne, *Charles Darwin*, vol. 2, 108.

"The Book . . . far": JH/Jenyns/26/1/60, quoted in Desmond and Moore, *Darwin*, 487.

"Darwin . . . out": JH/Jenyns/26/1/60, quoted in Desmond and Moore, *Darwin*, 488.

"the whole orang": Lyell/Huxley/17/7/1859, DCP-LETT-2469A.

"All . . . written": Kingsley/D/18/11/59, DCP-LETT-2534.

"a dozen . . . periodicals": D/Huxley/28/12/59, DCP-LETT-2611.

"As for . . . readiness": Huxley/D/23/11/59, DCP-LETT-2544.

"vagueness and incompleteness . . . short-sighted": *Edinburgh Review* 111 (1860).

"extremely . . . damaging": D/Lyell/10/4/60, DCP-LETT-2754.

"jealousy . . . bitter enemy": *Auto*, 105.

"very . . . explainer": D/Hooker/5/6/60, DCP-LETT-2821.

"She . . . herself": Quoted in Keynes, *Creation*, 237.

"a . . . shot": D/Gray/22/7/60, DCP-LETT-2876.

"no one . . . him!": D/Wyman/3/10/60, DCP-LETT-2936.

"I never . . . America": A/Gray/28/1/60, DCP-LETT-2665.

From his . . . Society": The quotes from Livingston and Marx in this paragraph are from Desmond and Moore, *Darwin*, 485.

By June . . . men": Wilberforce's review is in *The Quarterly Review* (July 1860).

"flatulent stuff": Hooker/D/2/7/60, DCP-LETT-2852.

There are many different, slightly diverging eyewitness accounts of the Oxford debate. These include an account in *The Athenaeum Magazine*, July 14, 1860. Hooker's letter to Darwin, Hooker/D/2/7/60, DCP-LETT-2852, gives his account. A letter (DAR106/7.36) of May 17, 1895, from G. Stoney to Darwin's son Francis details FitzRoy's involvement.

"poor . . . insanity": D/JH/16/7/60, DCP-LETT-2869.

"How . . . well": D/Huxley/5/7/60, DCP-LETT-2861.

"I . . . assembly": D/Huxley/3/7/60, DCP-LETT-2854.

CHAPTER TWENTY-FIVE   Natural Selection

"an odious spectre": D/Wallace/12/7/71, DCP-LETT-7858.

"Do the . . . mouth": D/Bridges/6/1/60, DCP-LETT-2640.

"I have . . . humanity": D/Gray/5/6/61, DCP-LETT-3176.

"just as much . . . mine": D/Wallace/28/3/64, DCP-LETT-4510.

"talking . . . disagreable": Wallace, *My Life*, vol. 1, 417.

"shy . . . society": Quoted in Browne, *Charles Darwin*, vol. 2, 199.

"a grand . . . times": D/Dana/7/1/63, DCP-LETT-3905.

"I shd. . . . vomiting": D/Hooker/23/4/61, DCP-LETT-3125.

"when . . . me": Quoted in Healey, *Emma Darwin*, 270.

"He was . . . so much": Quoted in Healey, *Emma Darwin*, 272.

"but . . . strength": D/Wallace/28/5/64, DCP-LETT-4510.

"I hope . . . child": D/Wallace/27/3/69, DCP-LETT-6684.

"I . . . rubbish": D/Hooker/18/1/74, DCP-LETT-9247.

"taunted . . . opinions": D/Muller/22/2/67, DCP-LETT-5410.

"It is . . . seal": Darwin, *Descent of Man*, vol. 1, 7–8.

"the same . . . degrees": Darwin, *Descent of Man*, vol. 1, 120.

"all . . . developed": Darwin, *Descent of Man*, vol. 2, 380.

"No other . . . red": Darwin, *Descent of Man*, vol. 2, 292.

"Ultimately . . . conscience": Darwin, *Descent of Man*, vol. 1, 165–166.

"man . . . woman: Darwin, *Descent of Man*, vol. 2, 328–329.

"I certainly . . . man": D/Kennard/9/1/82, DCP-LETT-13607.

"the subject . . . read it": D/Murray/21/9/61, DCP-LETT-3259.

"The chief . . . hands": Darwin, *Descent of Man*, vol. 2, 327.

"The careless . . . him": Darwin, *Descent of Man*, vol. 1, 174. Darwin is quoting from an
    article by William Rathbone Greg.

"The Fuegians . . . faculties": Darwin, *Descent of Man*, vol. 1, 34.

"I don't . . . view": Quoted in Healey, *Emma Darwin*, 292.

"I think . . . off": Quoted in Keynes, *Creation*, 259.

"I have . . . nation": D/Muller/28/8/70, DCP-LETT-7310.

"half-killed me": D/Sulivan/13/3/71, DCP-LETT-7579.

"The astonishment . . . ancestors": Darwin, *Descent of Man*, vol. 2, 404.

"the . . . England": D/Lankester/22/3/71, DCP-LETT-7612.

"evolution . . . fact": Hooker/D/26/3/71, DCP-LETT-7627.

"there . . . life": Quoted in Browne, *Charles Darwin*, vol. 2, 352.

"I . . . interesting": Quoted in Browne, *Charles Darwin*, vol. 2, 368.

"You will . . . lifetime": Huxley/D/12/9/68, DCP-LETT-6363.

CHAPTER TWENTY-SIX   "The Clerk of the Weather"

"I . . . grieve . . . as . . . later": FitzRoy to Sir Thomas Gladstone, April 14, 1852,
    Gladstone Library.

"He looked . . . House-keeping": D/P.G.King/21/2/54, DCP-LETT-1554A.

"he has . . . sad": Gribbin and Gribbin, *FitzRoy*, 254.

FitzRoy . . . rain: FitzRoy, *Barometer and Weather Guide*, 5, 14, 20.

"Papa . . . friendly . . . Mrs FitzRoy . . . . say": Quoted in Browne, *Charles Darwin*,
    vol. 2, 53.

FitzRoy . . . fine: Quotes in this paragraph are from *The (London) Times*, August 1, 1861.

"those pecuniary . . . market": Quoted Mellersh, *Fitzroy of the Beagle*, 278.

The Weather Book: In this FitzRoy wrote, "Dampier's admirable (but *now* too little
    appreciated) descriptions and intelligent explanations were text-books among the
    navigators of that and the following century (as Cook, La Perouse and Flinders
    showed in their works), being then the only systematic and reliable general
    account of winds, weather, and climate around the world."

"This . . . life . . . that . . . present": FitzRoy, *The Weather Book*, 15. Despite the pressures of authorship and the Meteorological Office, FitzRoy found time in this period to support the pioneering balloon ascents by the British meteorologist James Glaisher to explore and gather data on the upper atmosphere, on one occasion ascending over seven miles.

"The Doctors . . . there": Quoted in Mellersh, *Fitzroy of the Beagle*, 281.

"I was . . . character": D/Hooker/4/5/65, DCP-LETT-4827.

"FitzRoy's . . . sway": *Auto*, 73.

"Darwin contributed": Sulivan too contributed to the fund for FitzRoy's wife and according to a friend "moved heaven and earth" to get FitzRoy's services acknowledged by the authorities and some at least of his unauthorized expenditure incurred on government business refunded to his family. He was partially successful. Subsequently a grace and favor apartment in the royal palace of Hampton Court was allocated to FitzRoy's widow.

CHAPTER TWENTY-SEVEN "All Soon to Go"

"I shall . . . evolution": D/Wallace/27/7/72, DCP-LETT-8429.

"I often . . . time": Quoted in Browne, *Charles Darwin*, vol. 2, 460.

"I never . . . would": Quoted in Browne, *Charles Darwin*, vol. 2, 440.

"pleasing . . . appearance": Quoted in Desmond and Moore, *Darwin*, 626.

"I heartily . . . mankind": D/Marx/1/10/73, DCP-LETT-9080.

"spread . . . mankind": D/Bradlaugh/6/6/77, DCP-LETT-10988.

"Half . . . questions": D/R.Darwin/8/4/79, DCP-LETT-11982.

"he felt . . . alone": F. Darwin, *The Life of Charles Darwin*, 56.

"any benefit": D/Browne/18/12/80, DCP-LETT-12919.

"I know . . . vanities": D/GH Darwin/10/2/75, DCP-LETT-9851.

"I do . . . God": D/McDermott/24/11/80, DCP-LETT-12851.

"I can . . . doctine": *Auto*, 87.

"Science . . . probabilities": D/Mengden/5/6/79, DCP-LETT-12088.

"beyond . . . intellect": D/Doedes/2/4/73, DCP-LETT-8837.

"The mystery . . . us": *Auto*, 94.

"I may . . . mind": D/Fordyce/7/5/79, DCP-LETT-12041.

"as if . . . go . . . I should . . . shoulders": D/Hooker/25/2/75, DCP-LETT-9873.

"I am . . . problem": *Auto*, 141.

"During . . . age": F. Darwin, *The Life of Charles Darwin*, 327–328.

DARWIN'S LEGACY

"perpetuating . . . kind": Quoted in Weintraub, *The Link between the Rockefeller Foundation and Racial Hygiene in Nazi Germany*.

Balfour's speech at the Congress of Eugenics was printed in full in *The Times* on July 25, 1912.

"the evil . . . procreate": D/Gaskell/15/11/78, DCP-LETT-11745.

"Climate . . . checks": Darwin, *On the Origin of Species*, 1st ed., 68.

POSTSCRIPT "Weep for Patagonia"

"Plea for Patagonia": Quoted in Nichols, *Evolution's Captain*, 264.

"deaths . . . mission": Quoted in Nichols, *Evolution's Captain*, 270.

One . . . ladder?'": Quotes in this paragraph are from Snow, *A Two Years' Cruise off Tierra del Fuego*, vol. 2, 29–30.

Helped . . . gifts: Quotes in this paragraph are from Snow, *A Two Years' Cruise off Tierra del Fuego*, vol. 2, 33–34.

"five moons": Quoted in Hazlewood, *Savage*, 196.

"I . . . Island": Quoted in Hazlewood, *Savage*, 204.

"wild . . . season": Quoted in Hazlewood, *Savage*, 226.

"A canoe . . . away": Fell's diary, 2/11/59, quoted in Hazlewood, *Savage*, 250.

Soon . . . right to": Coles's account in these three paragraphs is given in CO78/42, UKPA.

Believing . . . man": Jemmy Button's account in this paragraph is given in CO78/42, UKPA.

"How . . . London": Quoted in Hazlewood, *Savage*, 326.

"Captain . . . on board": Footnote to p. 217 in Darwin, *Journal of Researches*, 2nd ed.

"little . . . girl": Quoted in Hazlewood, *Savage*, 327.

BIBLIOGRAPHY

PRIMARY SOURCES

Charles Darwin's Works Published in His Lifetime

All Darwin's works are available in each of their editions from the Darwin Online
Project, the world's largest collection of Darwin's writings, from his books to his
manuscripts, notebooks, and private papers. The editions, listed below in date order,
are those referenced in particular in this book.

- *Journal of Researches into the Geology and Natural History of the Countries Visited
  during the Voyage round the World of H.M.S. Beagle under Command of Captain
  FitzRoy, R. N. Vol. 3 of The Narrative of the Voyages of H. M. Ships, Adventure and
  Beagle, 1826–1836.* London: Henry Colburn, 1839.
- *The Geology of the Beagle, 1842–1846—Part I, The Structure and Distribution of
  Coral Reefs.* London: Smith, Elder & Co., 1842; *Part II, Geological Observations
  on the Volcanic Islands Visited during the Voyage of H.M.S. Beagle.* London: Smith,
  Elder & Co., 1844; *Part III, Geological Observations on South America.* London:
  Smith, Elder & Co., 1846.
- *The Zoology of the Voyage of H.M.S. Beagle under the Command of Captain FitzRoy.*
  (5 vols.). London: Smith, Elder & Co., 1838–1843.
- *The Voyage of the Beagle (2nd edition of Darwin's Journal of Researches into the Natu-
  ral History and Geology of the Countries Visited during the Voyage round the World
  of H.M.S. Beagle under Command of Captain FitzRoy, R. N.* (See above.) London:
  John Murray, 1845. Reprinted many times, including London: Marshall
  Cavendish, The Great Writers Library, 1987. Page references are to this 1987
  edition.
- *On the Origin of Species by Means of Natural Selection, or the Preservation of Favoured
  Races in the Struggle for Life.* London: John Murray, 1859.
- *On the Origin of Species by Means of Natural Selection, or the Preservation of Favoured
  Races in the Struggle for Life.* 2nd ed. London: John Murray, 1860.
- *On the Various Contrivances by Which British and Foreign Orchids Are Fertilised by
  Insects.* London: John Murray, 1862.

- *The Variation of Animals and Plants under Domestication* (2 vols.). London: John Murray, 1868.
- *The Descent of Man and Selection in Relation to Sex* (2 vols.). London: John Murray, 1871.
- *The Expression of the Emotions in Man and Animals.* London: John Murray, 1872.
- *The Formation of Vegetable Mould, through the Action of Worms with Observations on Their Habits.* London: John Murray, 1881.

Darwin's Other Writings, Published and Unpublished

- *The Autobiography of Charles Darwin, 1809–1882 with Original Omissions Restored*, ed. Nora Barlow. London: Collins, 1958.
- Autobiographical note, "Life," August 1838. Darwin Online, ref. DAR 91.56–63.
- *Beagle Diary*, ed. R. D. Keynes. Cambridge: Cambridge University Press, 2001.
- Pocket diary. Darwin Online, ref. DAR 158.

Darwin's scientific notebooks and diaries are available from Darwin Online, as well as in edited published versions. Those referenced in the text are:

- Geological diary from the *Beagle* voyage. Darwin Online, ref. DAR 32–38.
- Ornithological notes from the *Beagle* voyage transcribed in N. Barlow, *Bulletin of the British Museum (Natural History) Historical Series* 2, no. 7 (1963): 201–278. Darwin Online, ref. F1577.
- Zoology notes from the *Beagle* voyage transcribed in R. Keynes, *Charles Darwin's Zoology Notes and Specimen Lists from H. M. S. Beagle.* Cambridge: Cambridge University Press, 2000. Darwin Online, ref. F1840.
- Field notebooks from the *Beagle* voyage are on Darwin Online under the general heading "Darwin's *Beagle* Field Notebooks, 1831–1836," and then by individual geographical area. Field notebooks quoted in this book include: Australia, Banda Oriental Santa Cruz, Falklands, Galápagos, Port Desire, and Santa Fe. They are held by the English Heritage organization.
- *Red Notebook* transcribed in Sandra Herbert, *The Red Notebook of Charles Darwin. Bulletin of the British Museum (Natural History) Historical Series* 7 (1980): 1–164. Darwin Online, ref. F1583e.
- Transmutation notebooks B, C, D, and E, and notebooks M and N. Darwin Online, ref. DAR 121–126.
- Memorandum on marriage. Darwin Online, ref. DAR 210.8.1.
- First "pencil" sketch of species theory 1842. Darwin Online, ref. DAR 6. Transcribed in F. Darwin, *The Foundations of the Origin of Species.* Cambridge: Cambridge University Press, 1909. Darwin Online, ref. F1556.
- "Elevation of Patagonia," unpublished essay. Folios 49–52, folio 73, vol. 42, Darwin papers, Cambridge University Library.

CORRESPONDENCE

Darwin's surviving letters and those sent to him, in both cases, until 1882, are available online through the Darwin Correspondence Project.

The Beagle Letters, ed. F. Burkhardt, introduction by J. Browne. Cambridge: Cambridge University Press, 2009, is another very helpful source.

ARTICLES

Darwin, Charles, et al. "On the Tendency of Species to Form Varieties; and on the Perpetuation of Varieties and Species by Natural Means of Selection," by Charles Darwin Esq., F.R.S., F.L.S., and F.G.S., and Alfred Russel Wallace Esq. communicated by Sir Charles Lyell and J. D. Hooker Esq. Read to Linnean Society, July 1, 1858.

OTHER PRIMARY SOURCES

Cook, J. *Cook's Voyage to the Pacific Ocean*. London: Critical Review, 1784.

Covington, Syms. Journal, December 1831–September 1836, edited by V. Weitzel. asap .unimelb.edu.au.

Dampier, William. *Voyages—Dampier's Collected Works*, edited by John Masefield. London: Grant Richards, 1906.

Darwin, E. *Zoonomia; or the Laws of Organic Life*. London: J. Johnson, 1794.

FitzRoy, Robert. *Barometer and Weather Guide*. London: Eyre and Spottiswoode, 1859.

FitzRoy, Robert. Family correspondence. Cambridge University Library. GB 12 MS. Add 8853.

FitzRoy, Robert. *Remarks on New Zealand*. London: White, 1846.

FitzRoy, Robert. *The Weather Book: A Manual of Meteorology*. London: Longman, Green, Longman, Roberts & Green, 1863.

King, Philip Gidley. "Reminiscences of Darwin during the Voyage of the *Beagle*." Darwin Online Project and Mitchell Library, Sydney. MLS-FM4.6900.

King, Philip Parker, Robert FitzRoy, and Charles Darwin. *Narrative of the Surveying Voyages of His Majesty's Ships Adventure and Beagle 1826–1836*, 3 vols. plus appendix. London: Henry Colburn, 1839.

Martens, Conrad. *Journal of a Voyage from England to Australia aboard H.M.S. Beagle and H.M.S. Hyacinth, 1833–35*, transcribed by Michael Organ. Sydney: State Library of New South Wales Press, 1994.

McCormick, Robert. HMS *Beagle* diary, transcribed in E. Steel, *He Is No Loss—Robert McCormick and the Voyage of HMS Beagle*. British Society for the History of Science Monographs, no. 14 (2011).

McCormick, Robert. *Voyages of Discovery*. London: Sampson Low and Co., 1884.

Snow, William Parker. *A Two Years' Cruise off Tierra del Fuego, the Falkland Islands, Patagonia and the River Plate*. London, 1857.

Stokes, John Lort. *Discoveries in Australia*. London: T. and W. Boone, 1846.

Sulivan, N. H., ed. *Life and Letters of the Late Admiral Sir Bartholomew James Sulivan, KCB*. London: John Murray, 1896.

Thomson, Arthur S. *The Story of New Zealand, Past and Present, Savage and Civilized*. London: John Murray, 1859.

Wakefield, E. J. *Adventure in New Zealand*. London: John Murray, 1845.

Wallace, Alfred Russel. *My Life*. London: Chapman and Hall, 1905.

Wallace, Alfred Russel. "On the Law Which Has Regulated the Introduction of New Species." *Annals and Magazine of Natural History* 2, no. 16 (February 1855): 184–196.

COMPILATIONS OF PRIMARY SOURCES

Keynes, R. D., ed. *The Beagle Record*, extracts from writings of Darwin, Fitzroy, Martens, and others. Cambridge: Cambridge University Press, 1979.

Orel, H., ed. *Charles Darwin: Interviews and Recollections*. London: Macmillan, 2000.

Ralling, C., arranger. *The Voyage of Charles Darwin*. London: BBC, 1978.

Stanbury, David, ed. *Narrative of the Voyage of HMS Beagle* (extracts from FitzRoy's Narrative, logs, reports and letter; extracts from Darwin's diary and letters; notes by King and letters from Sulivan). London: Folio Society, 1977.

SECONDARY SOURCES

Armstrong, P. H. *Charles Darwin in Western Australia*. Nedlands, Western Australia: University of Western Australia Press, 1985.

Armstrong, P. H. *Charles Darwin's Last Island: Terceira, Azores, 1836*. Nedlands, Western Australia: Geowest No. 27, University of Western Australia Press, 1992.

Armstrong, P. H. *Darwin's Desolate Islands: A Naturalist in the Falklands, 1833 and 1834*. Chippenham, UK: Picton Publishing, 1992.

Armstrong, P. H. *Darwin's Luck*. London: Continuum Books, 2009.

Armstrong, P. H. *Darwin's Other Islands*. London: Continuum Books, 2004.

Armstrong, P. H. *Under the Blue Vault of Heaven: A Study of Charles Darwin's Sojourn in the Cocos (Keeling) Islands*. Nedlands, Western Australia: Indian Ocean Centre for Peace Studies, 1991.

Aydon, C. *Charles Darwin*. London: Constable, 2002.

Bowlby, J. *Charles Darwin*. London: Hutchinson, 1990.

Bowler, P. J. *Charles Darwin—The Man and His Influence*. Oxford: Basil Blackwell, 1990.

Browne, J. *Charles Darwin: Voyaging* (biography vol. I). London: Jonathan Cape, 1995.

Browne, J. *Charles Darwin: The Power of Place* (biography vol. II). London: Jonathan Cape, 2002.

Browne, J. *Darwin's Origin of Species—A Biography*. London: Atlantic Books, 2006.

Darwin, F. *The Life and Letters of Charles Darwin*. London: Senate, 1995.

Desmond, A., J. Moore, and J. Browne. *Charles Darwin*. Oxford: Oxford University Press, 2007.

Desmond, A., and J. Moore. *Darwin*. London: Michael Joseph, 1991.

Desmond, A., and J. Moore. *Darwin's Sacred Cause*. London: Penguin, 2010.

Fara, P. *Sex, Botany and Empire*. Cambridge: Icon Books, 2003.

Grant, P. R. *Ecology and Evolution of Darwin's Finches*. Princeton, NJ: Princeton University Press, 1986.

Green, T. *Saddled with Darwin*. London: Phoenix, 2001.

Gregor, A. S. *Charles Darwin*. London: Angus and Robertson, 1967.

Gribbin, J., and M. Gribbin. *FitzRoy*. London: Review, 2004.

Gribbin, J., and M. White. *Darwin—A Life in Science*. London: Pocket Books, 2009.

Hazlewood, N. *Savage—The Life and Times of Jemmy Button*. London: Hodder and Stoughton, 2000.

Healey, E. *Emma Darwin*. London: Headline, 2001.

Herbert, S. *Charles Darwin, Geologist*. Ithaca, NY: Cornell University Press, 2005.

Horwell, D., and P. Oxford. *Galápagos Wildlife*. Chalfont St. Peter, UK: Bradt Publications, 1999.

Hutton, J. *Theory of the Earth*. Edinburgh: Transactions of the Royal Society of Edinburgh, 1788.

Huxley, J., and H. B. D. Kettlewell. *Charles Darwin and His World*. London: Book Club Associates, 1975.

Jones, S. *Almost Like a Whale*. London: Doubleday, 1999.

Keynes, R. *Creation—The True Story of Charles Darwin*. London: John Murray, 2009.

Keynes, R. *Fossils, Finches and Fuegians*. London: Harper Collins, 2003.

Lyell, Sir Charles. *Life, Letters and Journals of Sir Charles Lyell*, ed. K. M. Lyell, vol 1. London: John Murray, 1881.

Lyell, Sir Charles. *Principles of Geology, volumes 1–3*. London: John Murray, 1830–1833.

Marks, R. L. *Three Men of the Beagle*. New York: Avon Books, 1991.

Mellersh, H. E. L. *Fitzroy of the Beagle*. London: Rupert Hart-Davis, 1968.

Moore, J. *The Darwin Legend*. London: Hodder and Stoughton, 1995.

Moore, P. *The Weather Experiment*. London: Chatto and Windus, 2015.

Moorehead, A. *Darwin and the Beagle*. London: Penguin, 1971.

Moorehead, A. *The Fatal Impact*. London: Reprint Society, 1967.

Nicholas, F. W., and J. M. Nicholas. *Charles Darwin in Australia*. Cambridge: Cambridge University Press, 1989.

Nichols, P. *Evolution's Captain*. London: Profile Books, 2003.

North, F. J. *Sir Charles Lyell*. London: Arthur Barker, 1965.

Pearn, A. M., ed. *A Voyage Round the World—Charles Darwin and the Beagle Collections in the University of Cambridge*. Cambridge: Cambridge University Press, 2009.

Preston, D. *Paradise in Chains*. London: Bloomsbury, 2017.

Preston, D., and M. Preston. *A Pirate of Exquisite Mind*. London: Transworld, 2004.

Rodger, N. A. M. *The Wooden World*. London: Folio, 2009.

Ruse, M., and R. J. Richards, eds. *The Cambridge Companion to the "Origin of Species."* Cambridge: Cambridge University Press, 2009.

Salmond, A. *Aphrodite's Island*. London: Penguin Viking, 2009.

Salmond, A. *The Trial of the Cannibal Dog*. London: Penguin, 2004.

Sobel, D. *Longitude*. New York: Walker Books, 1995.

Steel, E. *He Is No Loss—Robert McCormick and the Voyage of HMS Beagle*. London: British Society for the History of Science, 2011.

Thomson, K. *The Young Charles Darwin*. New Haven, CT: Yale University Press, 2009.

Thomson, K. S. *HMS Beagle—The Story of Darwin's Ship*. London: Phoenix, 2003.

Weiner, J. *The Beak of the Finch*. London: Jonathan Cape, 1994.

Wesson, R. *Darwin's First Theory: Exploring Darwin's Quest for a Theory of Earth*. New York: Pegasus Books, 2017.

Wilson, A. N. *Charles Darwin: Victorian Mythmaker*. London: John Murray, 2017.

Wilson, E. O. *Half-Earth—Our Planet's Fight for Life*. New York: Liverright Publishing, 2016.

Wyhe, J. van. *Darwin—The Story of the Man and His Theories of Evolution*. London: Andre Deutsch, 2008.

ARTICLES

Fernicola, J. C., S. F. Vizcaino, and G. de Iuliis. "The Fossil Mammals Collected by Charles Darwin in South America during His Travels on Board the HMS *Beagle*." *Revista de la Asociacion Geologica Argentina* 64, no. 1 (2009): 147–159.

Gelber, D. "The Origins of Rio de Janeiro." *History Today* 66, no. 8 (August 2016).

Gillham, N. W. "Cousins: Charles Darwin, Sir Francis Galton and the Birth of Eugenics." *Royal Statistical Society* 6, no. 3 (2009): 132–135.

Mayr, E. "Darwin's Influence on Modern Thought." *Scientific American* 283, no. 1 (July 2000): 78–83.

Melnick, D., M. Cisternas, M. Moreno, and R. Norambuena. "Estimating Coseismic Coastal Uplift with an Intertidal Mussel: Calibration for the 2010 Maule Chile Earthquake (Mw = 8.8)." *Quarterly Science Reviews*, no. 42 (2012): 29–42.

Ryding, S., M. Klaassen, G. J. Tattersall, J. L. Gardner, and M. R. E. Symonds. "Shape-Shifting: Changing Animal Morphologies as a Response to Climatic Warming." *Trends in Ecology and Evolution* 36, no. 11 (November 2021): 1036–1048.

Slatta, R. W. "The Demise of the Gaucho and the Rise of Equestrian Sport in Argentina." *Journal of Sport History* 13, no. 2 (Summer 1986).

Sloan, P. "Evolutionary Thought before Darwin." *The Stanford Encyclopedia of Philosophy* (Winter 2019). https://plato.stanford.edu/archives/win2019/entries/evolution -before-darwin/.

Strelin, J., and E. Malagnino. "Charles Darwin and the Oldest Glacial Events in Patagonia: The Erratic Blocks of the Rio Santa Cruz Valley." *Revista de la Asociacion Geologica Argentina* 64, no. 1 (2009): 101–108.

Thomson, K. "Darwin's Enigmatic Health." *American Scientist* 97, no. 3 (May–June 2009): 198.

Thomson, K. "HMS *Beagle* 1820–1870." *American Scientist* 102, no. 3 (May–June 2014): 218.

Weintraub, L. "The Link between the Rockefeller Foundation and Racial Hygiene in Nazi Germany," https://dl.tufts.edu.

Wilson, M. F., and J. J. Armesto. "The Natural History of Chiloe: On Darwin's Trail." *Revista Chilena de Historia Naturel* 69 (1996): 149–161.

Winn, P. "British Informal Empire in Uruguay in the Nineteenth Century." *Past and Present*, no. 73 (November 1976): 100–126.

Zappettini, E. O., and J. Mendia. "The First Geological Map of Patagonia." *Revista de la Asociacion Geologica Argentina* 64, no. 1 (2009): 55–59.

IMAGE CREDITS

118 Guanaco in Patagonia, February 2010. (Photograph by M. and D. Preston)

128 Button Island near Woollya, by C. Martens, engraved by T. Landseer. (Courtesy of the Wellcome Trust; Wellcome Images)

136 (c) Martin Lubikowski, ML Design, London

138 Settlement at Port Louis, Falkland Islands, by C. Martens, engraved by J. W. Cook. (M. and D. Preston's collection)

156 General Juan Manuel de Rosas, by F. G. del Molino. (Courtesy of Museo Histórico de Argentina)

163 Argentinian gaucho, 1868. (Courtesy of Library of Congress)

175 Port Desire, by C. Martens, engraved by S. Bull. (Courtesy of the Wellcome Trust; Wellcome Images)

177 A Rhea Petise (*Rhea darwinii*). (M. and D. Preston's collection)

180 A Patagonian, by P. P. King, engraved by T. Landseer. (Courtesy of Wikimedia Commons)

189 A Falkland fox. (Courtesy of J. G. Keulemans)

193 The *Beagle* laid ashore, River Santa Cruz, by C. Martens, engraved by T. Landseer. (Courtesy of the British Library)

197 Basalt glen, Santa Cruz River, Patagonia, by C. Martens, engraved by T. Landseer. (Courtesy of the Wellcome Trust; Wellcome Images)

202 Fuegian wigwams, Magdalena Channel, by P. P. King, engraved by S. Bull. (Courtesy of the Wellcome Trust; Wellcome Images)

204 San Carlos (Ancud), Chiloe, by C. Martens, engraved by S. Bull. (Photograph by M. and D. Preston)

206 Valparaiso harbor, watercolor, by C. Martens. (Courtesy of the National Library of Australia)

215 Osorno Volcano. (Photograph by M. and D. Preston)

225 Spanish fort at Valdivia, March 2020. (Photograph by M. and D. Preston)

228 Remains of the cathedral at Concepcion, ruined by the great earthquake of 1835, by J. C. Wickham, engraved by S. Bull. (Courtesy of the Wellcome Trust; Wellcome Images)

238 Fossilized trees near the Uspallata Pass, March 2020. (Photograph by M. and D. Preston)

239 The so-called Inca Bridge, actually a natural formation and now a tourist attraction, March 2020. (Photograph by M. and D. Preston)

241 Darwin's geological cross-section of the Uspallata Pass. (Courtesy of the Wellcome Trust; Wellcome Images)

248 Bernardo O'Higgins. (Courtesy of Galería Nacional o Colección de Biografías i Retratos de Hombres Celebres de Chile)

261 Mockingbird, Floreana (Charles Island), Galápagos, January 2022. (Photograph by M. and D. Preston)

263 Charles Island (Floreana), Galápagos, by P. G. King, engraved by S. Bull. (Courtesy of the Wellcome Trust; Wellcome Images)

269 Marine iguanas, Albemarle Island (Isabela), Galápagos, January 2022. (Photograph by M. and D. Preston)

273 View of Tahiti by C. Martens, engraved by T. Landseer. (Courtesy of the Biodiversity Heritage Library)

280 Missionary chapel, Tahiti, by P. G. King, engraved by T. Landseer. (Courtesy of the Wellcome Trust; Wellcome Images)

284 Maori men, sketched by R. FitzRoy. (M. and D. Preston's collection)

295 First known sketch of duck-billed platypus, 1799. (Courtesy of Wikimedia Commons)

307 Coconut (robber) crab. (Courtesy of Wikimedia Commons)

316 Napoleon's Tomb, St. Helena. (Courtesy of Library of Congress)

323 Vineyard lava stone protective walls, Terceira, Azores, October 2021. (Photograph by M. and D. Preston)

336 Beaks of Galapagos finches by J. Gould. (M. and D. Preston's collection)

340 Charles Darwin's "Tree of Life." (M. and D. Preston's collection)

345 Jenny the Orangutan. (Courtesy of Wikimedia Commons)

350 Charles Darwin, 1840, by G. Richmond. (Courtesy of Wikimedia Commons)

352 Emma Darwin, 1840, by G. Richmond. (M. and D. Preston's collection)

356 Joseph Hooker. (M. and D. Preston's collection)

361 Charles Darwin's study, Down House. (Courtesy of the Wellcome Trust; Wellcome Images)

378 Charles Darwin in middle age. (M. and D. Preston's collection)

379 Thomas Huxley. (M. and D. Preston's collection)

380 Asa Gray. (M. and D. Preston's collection)

383 Alfred Russel Wallace. (Courtesy of the Wellcome Trust; Wellcome Images)

395 *Vanity Fair* cartoon of Samuel Wilberforce, Bishop of Oxford. (Courtesy of the City College of New York libraries)

409 Cartoon of Charles Darwin as "A Venerable Orang-outang," published in *The Hornet*, March 1871. (Courtesy of University College London)

415 Robert FitzRoy. (M. and D. Preston's collection)

420 Charles Darwin in old age, by J. Collier. (M. and D. Preston's collection)

424 Charles Darwin's funeral, Westminster Abbey. (Courtesy of the Wellcome Trust; Wellcome Images)

INDEX